Handbook for Prospectors

Richard M. Pearl

Professor of Geology, Colorado College
Colorado Springs, Colorado

FIFTH EDITION

A revision of *Handbook for Prospectors and Operators of Small Mines* by M. W. von Bernewitz

McGRAW-HILL BOOK COMPANY

New York St. Louis San Francisco Düsseldorf Johannesburg Kuala Lumpur London Mexico Montreal New Delhi Panama Rio de Janeiro Singapore Sydney Toronto

Library of Congress Cataloging in Publication Data

Pearl, Richard Maxwell, date
Handbook for prospectors.

"A revision of Handbook for prospectors and operators of small mines, by M. W. von Bernewitz."
Includes bibliographical references.
1. Prospecting. I. Von Bernewitz, Max Wilhelm, 1878–1940. Handbook for prospectors and operators of small mines. II. Title.
TN270.P4 1973 622'.1 72-11749
ISBN 0-07-049025-2

234567890 MUBP 76543

The editors for this book were Harold B. Crawford, Robert E. Curtis, and Lydia Maiorca Driscoll, the designer was Naomi Auerbach, and its production was supervised by George E. Oechsner. It was set in Caledo by University Graphics, Inc. It was printed by The Murray Printing Company and bound by The Book Press.

To Meredith and Verna Alair

Contents

Preface vi

PART 1

1. The Art and Science of Prospecting 3
2. Aids to Prospectors 27
3. Transportation 48
4. Prospecting Equipment 58
5. Camping, Clothing, Cooking 85
6. Health and Safety 96

PART 2

7. Basic Prospecting 102
8. Prospecting for Radioactive Minerals 119
9. Prospecting for Luminescent Minerals 129
10. Geophysical Prospecting 137
11. Geochemical Prospecting 149
12. Sampling, Testing, Assaying 157
13. Staking a Mining Claim 167

PART 3

14. Studying the Earth 207
15. Elementary Mineralogy 215
16. Rocks: Elementary Petrology 235
17. Elementary Geology 250

PART 4

18. Precious Metals 287
19. Nonferrous Metals 298
20. Iron and Ferroalloy Metals 309
21. Lightweight Metals 324
22. Radioactive Metals 330
23. Other Major Metals 338
24. Minor Metals 342

PART 5

25. Gemstones 349
26. Fertilizers 357
27. Abrasives 360
28. Ceramic Materials 364
29. Metallurgical Materials 367
30. Chemicals 370
31. Building Materials 374
32. Manufacturing Materials 378

PART 6

33. Opening a Mine 385
34. Mineral Processing 408

PART 7

Data Lists and Tables 419

PART 8

Glossary of Terms 429

Index 449

Preface

In the original *Handbook for Prospectors and Operators of Small Mines,* the late Max W. von Bernewitz created a classic volume, which has become the standard treatise on this intriguing subject. After his death in 1940—following a broad career in British Commonwealth countries, the United States, and elsewhere—a 4th edition was edited and revised by Harry C. Chellson, likewise a man of worldwide experience in mining.

The present book, newly titled to emphasize prospecting and the mineral industries for the layman—even the weekend rockhound-prospector and his wife—has been entirely rewritten. Not one sentence from the former books has been retained; yet, it is strongly hoped that the spirit and enthusiasm of Von Bernewitz and Chellson remain evident in these pages. For here is an exciting as well as an important subject!

This book has been brought up to date in coverage, enlarged to take in more of the hobby aspects that have become so conspicuous in recent years, and expanded to serve as a general guide to the sources of help that serve well the prospector-rockhound.

Although this book deals with the whole globe, as befits a handbook, emphasis has properly been placed on prospecting in the United States, Canada, Mexico (where rockhounds are going in increasing numbers),

Australia, and the other countries in which prospecting and mineral collecting are encouraged or at least permitted and where they can be done on a basis that is reasonably congenial to the individual who is familiar with the mining customs and laws of a free-enterprise system. Opportunities seem best today in the United States, Canada, and Australia.

Part 1 covers the general aspects of prospecting in a survey orientation, which will, it is hoped, prove of interest to the person who does not know a rock from a mineral but who is curious as to what the fuss is all about: What are so many otherwise sensible people doing? In these chapters, you will find a discussion of prospecting as both an art and a science.

Part 2 deals with the actual methods of prospecting, beginning with the basic procedures that use a gold pan, continuing with prospecting for radioactive and luminescent materials and the application of geophysical and related techniques, and telling how to stake a mining claim and obtain other mineral rights in the United States. This section concludes with a reference to the laws that pertain to mining throughout the rest of the world.

Parts 3, 4, and 5 are the heart of the book, its meat and potatoes. Part 3 explains the fundamental mineralogy, petrology, and geology of mineral deposits simply but thoroughly. Part 4 describes the metallic minerals (the ores) that are of most value. Part 5 covers the nonmetallic minerals and rocks, which doubtless offer the lone prospector his greatest opportunities. These chapters tell where to look for them, right down to the basic principles that must guide the seeker for mineral wealth. Prospecting for fuels is not emphasized in this book.

Part 6 has to do with what comes next: sampling and testing or assaying, developing a prospect into a mine (at least in its preliminary stages), dressing and treating ore, and finally disposing of the property when the urge comes to start over again and find another mine.

Part 7 contains tables of the chemical elements, weights and measurements, and other useful memoranda.

Part 8 is a glossary of terms, prepared afresh for this book; it does not duplicate any entry of the extensive index that follows it.

Alternate names of minerals are given in the order in which they were used historically. In the lists of references, the names of authors are given exactly as they appear on the title pages of the books. For convenience in purchasing or inquiring, however, the names and addresses of publishers are given in the business forms that they use today. The references are to subjects that are directly related to the technical problems of prospecting rather than to general subjects such as camping and boating, for which new books are published frequently.

To write this book, I have reexamined the major literature, including the 1971 edition of the excellent Canadian government book on prospecting by A. H. Lang (which is highly recommended to all readers who intend to work in Canada) and the Swiss book (written in German and published in Austria) by Gunter Zeschke; attended the Uranium Workshop of the U.S. Atomic Energy Commission, in Grand Junction, Colorado, in 1970; and held the final revision until the publication in 1971 of the 4th edition of the *U.S. Bureau of Mines Bulletin 650* (dated 1970).

Appreciation is hereby expressed to the late Dr. Alan M. Bateman, of Yale University, whose textbooks and ideas I have used for more than 25 years to teach my own thousands of students. The drawings by Mignon Wardell Pearl illustrate geologic relationships that are best shown in this way. Photographs, which are credited where printed, are used to make clear other aspects of the subject.

Richard M. Pearl

Part 1

1

The Art and Science of Prospecting

Prospecting is both an art and a science. Being observant (an art) is helpful; knowing what to look for (the science) is necessary; and these assets when combined to a high degree produce the most successful finder of mineral wealth. Close and careful observation can be cultivated, even though it is often a natural talent, made more congenial by one's possession of keen vision. Knowledge, however, must be acquired and is to a large extent equally available to everyone who is interested in a given subject. Knowledge, in fact, increases one's interest in a subject, and both grow rapidly together.

Knowing where to search is another aspect of knowledge, and a most valuable one, for the odds greatly favor the prospector or collector who is in the right place at the right time. Some prospectors, who have much of the beachcomber's attitude toward life, may be on the constant lookout for almost anything that may be used or turned into cash. Most prospectors and collectors, however, usually have something specific in mind as they examine the countryside. The typical oldtime prospector was interested only in gold and silver and overlooked nearly everything else. The uranium prospector of the 1950's was equipped with a Geiger counter or scintillation counter, which was able to detect only radioactivity (usually of uranium or thorium), and he may not have been able to recognize a vein of native gold if he had camped on it for the night. The rockhound may be more interested in pretty agates than in copper ore—and the agates may even prove to be the more valuable.

It is the purpose of this book to enable you to develop the art of prospecting while increasing your knowledge of the most worthwhile materials to search for and the most likely places in which to look. A vast amount of knowledge is not needed: simply the ability to recognize the common minerals, metallic and nonmetallic, and the common rocks and geologic formations in which they generally occur. But, for knowledge as for other things, every little bit helps!

The most fundamental question in modern prospecting is probably this: Have conditions changed so much in recent decades that the lone prospector can no longer hope to justify his efforts? The outstanding authorities in mining are in complete disagreement. It seems only fair, therefore, to present here in condensed form the main evidence on both sides and then try to give a balanced (not merely a compromise) viewpoint.

After all is said, the proof of the pudding is in the eating, and if prospectors are still finding valuable minerals, their opportunities must obviously still exist. Repeated failure would not necessarily make the reverse situation true, but it would certainly be discouraging. On the other hand, the hobbyist is seldom concerned with making his efforts pay: the weekend fisherman, for example, does not contrast his expenses with the price of fish in the market.

It must first be recognized that certain conditions have most definitely changed. The large bodies of low-grade ores are now best left in the hands of the big mining companies, who can afford the great investment in exploration—beginning with aerial photography and aerial survey—and in development, which may take many years from discovery to production. Even in the old days, nevertheless, it was usually customary for the individual or group prospector to sell out to the companies (or speculators or investors) that had or could get the necessary capital, and they in turn carried on the mining operation. Only in California, as Rodman W. Paul, of the California Institute of Technology, has pointed out, were geologic conditions favorable for the individual miner to continue working, by simple machinery and primitive methods, the geologically recent placer deposits of gold that he had discovered. This book will tell you which kinds of minerals the lone prospector has the best chance of finding, and which kinds offer him (at least today) little hope of profitable exploitation. Fools are here suffered gladly, but fuels are not covered at all.

Whereas the mainstay of the oldtime prospector was a gold deposit—the placer-gold miner with his pick, shovel, and burro (the "beard, burro, and bourbon" type) is virtually the picture of the field prospector—this simple way of life now offers almost the least satisfactory variety of prospecting and mining. The current price of gold is not high enough, even the $38 an ounce put in effect in February 1972. It may, indeed, double or treble in the near future as a result of inflation or a counterattack on deflation, or it may be reduced in the same proportion as a result of the invention of techniques that are being investigated to recover the metal from low-grade ores at much lower cost. (A process that was announced by the U.S. Bureau of Mines in 1970 lowers the profitable limit of gold ore from 0.06 to 0.02 ounce per ton.) Regardless of this, however, it seems that practically all the world has already been searched rather thoroughly for gold. The two large new American gold mines of late years, the Carlin and Cortez mines in Nevada, were opened only after intensive and expensive scientific and technical research beyond the means of the ordinary prospector.

This typical oldtime prospector operated out of Cripple Creek, Colorado, around the turn of the century. *(Library, State Historical Society of Colorado)*

How is one to say whether prospecting pays or not? Some prospectors have done well, others have not; so also have some businessmen, stock market investors, and real estate buyers, and perhaps even some gamblers have prospered. Harry C. Chellson, who revised the previous edition of this book, calculated that the proportion of properties that were developed to become profitable mines was one in three or four hundred. The probability of success is thus statistical and cannot predict what will happen to John Doe, Richard Roe, or Joe Doakes.

Even though your chance of getting rich is small, it does exist. A former student of the author, who was inspired partly by the Von Bernewitz-Chellson book (and, it is hoped, by his professor's lectures), gathered about $75,000 in gold nuggets during one summer in Alaska, which gave him the capital to drill for oil in Kansas, which in turn provided money for large-scale land investments in Colorado. This example, incidentally, was used in the forword prepared by the author for *Simple Methods of Mining Gold,* a paperback booklet, by Terry Faulk, probably the first book of this kind ever written by a person who was confined to a state penitentiary (Colorado), where the use of pick and shovel would have been frowned on.

To some people, prospecting is eminently satisfying; others will not follow the lure. It is usually at least a healthful activity, in spite of the occasional disaster from a bouncing rock, a rattlesnake bite, or a fall from a cliff. You cannot tell about prospecting until you try it. Not only is the invitation freely extended here, but the author hopes that those who read this book will have the greatest success! That is certainly his intention.

The true record of the past is clear. The majority of American metal mines were discovered by qualified prospectors who were searching for valuable minerals at the time, as E. D. Gardner, of the U.S. Bureau of Mines, said in 1935. Yet, largely because of their use of Geiger and scintillation counters—which do much of (but not all) the thinking—83 percent of the major discoveries of uranium and other radioactive ores in the United States were made by amateur prospectors, according to Henry C. Mulberger, a former manufacturer of field instruments.

And what then of the future? In 1942, Harry C. Chellson, editor of the *Mining Congress Journal,* stated that the rate of mineral discovery in North America had slowed down since the golden age of surface prospecting that marked the half century after the discovery of the yellow metal in California in 1848. Nevertheless, enough valuable finds have been made since then throughout the world—consider Canada, Australia, Africa, the Philippines, New Guinea—to serve as an encouragement to even the most cautious person. When the staking of the millionth claim in British Columbia was celebrated in June 1969, Frank Richter, the provincial minister of mines, pointed out that the 1969 rate of staking (and hence presumably of mineral finding) was double that of 1968.

Thomas B. Nolan, director of the U.S. Geological Survey, predicted in 1961 that perhaps 10 times as many new mining districts remained to be discovered in the United States as those then known, and that they ought to include as many major districts. He based his prediction—which indicates $1 trillion worth of ore—on studies that show a statistically sound relationship between the number and size of productive mining districts and the surface extent of geologically similar areas. This is a perfectly logical assumption, but it will probably include the better recognition of hidden deposits, mainly by the geophysical and related methods that are described in chapters 10 and 11.

Large areas, even in the settled parts of the United States, have scarcely been examined, as was shown during the uranium rush of the first half of the 1950's. Much of this land remains, of course, on property that is owned by railroads and large ranchers, and it will be explored by company geologists or on a lease arrangement; the lone prospector has little opportunity here unless he is a professional. Some prospectors have indeed made a place for themselves in industry by rendering a service that is not readily obtained elsewhere.

The long interval that is often needed to bring a discovery into production can be illustrated by such examples as the Yerrington copper mine in Nevada, which was known in 1918 but not developed until 1940 and not put into production until 1953; and the Lakeshore copper deposit in Arizona, which was discovered more than 85 years before it went into operation in 1968.

On the one hand, it is likely true that most of the easily accessible parts of the earth have been looked at at least casually by prospectors and that future discoveries may need to be made in those regions where exploration is difficult and hazardous: excessively cold or hot places, deserts or swamps or jungles, the habitats of unfriendly or savage people, and areas covered by ice, lava, lakes, much gravel, or thick vegetation. A large part of the world is still like this.

On the other hand, access to such uninviting regions has been made vastly easier by modern means of transportation, discussed in chapter 3. The airplane may still remain chiefly in the hands of large companies, although groups of would-be prospectors, especially those having military connections or a background of flying, have used airplanes successfully in recent years. Ground vehicles that will go almost anywhere on any kind of terrain where they are allowed—the ATV, or all-terrain vehicles—are readily available to nearly everyone.

Equipped with gravity meter, this helicopter represents the modern approach to prospecting. *(Standard Oil of California)*

Said Hollis M. Dole, Assistant Secretary of the U.S. Department of the Interior, in 1969:

> Sometimes we have heard it said that—just as the mom and pop grocery and the independent corner drugstore are going out of the picture—the day of the individual prospector and the small mining operator is past. Undoubtedly, the sophisticated and costly technological developments in recent times tip the balance toward larger mining companies able to command sufficient capital and technical skills, but I suggest to you that ways should be considered to keep the individual prospector and the small operator in business. I am not convinced that the large open pit is "the only way to fly."

In Australia in late years, Ken Shirley, a prospector, discovered the Poseidon mine, even though the Western Mining Corporation had two teams of geologists in the area. The Scotia nickel deposit was found by

John Jones, a 24-year-old prospector, in an area that had previously been examined by geologists. Jones became a millionaire, so too did Colin Campbell, another nickel prospector; but both men continue to prospect extensively. Two prospectors named Morgan and Cowcill discovered the Kambalda nickel deposit.

Altogether, however, prospecting has become a more complex, better

Prospecting can even be done under water.

organized, and more skilled occupation than before. Even the hobby aspects are different, if only because so many mineral collectors are involved that group participation has become commonplace. The amateur prospector has in general become much better informed than previously. Many amateurs, of course, do not know how to proceed, and their uncertainty and their expectation of private assistance from busy professionals and government agencies tend to clutter up the channels of information. One of the purposes of this book is to help them find their way in this interesting activity.

Chance plays its part, too, in the discovery of mineral deposits. W. R. Cross once listed 130 gold and silver mines in the United States, of which 11 were found in this way. Miners rounding up burros, ranchers picking up a stray rock while chasing stray cows, hunters looking for game, campers occupying the right spot—these are some of the events that illustrate the factor of luck. Ore bodies have accidentally come to light while cuts were made for roads, trails, and railroads and in excavating for buildings. Burrowing insects and larger animals have brought ore to the surface where it could be found. Ants in the Atacama Desert of Chile are said to bring up pieces of gold and copper ore, which have indicated places to prospect. Herodotus, the Greek historian, referred to "gold-digging ants," probably anteaters, in India.

In 1810, well before the California gold rush, Thomas Pennant wrote that "chance was the general detector of metallic riches in early times," and he mentioned the discovery of the gold mines of India as a result of seeing the bright-yellow metal in anthills and the silver mines of Spain by the casual burning of wood. At Llangynnog, in Wales, a woman slipped while climbing a hill, and a vein of rich ore was revealed at her foot. Even earlier—much earlier—the Greek historian Diodorus Siculus related that "in the country of the Paeonians, the farmers, cultivating the soil, often turned up bits of virgin gold with the plough." In this way, also, were found the gold mines of Galicia (in Spain). The first metals that were used—gold and copper—were doubtless picked up in native form from streams by primitive man.

In a real sense, the accidental discovery of mineral deposits might be said to have been virtually the sole means prior to the development of geophysical methods. The list of such discoveries would therefore be endless and eventually tedious. This chapter, therefore, avoids them and continues with a brief summary of some of the famous mineral deposits that have been brought to light by virtue of an unusual combination of circumstances.

A story persists that the telluride minerals of Cripple Creek were first recognized when a miner made a camp fireplace with pieces of rock that released the tellurium as fumes, leaving globules of gold behind. This episode does not denigrate the shrewd observations of Richard Pearce or Winfield Scott Stratton. The "sweating" of gold still yields spectacular specimens from Cripple Creek ore.

Another fire, a tragic one, in 1911 in the Porcupine district of Ontario, swept clear the underbrush and exposed rock surfaces that were to reveal gold ore to the prospectors in eastern Canada.

The history of Africa was transformed in 1867 when Schalk van Niekerk was attracted by an unusual stone lying on the floor of a farmhouse where he was visiting. He was laughingly told that it was only a child's plaything picked up in the field. When the "worthless" stone was shown to a mineralogist, it was identified as a diamond worth about £500. The pros-

Cripple Creek as it appeared on July 14, 1892. *(Library, State Historical Society of Colorado)*

The Wesselton diamond mine, near Kimberley, Republic of South Africa, is relatively small but noted for the very fine quality of its stones. *(N. W. Ayer & Son, Inc.)*

pecting that followed culminated in the finding of the Star of South Africa, a magnificent diamond, by a shepherd boy in 1869. The African mineral industry dates from this time. The Republic of South Africa is still regarded as a prime candidate for mineral exploration.

Elisha Holmes took up land in Divine Gulch, in California, which he homesteaded. Years later, his grandson, working in the family garden, uncovered a gold nugget that was worth more than $1,000, which was only the beginning of the afternoon's output.

A straying burro, which took shelter under an outcrop of dark rock, is held responsible for the discovery by Jim Butler of the Tonopah mine in Nevada. The vein in the rock ultimately yielded about $150 million in gold and silver.

Even more observant was the white donkey of Noah Kellogg, prospecting in the Coeur d'Alenes, in Idaho. Wandering off during the night, into a canyon and up a slope, this intelligent animal was found standing on an outcrop of ore, gazing "in apparent amazement" across the canyon at another body of ore "reflecting the sun's rays like a mirror." Thus began the Bunker Hill mine, a classic property in the history of lead and silver mining. T. A. Rickard, historian of mining, says that the story about the "gentle jackass" is "pure moonshine," but Jim Wardner, who developed the mine, left the account for us in his memoirs, and who can doubt what a mine promoter says? The legend of Kellogg has been portrayed frequently since 1935 by Norman McLeod and a burro.

The Big Blue mine, which began the boom at Quartzburg, California, was discovered when "Lovely" Rogers threw a rock at his wandering mule and struck a piece of gold-rich quartz instead.

Angel's Camp, famous in California history, was renewed as a mining town when a miner known as Raspberry fired his gun in order to free it of a stuck ramrod; the metal hit a rock, revealing its content of gold.

Completely untrained and uneducated men have in such ways discovered vast riches. Nevertheless, it is true that Lady Luck usually favors the trained and the diligent. Many ore bodies have been found by experienced prospectors after hundreds of untrained men had already passed over the ground. The author's wife discovered, in the summer of 1969, a new field of gem jasper, small but worth more per pound than most gold ore, by following her curiosity beyond a place in Colorado where numerous picnic parties must previously have stopped; but she recognized what she had seen. The prospector might consider Abraham Lincoln's early promise: "I will study and get ready and some day my chance will come."

Perhaps we have taken it for granted that prospecting is worth doing for its own sake and our own. The welfare of one's country and of society in general might also be considered. How important are the products of the mineral kingdom?

It seems unnecessary to try to justify their importance in today's world.

Entirely apart from the benefits and pleasures of the mineral-collecting hobby, we recognize the extent to which our industrial civilization is based on the abundance of the resources of the earth's crust: metallic, nonmetallic, and fuel. Without an adequate supply of them, we should have to return to a more primitive mode of existence, perhaps even to the nomadic life. And, as stated at the beginning of the author's book *Successful Mineral Collecting and Prospecting,* "The prospector, from the Old and New Stone Ages right down to the present day, has made his contribution to almost every material advance of civilization."

Prospectors have contributed enormously to the development of the United States, Canada, Australia, South Africa, and certain other countries. From the time of the earliest European settlement in North America, the need for iron for tools and guns, lead for bullets, and copper for utensils prompted a continuing search for sources of these metals. The lure of gold and silver provided the impetus for much of western development between 1850 and 1910. The demands for a wider variety of metals took the prospector on the search for them, climaxing in the great uranium boom that subsided, at least temporarily, in 1955.

The current production of ore (the raw material that yields metals) amounts to nearly half a billion tons yearly in the United States. Minerals are relatively even more important to the well-being of Canada and Australia, and the rate of consumption of many of the key mineral materials in the rest of the world is rising even faster than it is in the United States, which has used more mineral resources during the past 10 years than the entire population of the world did in all previous history. Even more ore must be found if industry is to expand to satisfy growing populations and flourishing economies. The uninterrupted need for all kinds of minerals justifies the remark of Howard A. Meyerhoff, executive director of the Scientific Manpower Commission, that in spite of the urgent requirements of conservation, "the United States has become too concerned about how much of our mineral resources have been used and has paid too little attention to how much more we can find." This admonition applies to the entire world.

Prospecting probably began a long time ago. After the birth of civilization, when specialized labor took form and metallurgy became a vital part of this new way of human life, those few observant men whose talent was finding minerals, rather than mining or preparing them for use, became prospectors.

The earliest history of prospecting is even more obscure than the oldest history of mining, which is less well known than the development of the mineral industries and metallurgy. The last subject has been carefully investigated with relation to the beginnings of civilization, for metallurgy (and the specialization of labor that was associated) is one of the principal factors in it. The history of mining is somewhat known, for it was an obvious group activity and played a significant part in ancient migrations and military activities. But prospecting seems to have received little attention.

According to Sydney H. Ball, a Captain Haroernis led the first party of Egyptian prospectors and miners into the Sinai Peninsula and the Sudan in about 2000 B.C. After 3 months of prospecting, they discovered and mined large amounts of turquoise.

This new occupation developed extensively in central Europe after the Middle Ages. Georg Bauer (called Agricola), who in 1556 wrote the first comprehensive book on mining, said:

> A miner must have the greatest skill in his work, that he may know first of all what mountain or hill, what valley or plain, can be prospected most profitably, or what he should leave alone; moreover, he must understand the veins, stringers and seams in the rocks. Then he must be thoroughly familiar with the many and varied species of earths, juices, gems, stones, marbles, rocks, metals, and compounds.

The tin mines of Cornwall, England, worked since the days of the Phoenicians, sent forth a host of experienced prospectors and miners who carried their knowledge and skills to the far corners of the expanding British Empire. Prospectors and miners from other parts of Europe, as well as the Chinese, brought special techniques that helped the mineral industries to grow throughout the newly awakening regions of the world.

America's most romantic and dramatic experience was the winning of the West—and the prospector was in the heart of it. The story of the mountain man has endured as saga, and that of the cowboy has been transformed into the Great American Legend; likewise, that of the prospector is only partly true.

Many a prospector took to the field to escape responsibilities at home, or the law, or to find adventure rather than mineral. After enduring the almost incredible hardships of the 1898 trek to the Yukon, for example, fewer than half of the participants even bothered to look for gold. To have accomplished the trip itself was satisfaction enough. Or take M. Gordon Fowler, who recently died in Moab, Utah, at age 80, the last of the old-timers of Miners Basin. He packed in supplies by mule train for 17 years, until he built a road. His chief interest was in proving his theory that the heart of the La Sal Mountains contains veins of metal. Perhaps it does.

Some prospectors became miners when they made a discovery. Others sold out and turned their search elsewhere, unwilling to settle down, because of restlessness or high hope. Except notably in California, where the abundance of placer gold made individual mines possible, and in Utah, where Mormon influence effectively delayed the mining fever until it burst out "with truly Gentile zeal," most western mining properties were sold to large operators, who worked them. And, except for the small number of men who became very rich in mining—their names are well known in their home country: Cripple Creek had 28 "authentic millionaires"—most of the really substantial wealth produced through western mining came to those who succeeded in the businesses of smelting, transportation, merchandising, building, real estate, or banking. The business and professional man in remote regions—such as the interior of Alaska—is now largely being bypassed by the airplane, which brings supplies and services from distant trading centers.

Central City, Colorado, was one of the famous mining camps of the Rocky Mountains. *(Colorado Advertising and Publicity Department)*

Prospectors were responsible for the earliest system of law in much of the West, that of the mining districts. Lacking local government and recognizing the weakness of territorial administration when it came, the first settlers sometimes established their mining districts even before

Ouray, in the San Juan Mountains of Colorado, is headquarters for the historic Camp Bird mine. *(Colorado Advertising and Publicity Department)*

arriving at their new home. It has been shown that one of their purposes was to aid in land speculation. Later mining districts superseded the government of the individual camps.

The sustained development of western Canada followed the mineral discoveries of the Cariboo in 1860, the Klondike in 1897, and other fields. Eastern Canada came into mining prominence with the discoveries at Sudbury in 1883 and Cobalt in 1903, as railroads were constructed in this region. The northern frontier stretching from Labrador to the Yukon is now developing at an unforeseen pace for the same reason.

Prospectors and miners carried forward not only civilized society as they advanced from frontier to frontier but also methods for extracting metal, each new region presenting its own geologic conditions and technologic problems.

The oldtime prospectors thus made a contribution to the opening up of western America, both the United States and Canada, that is just becoming appreciated. Most of them had a fair knowledge of their craft, but the fact that they were pathfinders in a new country gave them a tremendous advantage, for they were often the first to appear on a given scene with prospecting in mind. As A. H. Lang, of the Geological Survey of Canada, says: "Their outstanding characteristics were ability to travel and live under pioneer conditions, buoyant optimism, dogged perseverance, and open-handed hospitality. They were adventurers willing and often eager to undergo hardship for the opportunity to lead an independent and roving life and for the chance of 'striking it rich.'" Some prospectors today direct their search for "lost mines" of fact and fiction. Most of these mines never existed, as even some of their most enthusiastic advocates cheerfully admit.

The methods of the oldtimer are still the basis of all ordinary prospecting. These methods begin with developing a sharp eye for metallic minerals and the "signs" that often accompany them, as discussed in chapters 18 through 24. We shall emphasize in chapters 25 through 32 the nonmetallic minerals and rocks, both common and unusual, which today may offer much better opportunities for profit. (In fact, as brought out in 1970 by Robert M. Dreyer, of Western Mineral Associates, the disproportion between nonmetallic and metallic minerals becomes greater as a nation grows into economic maturity. This is because nonmetallic minerals are basic raw materials for the construction and chemical processing industries, both of which are essential for national economic growth.) The fragments of valuable minerals that may be seen are traced to their source, using the so-called gold pan to separate grains of heavy minerals in crushed rock or loose gravel or sand. When promising bedrock is found, it is scraped, pitted, or trenched to expose it for testing. These techniques are all explained in this book.

The winds of change in prospecting began at about the opening of the Second World War. This war, and the rapidly expanding economy that marked the end of the Great Depression, revealed the coming depletion of many of the world's important mineral deposits. Mining and government officials realized that new deposits must be found, and that the prospectors had to be the ones to find them, either alone or in crews, either with or without electronic and other elaborate equipment. The uranium boom of the 1950's, which saw the most concentrated search for minerals ever known, proved that the prospector could produce what was needed if encouraged to do so.

The government of Canada, and more recently of Ireland, seeks to stimulate private exploration and development of mining by means of tax concessions and subsidies. On the other hand, the governments of Sweden, Norway, Greece, Turkey, and France are actively engaged in prospecting. The Soviet Union carries on what might be called saturation prospecting, leaving literally no stone unturned. In Leningrad is the Chernyshov Museum of Geological Prospecting.

Accompanying the wartime and postwar need for minerals was the revolution in transportation, discussed in chapter 3. This ranged all the way from float-equipped aircraft to trail buggies. Geophysical and related techniques, discussed in chapters 10 and 11, have vastly expanded the scope of prospecting. Some of these methods are familiar to the ordinary prospector: Geiger counters and scintillation counters, though electronic in nature, are included among the tools of the ordinary prospector (see chapter 8) because they can be used rather successfully by the untrained person. Another contributing factor is a remarkable improvement in diamond drilling, which makes it possible to test quickly the favorable indications suggested by observation or instrumental evidence.

Speaking of Alaska, Ernest N. Wolff, of the University of Alaska, said, "Sixty years of history makes men less optimistic about making a fortune." Also, he notes that "because of improved standards of living, modern man is unused to hardships." These attitudes must be taken into account.

Let us look at a few examples of recent mineral discoveries and see if they follow any sort of pattern. These are well documented in print, and the references from which they are taken (not necessarily the original publication) are given here.

"One of Montana's greatest ore discoveries in many years" was reported by John McGee in *The Mining Record* of September 10, 1969. A mammoth vein of tungsten was uncovered by Benny Dickson, a bulldozer operator, while he was helping Dee Alexander, 56, his brother John, 54, and son Bill, 27, who have combined prospecting with their varied logging operations. An offer of $3 million by a Canadian firm for the claims has been reported. The first find was thought to be gold ore. Note that the veins are exposed at the surface and, according to Dee Alexander, "The whole mountain above Snider glows at night under the black light." Any outdoorsman who was equipped with a portable ultraviolet lamp could have made the same discovery. (This method is described in chapter 9.)

Gordon Milbourne, a Canadian prospector, discovered copper mineralization on Vancouver Island and sold his property to the Utah Construction and Mining Company, retaining an interest in the profits for himself. Development of an open pit expected to yield 280 million tons of copper and molybdenum ore was announced in *The Mining Record* of June 18, 1969.

The first major uranium discovery in New Mexico was made in 1950 by a part-Navajo Indian named Paddy Martinez, who had met two prospectors as they were talking about the subject in a hotel lobby in Grants. The prospectors showed Martinez a sample of yellow sandstone, which he remembered when he found rock like it a few weeks later near Haystack Mesa, where he had a sheep camp. For his discovery, the Santa Fe Railroad paid him $200 a month for the rest of his life—$400 an hour for the time it actually took him to collect his pay as a "uranium scout."

Faith that a prospector was right, that deep tests would disprove the

A former Laguna governor, Walter Sarracino, was an early-day Indian prospector near Grants, New Mexico. *(Grants and Western Valencia County Chamber of Commerce)*

discouraging assay reports of surface samples in the Blind River area of Ontario, led Joseph Hirschhorn, of Brooklyn, New York—hardly the mining center of the world—to gamble $30,000 on diamond drilling. Two months later, two men, prospector and speculator (investor?), found a $31 million uranium mine. *Fortune* told the story.

Mark Twain reported a somewhat different experience. In a letter sent in 1869 to a group of California pioneers, he wrote as follows:

I purchased largely in the Wide West, the Winnemucca, and other fine claims, and was very wealthy. I fared sumptuously on bread when flour was $200 a barrel, and had beans every day when none but bloated aristocrats could afford such grandeur. But I finished by feeding batteries to a quartz mill at $15 a week, and wishing I was a battery myself and had somebody to feed me. My claims in Esmeralda are there yet.

I suppose I could be persuaded to sell. I went to the Humboldt district when it was new. I became largely interested in the Alba Nueva and other claims with gorgeous names, and was rich again in prospect. I owned a vast mining property there. I would not have sold out for less than $400,000 at that time, but I will now. Finally I walked home—some 200 miles—partly for exercise and partly because stage fares were expensive.

Next I entered upon an affluent career in Virginia City and by judicious investment of labor and capital of friends, became the owner of about all the worthless wildcat there in that part of the country.

Assessments did the business for me there. There were 117 assessments to one dividend, and the proportion of income to outlay was a little against me. My financial thermometer went down to 32°F. and the subscriber was frozen out. I took up extensions on the mainland—extensions that reached British America in one direction and the Isthmus of Panama in another—and I verily believe I would have been a rich man if I ever found those infernal extensions. But I didn't. I ran tunnels until I broke through the roof of perdition, but those extensions turned up missing every time.

I paid assessments on Hale & Norcross until they sold me out and I had to take in washing for a living and the next month that infamous stock went up $7,000 a foot. I own millions and millions of feet of affluent silver leads in Nevada—in fact, I own the entire undercrust of that country and if Congress would move that state off my property so that I could get at it, I would be wealthy yet. But no, there she squats—and here am I. Failing health persuades me to sell. If you know of anyone desiring a permanent investment, I can furnish him one that will have the virtue of being eternal.

Only a few years ago, Harry V. Warren, professor of geology at the University of British Columbia, stated that basic prospecting is much the same as when men fought their way to the Klondike in 1898. "The main tools in searching for wealth in the ground," he said, "are a need, knife, and magnifying glass." On the same occasion, Thomas Elliott, secretary-manager of the Yukon Chamber of Commerce, affirmed that "virtually every major discovery in British Columbia and the Yukon" was made by the everyday type of prospector. In Canada, 46 important new mines were brought into production or nearing production during the first 10 years after the Second World War (1945–55); of these, 22 were ascribed to conventional prospecting by men who were not scientists or engineers. Drilling according to geologic hunches found 17 mines, and 7 were found by physical prospecting. During the decade 1956–66, 55 new mines were discovered in Canada, of which 18 were credited to conventional methods, 15 were geologic, and 20 were geophysical.

Mining Engineering, in September 1970, listed the following new mines in British Columbia that were originally discovered by prospectors; note that the value of the orebodies totals nearly $9 billion.

Property	Size of ore body (thousands of tons)	Gross value of ore body (millions of dollars)	Man-years of work to mine ore body
Lornex (Highland Valley)	300	1,200	10,300
Bethlehem Copper (Highland Valley)	70	350	4,900
Brenda (Peachland)	146	360	5,500
Granduc (Stewart)	50	1,000	12,200
Granisle (Babine Lake)	25	100	1,600
Wesfrob (Queen Charlotte Islands)	40	250	2,900
Utah Construction (Port Hardy)	270	1,350	9,300
Anvil Mining (Ross River, Y.T.)	70	1,750	7,900
Ingerbelle Mines (Princeton)	65	315	4,300
Hudson Bay Mountain (Smithers)	75	300	7,000
B.C. Molybdenum (Alice Arm)	40	280	2,700
Cassiar Asbestos (Cassiar)	25	500	18,000
Wellgreen Nickel Mines (Kluane, Y.T.)	750	50	400
Endako Mine (Fraser Lake)	200	800	10,000
	2,126	8,605	97,000

Geology is the foundation of prospecting. Yet, as A. H. Lang has so clearly brought out, geologists do not in general make the best prospectors!

This seeming incongruity is partly due to the fact that geologists are usually trained to cover the ground rapidly—"reconnaissance," it is termed—whereas the prospector may need to examine carefully every square foot of an outcrop and then scrape it or even dig into it. Another reason is that the geologist is readily diverted by the scientific aspects, neglecting the economic.

The example of John T. Williamson indicates something different. A geology graduate of McGill University, in Canada, he followed his long quest, backed by a firm conviction that his scientific reasoning was right, until he discovered the world's largest diamond mine in 1940 at Mwadui, Tanganyika. Six years of privation, hardship, and searching amidst malaria, sunstroke, hunger, thirst, and ridicule paid off handsomely, and Williamson is said to have died the richest bachelor on earth.

Having a geologist as a friend, therefore, may be good policy, and geologists are certainly necessary to supervise a prospecting program, to interpret assays, and to prepare maps and reports. A geologist might make a good partner for the prospector. That is, unless the elementary-school pupil in Calgary, quoted by the *Western Miner and Oil Review,* is correct when he wrote:

> All the mines are found by a kind of geology labourer called prospectors. These poor prospectors have no book learning, but make use of their thumb in a secret way called the rule of thumb. When they discover a good thing the geologists and their pals called promoters swindle him out of it. This kind of swindling is supposed to be fair game and is called litigation or something like that. My pop wasn't sure.

The attitude of society toward the prospector and miner has in recent years undergone a modification as extensive as the changes in techniques and equipment. Prospectors and collectors should be aware of the disposition of mind that generally prevails and should accommodate themselves to it, partly to make their activities more pleasant, and partly even to make them possible.

Trespassing on private property has always been regarded as a serious offense. The pressures of a growing population may in time modify, if not eliminate entirely, the concept of private property. Until that time, however, it is unwise and unsafe to commit trespass, and every precaution should be taken to avoid doing so.

Conservation has become more and more desirable, if not absolutely necessary, in the eyes of the American people. Waste and destruction are less lightly tolerated than formerly, and the prospector and collector should use all due care not to become obnoxious when in the field.

The Code of Ethics of the American Federation of Mineralogical Societies, presented below, will call to your attention the sort of conduct that should be maintained in order to preserve the goodwill that will enable the search for minerals to continue.

I WILL respect both private and public property and will do no collecting on privately-owned land without the owner's permission.

I WILL keep informed on all laws, regulations, and rules governing collecting on public lands and will observe them.

I WILL, to the best of my ability, ascertain the boundary lines of property on which I intend to collect.

I WILL use no firearms or blasting material in collecting areas.

I WILL cause no willful damage to property of any kind—fences, buildings, signs, etc.

I WILL leave all gates as found.

I WILL build fires in designated or safe places only and will be certain they are completely extinguished before leaving the area.

I WILL discard no burning material—matches, cigarettes, etc.

I WILL fill all excavation holes which may be dangerous to livestock.

I WILL NOT contaminate wells, creeks, or other water supplies.

I WILL cause no willful damage to collecting material and will take home only what I can reasonably use.

I WILL support rockhound project H.E.L.P.—Help Eliminate Litter, Please, and will leave all collecting areas devoid of litter, regardless of how found.

I WILL cooperate with field trip leaders and those in designated authority in all collecting areas.

I WILL report to my club or federation officers, Bureau of Land Management, or other proper authorities, any deposit of petrified wood or other material on public lands which should be protected for the enjoyment of future generations and for public educational and scientific purposes.

I WILL appreciate and protect our heritage of natural resources.

I WILL observe the "Golden Rule," will use "Good Outdoor Manners," and will at all times conduct myself in a manner which will add to the stature and "Public Image" of rockhounds everywhere.

The Colorado Mining Association adopted in 1970 the following Code of Exploration Practices:

1. Know and comply with the mining laws relating to exploration in Colorado.

2. Know and comply with Forest Service, Bureau of Land Management, or other appropriate government agency's rules and regulations.

3. Establish and maintain cordial relations with land owners and/or lease holders in the area of activity. These may be owners of private lands or administrators of public lands such as Forest Service, Bureau of Land Management, or state agencies.

4. Avoid harmful impacts upon the environment. For example:

a. Deposit lunch sacks, cans, or other litter in containers provided by U.S.F.S. or other agencies or dispose of such debris in a manner consistent with sound conservation and environmental practices.

b. Use all reasonable efforts to avoid spillages of petroleum products and other noxious materials in connection with drilling activities.

c. Plug or cap drill holes upon completion and abandonment.

d. Inspect each drill site upon abandonment to make sure it has been properly cleaned up.

e. Remove plastic flagging and/or aluminum foil used in geochemical, geophysical, or other surveys when no longer needed.

5. Keep excavation (roads, drill sites, and cuts) at a minimum. When no longer needed, reclaim to the extent practicable, including efforts to re-establish vegetation.

6. Conduct all activities in the manner which minimizes danger of man-caused fires either in timber lands or grass lands.

7. Drive vehicles and conduct other activities in a manner such as to minimize disturbances of people and livestock in the area, keeping to established roads or trails when possible. Leave stock control gates in the position found (closed or open), unless a sign advises to the contrary or other advice is obtained from property owners.

8. Do not hunt or fish on private lands except with full permission of the land owner or authorized agent and then only when properly licensed.

9. Use no timber, water, or other resource on private land without permission of the owner.

10. Promptly initiate negotiations with the owner for settlement of any claims for damages resulting from activities of the exploration group or contractors.

And what of the far future, beyond our own time? Harrison Brown, professor of geochemistry at the California Institute of Technology, expressed the belief some years ago that future world requirements for minerals will be so great that it will be necessary to obtain them by breaking down common rock. One hundred tons of ordinary rock would thus yield about 12 pounds of copper, 4 pounds of lead, and 17 pounds of zinc. Mining would then no longer exist as such, having been absorbed into the chemical industry: there would simply be rock factories. This "cost concept" view has been vigorously disputed in recent months by such mineral economists as Peter T. Flawn, of the University of Texas, and Charles F. Park, of Stanford University, who insist that special concentrations of useful substances, which have furnished all our mineral resources to date, must continue to be relied on in the future, or else the economic and social effects will be disruptive.

This may well mean that the development of strategic minerals is the "highest and best use" of land, as real estate people term it, because of the great value obtained from small areas. This view is known as the "orebody concept." Almost all the world's supply of metallic minerals to date has come from only about 1,000 square miles of the earth's surface. Custodial management, as stated by William W. Porter II, is more meaningful than simple conservation.

LITERATURE ON PROSPECTING

Virtually a classic, the predecessor of the present book, titled *Handbook for Prospectors and Operators of Small Mines,* by M. (Max) W. von Bernewitz, went through four editions between 1926 and 1943, the last having been edited and partly rewritten by Harry C. Chellson. It was published by McGraw-Hill Book Company, New York.

Much information on prospecting and exploration is given in Section 10 of *Mining Engineers' Handbook,* 3d edition, edited by Robert Peele (John Wiley & Sons, Inc., New York, 1941. 2 volumes).

The first book "of general distribution that combines the related interests of prospectors and hobbyists in the worldwide search for minerals" was *Successful Mineral Collecting and Prospecting,* by Richard M. Pearl. It was published in 1961 by McGraw-Hill Book Company, New York, and two reprints have been issued by other publishers.

A small paperbound book, *Popular Prospecting,* by H. C. Dake, was published in 1955 by The Mineralogist Publishing Company, Portland, Oregon, in order to meet the needs of the fast-growing number of uranium seekers, but it emphasizes general prospecting. Approaching the subject from the opposite direction, most of the books on uranium prospecting (see pages 127–128) have some information on prospecting in general.

Prospecting is referred to in almost every book on mining, and the historical development is discussed especially in such books as *Man and Metals,* by T. A. Rickard (McGraw-Hill Book Company, New York, 1932); *The Romance of Mining* by T. A. Rickard (The Macmillan Company, New York, 1944); *A Pictorial History of American Mining,* by Howard N. and Lucille L. Sloane (Crown Publishers, Inc., New York, 1970); *Gold and Silver in the West,* by T. H. Watkins (American West Publishing Company,

Palo Alto, California, 1971); and such specialized articles as "Historical Notes on Gem Mining," by S. H. Ball (*Economic Geology*, vol. 26, 1931, pp. 681–739).

The varied aspects of mineral economics are well presented in:

Mineral Resources: Geology, Engineering, Economics, Politics, Law, Peter T. Flawn, Rand McNally and Company, Chicago, 1966.

Affluence in Jeopardy: Minerals and the Political Economy, Charles F. Park, Jr. Freeman, Cooper and Company, San Francisco, 1968.

Other recommended books in English on general prospecting, even though certain of their titles may suggest too specialized an approach to alluvial methods or limited areas, include the following:

The Prospectors' Handbook, 4th edition, W. L. Goodwin, National Business Publications, Ltd., Quebec, 1956.

Prospecting in Canada, 4th edition, A. H. Lang, Geological Survey of Canada, Economic Geology Report No. 7, Ottawa, 1970. An extensive book.

Handbook for the Alaskan Prospector, revised edition, Ernest N. Wolff, University of Alaska, College, Alaska, 1969.

The Alaska Prospector's Short Course in Introductory Prospecting and Mining, revised edition, Leo Mark Anthony, College of Earth Sciences and Mineral Industry Publication Bulletin 4, College, Alaska, 1969.

Alluvial Prospecting and Mining, 2d edition, S. V. Griffith, Pergamon Press Inc., New York, 1960.

Placer Examination Principles and Practice, John H. Wells, U.S. Bureau of Land Management Technical Bulletin 4, Washington, 1969.

Alluvial Prospecting, C. Raeburn and Henry B. Milner, Thomas Murby & Company, London, 1927.

Prospecting for Gold and Silver, Eric M. Savage.

Prospecting for Gold . . . , Ian L. Idriess, Angus and Robertson, Sydney, 1931.

Let's Go Prospecting, Edward Arthur, Joshua Tree, California, 1970.

Prospecting and Operating Small Gold Placers, 2d edition, William F. Boericke, John Wiley & Sons, Inc., New York, 1936.

Prospecting for Mineral Ores in Europe, James F. McDivitt, Organization for Economic Co-operation and Development, Paris, 1962.

2

Aids to Prospectors

Many aids are available to help the prospector. As described in this chapter, they run a fairly wide gamut, including actual financial assistance, information and advice, courses of study, testing and identification services, and maps and literature of all sorts. Laboratories and libraries exist in public agencies and other organizations for your benefit.

GRUBSTAKING

A time-honored means of raising funds for prospecting, or of being backed by equipment and supplies, grubstaking has more or less gone out of fashion. Yet, presumably, it can be resorted to as readily as any other kind of partnership, although it is technically a "tenants in common" arrangement. By pooling aid, labor, and expenses, two or more persons agree to locate claims on public land and to share in the interest, usually to an equal extent. Because the price of labor has increased much more than the cost of goods, a 50-50 division is no longer equitable, and some money should be given to the prospector in addition to supplies.

Grubstaking has as its basis the trust of the lender in the prospector. A chief reason for the failure of grubstaking to satisfy both parties is that the terms are not always spelled out clearly and carefully enough, rather than because of the dishonesty of the prospector. Therefore, grubstaking agreements must be recorded, in California at least, as legal documents.

A good many of the mines of history, especially gold mines, have been found as a result of grubstaking. Gold was favored, owing to the presumed chance of a rich strike; but Von Bernewitz believed that "this should not be the practice, because most minerals have a market."

H. A. W. Tabor, postmaster and storekeeper at Leadville, grubstaked two Colorado prospectors named August Rische and George F. Hook.

Horace A. W. Tabor got his start as the silver king of Colorado by grubstaking two prospectors. *(Library, State Historical Society of Colorado)*

Among their provisions was a jug of whiskey, which tempted them to rest awhile on Fryer Hill. The more they drank, the better the place looked to them, until they decided to sink a shaft right there. At a depth of 30 feet, they reached the ore body that became the Little Pittsburg mine. A goodly share of it, of course, went to Tabor.

Less happy were the results of the discovery of the Bunker Hill property, alleged to have been the doing of a burro, as described in chapter 1. Grubstaked by the firm of Cooper and Peck, Noah Kellogg received a donkey and $18.75 worth of supplies and provisions. After the provisions were depleted and Kellogg thought that the agreement had ended, the Bunker Hill was located. Cooper and Peck, however, won the case in court.

Howard Balsley, a pioneer in the uranium fields, where his experience goes back to the days of Madame Curie's interest in Colorado's sources of radium, told in 1969 how he had grubstaked a man on the strength of a dream. Charles Snell, he related, "came to me and stated that he had just had a dream, in which he had plainly seen a yellow circle in a block of sandstone up in the Upper Cave Springs Wash area, and he was sure that it was uranium and that he could find it." Balsley did not share much faith in dreams, but Snell was known to be honest and truthful, so he was grubstaked. "After perhaps 10 days," Balsley continued, "this prospector came riding in and stated that he actually had found the yellow circle in the block of sandstone." The five claims staked here brought in more than $1 million.

The above experience indicates that a person of proved integrity and preferably some ability can often interest a business or professional man or group in backing him as a prospector. Some small syndicates also do grubstaking, this having become again a more widespread procedure during the uranium rush of the 1950's. Prospectors' associations exist in several countries, furnishing both aid and advice. Again in 1970, for the first time in years, an advertisement that solicited grubstake funds appeared in a mining periodical. In 1971, the Rio Algom–Rio Tinto organization advertised its interest in new mining prospects in Canada, the United States, "or anywhere." "Hustle your burro over to Rio Tinto," the reader was advised. The (Alaska) State Division of Mines and Minerals (in Juneau) is authorized to reimburse prospectors for certain expenses when a venture seems worthy of help, in return for certain information, which is kept confidential for a specified period of time. Other information on government aid in Alaska can be obtained from the Assistant Director for Alaska, U.S. Bureau of Mines, Juneau 99801.

The following organizations are informal groups of prospectors; both include subscriptions to their publications with memberships:

Prospectors Club International
P.O. Box 548
Midland, Texas 79701

United Prospectors Inc.
5665 Park Crest Drive
San Jose, California 95118

In Canada, grubstaking schemes are in effect in the provinces of British Columbia and Saskatchewan. Grubstaking in the territories is provided under certain circumstances by the Department of Indian Affairs and Northern Development.

Canadian prospecting syndicates are not uncommon. Some of the provinces provide for limited-liability companies that are designed to protect such syndicates from excessive lawsuits that would injure the participants. The amount of money that is needed by the syndicate is typically divided into units of a certain value, perhaps $100 each. Each member subscribes for a number of units, and each prospector is allotted a percentage. After a discovery is made, a company may be formed, with shares substituting for the units; or the claims may be sold, and the money divided among the owners of the units. Some individuals and groups may be on the quiet lookout for a qualified prospector.

Employment by an established mining company is a different matter. In Canada, jobs for prospectors are occasionally listed with the Canada Manpower Centres. Such a listing almost never appears in the United States. The way to secure such employment is similar to that for any other kind of job, with special attention given to the mining periodicals that are listed on pages 405–406 of this book. A combination of employment and open prospecting has been used in Canada under certain circumstances. Organized prospecting teams, assembled in various ways, are increasing in popularity in Canada.

In Australia, a shortage of prospectors existed in 1971. Some mining companies are willing to train young men as geologic assistants.

COURSES

Prospecting courses are given from time to time by vocational schools, YMCA's, and other institutions in the western United States; a majority of the courses in recent years have been either sponsored or at least encouraged by rockhound clubs, whose members constitute most of those taking the instruction. Other courses of the same kind are given on club premises; announcements are likely to appear in the hobby magazines listed on pages 233–234 of this book. Some of the state schools of mines have given prospecting courses of two to five weeks' duration. Several of the uranium-prospecting courses given during the 1950's, including one given (with several repetitions) by the author at Phoenix College, in Arizona, had huge attendances. A new course was begun in 1971 by Richard K. Rinkenberger for the University of Colorado Division of Continuing Education, in Denver.

Apart from taking a specifically oriented prospecting course, the person contemplating this activity can well afford to learn more about mineralogy and geology, as further emphasized in later chapters. Day or evening courses in these interesting subjects are available in nearly every corner of the United States, Canada, and numerous other countries.

The New Mexico Institute of Mining and Technology, in cooperation with New Mexico Technical Services, offered in 1969 a short course in their School of Advanced Mineral Exploration. From time to time, the Colorado School of Mines and other institutions provide group instruction on this level. Such advanced courses could serve only experienced mining or petroleum engineers, especially the latter, who are interested in applying computer techniques to the search for mineral and oil deposits.

Prospecting courses, usually advertised in the mining or rockhound periodicals, are occasionally offered by mail. The University of Alaska Division of Statewide Services conducts a mining extension course, given without charge throughout Alaska; one-half of the entire course, incidentally, is devoted to identifying minerals.

Correspondence courses relating to prospecting, geology, and metal mining are given in Canada by the Welfare Services Branch of the Department of Veterans Affairs, Ottawa. The extent of the coverage can be seen from the titles of the booklets: *Geology and Mineralogy, Prospecting, Practical Mining, Business of Prospecting and Mining.* These were prepared by the Geological Survey of Canada for the Canadian Legion Educational Services. The course is open to Canadian military veterans, federal civil servants, and penitentiary inmates.

The Secondary School Correspondence Division of the British Columbia Department of Education, 546 Michigan Street, Victoria, B.C., offers a 20-lesson correspondence course in geology and prospecting for adults and other students who have Grade X standing. Residents of other provinces can usually take it also.

The appreciation that Canadians have for the mineral industries and what they can contribute to the growth of the country is shown by the extent to which good prospecting courses are offered to the public. The faith behind them—and it is well justified—is that education pays, and that even the amateur or hobbyist who intends to go prospecting should be willing to devote as much attention and practice to it as he would if making an intelligent approach to any other occupation, sport, or hobby.

Annual classes for prospectors are given in several of the provinces as well as in both territories. These are usually free and are held evenings (or afternoons and evenings) for a few weeks during winter, either in a convenient city or at several towns. The students are of all ages, and the "graduates" have over the years found hundreds of millions of dollars in mining claims.

In Alberta, prospecting classes are held jointly by the University of Alberta, the Alberta and Northwest Chamber of Mines and Resources, and the Edmonton Branch of the Canadian Institute of Mining and Metallurgy. The course is given in Edmonton in the winter. The address is 10060 100th Street, Edmonton.

In British Columbia, the British Columbia and Yukon Chamber of Mines combines with the Adult Education Department of the Vancouver School Board, the University of British Columbia, and the British Columbia Department of Mines and Petroleum Resources, offering courses in Vancouver. Even inmates at the British Columbia penitentiary have taken a prospecting course that was given by Robert M. Thompson.

In Manitoba, courses are given by the Evening Institute of the University of Manitoba, in Winnipeg.

In Nova Scotia, two cities have courses that are sponsored by the Mining Society of Nova Scotia and the Nova Scotia Department of Mines.

In Ontario, the Prospectors and Developers Association collaborates with the Ontario Department of Mines, as well as offering special instruction during some of its annual meetings. The Ontario Department of Mines also gives its own classes (including a field course) and unites in others with the federal Department of Indian Affairs and Northern Development.

In Quebec, the Quebec Department of Natural Resources holds classes in French or English. Longer courses have been given at Laval University, in Quebec, and at the École Polytechnique of the University of Montreal (in both languages).

In Saskatchewan, the Department of Mineral Resources has had brief courses, including one at Prince Albert Penitentiary.

In Yukon Territory, instruction has been offered by the Yukon Chamber of Mines.

OFFICE OF MINERALS EXPLORATION

The only direct financial aid available to prospectors in the United States, other than private arrangements discussed above, is through the program of the Office of Minerals Exploration (O.M.E.), an agency of the U.S. Geological Survey in the Department of the Interior.

To encourage exploration for certain mineral resources (excluding organic fuels) in the United States and its territories and possessions, money is available to private persons and companies on a participating basis. This is not for grubstaking or to finance prospecting expeditions, but to support investigations on already established claims, leases, or property that is owned by the applicant. If production results, the government is repaid at the rate of 5 percent royalty. If there is no production, no repayment is required. Therefore, this arrangement is much better than a loan, except that the applicant must begin the work at his own expense (until he submits a report and gets his monthly check), and he has to contribute a percentage of the money needed (25 or 50 percent, according to which mineral commodity is being sought).

The purpose of the law is to help those who are not likely to be able to do the work at their own expense and who are unable to secure commercial financing on reasonable terms. Either surface or underground exploration is acceptable, using any standard methods, which may include geophysical and geochemical techniques (see chapters 10 and 11). Exploration (including development) rather than mining is the goal.

Title 30, Chapter II, Part 229 of this program is outlined below. Statements are followed by questions most often asked by applicants and by the informal answers to them. This information is taken, with minor modifications that have the new prospector in mind, from official literature of the O.M.E.

229.1 *Purpose: stated above*

229.2 *Definitions: explained here where necessary*

229.3 *Eligible minerals or mineral products*

The following are eligible for government financial assistance of 50 percent of the allowable costs of exploration:

Asbestos
Bauxite
Beryllium
Cadmium
Chromite
Cobalt
Columbium
Copper
Corundum
Diamond (industrial)
Fluorspar
Graphite (crucible flake)
Iron ore
Kyanite (strategic)
Manganese
Mica (strategic)
Molybdenum
Monazite
Nickel
Quartz crystal (piezoelectric)

Rare earths
Selenium
Sulfur
Talc (block steatite)
Tellurium
Thorium
Uranium

The following are eligible for government financial assistance of 75 percent of the allowable costs of exploration:

Antimony
Bismuth
Gold
Mercury
Platinum group metals
Rutile
Silver
Tantalum
Tin

Combinations of the minerals or mineral products that are named above may be eligible for government financial assistance of 62.5 percent of the allowable costs of exploration.

Q. How can I become eligible for exploration assistance?

A. To be eligible for exploration assistance, you must

(1) Have a sufficient interest in property that can qualify for exploration for one or more of the eligible minerals or mineral products;

(2) Furnish evidence that funds for the exploration are not available from banking institutions or other commercial sources of credit on reasonable terms; and

(3) Certify that you would not ordinarily undertake the proposed exploration under current conditions and circumstances at your sole expense.

229.4 *Operator's property rights*

The operator must have and preserve the right to possession of the land (as owner, lessee, or otherwise) for a term that is at least sufficient to complete the exploration work. The operator shall devote the land and all existing improvements, facilities, buildings, installations, and appurtenances that are necessary to the purposes of the application.

Q. What is meant by a "sufficient interest"?

A. A "sufficient interest" gives you the right of possession of a property for the length of time that is required to complete the exploration and to protect the government's interest thereafter. Obviously, unencumbered ownership is sufficient. A leasehold, preferably with renewal rights, or a located claim may also be sufficient.

Q. If I am leasing the property, what must I obtain from my landlord?

A. You should obtain a lien agreement from your landlord, using Lien and Subordination Agreements (MME Form 52) supplied by the O.M.E. If the agreement cannot be obtained, you should provide O.M.E. with a copy of the letter of refusal. The O.M.E. may then accept, in lieu of the agreement, a performance bond (using Standard Form 25 supplied by O.M.E.) that is executed by an approved corporate surety or two responsible individual sureties.

Q. If the property is subject to a mortgage, lien, or other encumbrance, what document should I provide to the government?

A. You must obtain a subordination agreement from the mortgagee (using

MME Form 52) or provide a performance bond before an exploration contract can be executed.

Q. Why must I deal with my landlord if I am leasing or with my mortgagee if my interest is subject to a mortgage?

A. The lien agreement or subordination agreement, or performance bond in lieu thereof, are required to secure the payment of royalty to the government on any production during the exploration work or after certification if a certificate of possible production is issued.

Q. Would I be eligible under this program if I have no interest in a property and wish to purchase or lease the property only if I can get aid from the government?

A. No, but you could become eligible by entering into an agreement to purchase or lease the property subject to obtaining an exploration contract.

229.5 *Form and filing*

An application for federal financial assistance must be submitted in duplicate on forms which may be obtained from and filed with either of the following: Office of Minerals Exploration, Geological Survey, Department of the Interior, Washington, D.C. 20242, or Field Officers, Office of Minerals Exploration, Geological Survey.

Q. How may I obtain financial assistance for exploration?

A. You file an application for assistance on MME Form 40 in duplicate (in triplicate for Alaska), answering all questions to the best of your ability. Application forms may be obtained by writing to the Washington office or to the nearest O.M.E. field office.

229.6 *Information required*

(a) Each application shall fully describe the proposed exploration, and shall include all detailed data called for by the application form. The Secretary (of the Interior) may require the filing of additional information, including financial statements, reports, maps or charts, and exhibits and such physical on-site examinations as he deems necessary.

(b) The application must include evidence that funds for the exploration work are unavailable on reasonable terms from commercial sources. The evidence shall include information as to the commercial sources to which applications were made, the amounts requested, and the reasons why loans were not obtained.

(c) The application must include a certification by the applicant that he would not normally undertake the exploration at his sole expense under current conditions or circumstances.

Q. What information is required in my application?

A. Your application must include the following information:

(1) Evidence of your financial eligibility
(2) A legal description of the property to be explored
(3) A statement of your interest in the property
(4) A description of the pertinent geology of the property
(5) An explanation of your reasons for expecting to find ore

(6) An explicit statement of the proposed exploration work
(7) A listing of cost estimates for the labor, materials, and equipment that will be required for the work
(8) A detailed estimate of the cost of each type of work
(9) Maps showing land boundaries, existing workings, and proposed work

Q. What evidence must I give that funds are not available on reasonable terms from commercial sources?

A. You must furnish evidence of your efforts to obtain credit from two banking institutions or other commercial sources of credit. This evidence must include true copies of your letters to the banks showing the amount and terms requested and proposed use of loan funds and true copies of the replies showing why a loan was not granted.

Q. Will any other evidence of financial eligibility be required?

A. You must give assurance that your share of the exploration costs can be furnished. You may be required also to furnish additional information including a financial statement.

Q. Must I be an experienced operator or miner?

A. Not necessarily, but you will be responsible for securing experienced and competent people to supervise the exploration if an exploration contract is executed.

Q. How can I obtain help to gather this information and prepare the application?

A. You may consult with the O.M.E. field officer in your region, or the Washington office, on any question that may arise. If you are not experienced in exploration or mining, you may need a consulting mining engineer or geologist to help you with some of the information. The application form has full instructions. The general instructions on the front of the form tell you what to do, and each item of information is described in detail on the back of the form.

Q. May the work include blocking out ore for production?

A. No. The regulations state: "The work shall not go beyond a reasonable delineation and sampling of a mineral deposit, and shall not be conducted primarily for mining or preparation for mining."

229.7 *Criteria*

Q. What action does O.M.E. take upon my application?

A. The application is reviewed to determine if the information requested on the application form has been submitted. If not, the applicant may be asked to furnish the necessary information. When the application is complete, and if the facts warrant, a field examination of the property may be made by the government, usually with the applicant, before a final decision regarding the application is reached. A field examination is not made by the government at the request of the applicant.

The following factors will be considered and weighed in passing upon applications:

(1) The geologic probability of a significant discovery being made
(2) The estimated cost of the exploration in relation to the size and grade of the potential deposit
(3) The plan and method of conducting the exploration

(4) The accessibility of the project area
(5) The background and operating experience of the applicant
(6) The applicant's title or right to possession of the property

229.8 *Approval*

If the application is approved, the government may enter into an exploration contract with the applicant upon terms and conditions which the Secretary (of the Interior) deems necessary and appropriate as set forth in the contract form furnished by the government.

229.9 *Government participation*

Q. What are the principal elements of an exploration contract?

A. The contract includes the following:
(1) A description of the land upon which the work will be performed
(2) A detailed statement of the work to be performed
(3) A time limit within which the work must be performed (generally not more than 2 years)
(4) An estimate of costs
(5) Provisions for the government's contribution (not more than the percent specified or $250,000) to costs and for repayment with interest to the government in the event of production
The contract also includes the standard provision relating to nondiscrimination, settlement of disputes, 8-hour law, and rebate of wages.

Q. Is more than one form of contract used?

A. Yes. Two contract forms are presently used. Under the short form (MME Form 51), the government's contribution is a percentage of unit costs of the work, agreed upon in advance and fixed by the contract. Under the long form (MME Form 50), the government's contribution is a percentage of the allowable actual costs of the work as incurred or a combination of actual and fixed unit costs.

Q. Explain the fixed unit form of contract. What is meant by "fixed unit costs"?

A. The fixed unit costs of the work (per foot of drilling, per foot of drifting, per cubic yard of materials moved, etc.) are those agreed upon and stated in the contract. The government's contribution is a percentage of these fixed unit costs of work performed, verified by the representatives of the O.M.E. as having been completed and found acceptable under the specifications of the O.M.E. contract. The government's contribution will be based on the amount stated in the contract, whether the actual cost is greater or less.

Q. Am I required to complete all units of work specified by a fixed unit cost contract, regardless of the actual costs per unit?

A. Yes, because that is what you agreed to do under that form of contract. However, under certain circumstances, the number of units may be reduced by mutual agreement.

229.10 *Allowable costs*

Q. What is meant by "allowable costs" in the actual cost contract?

A. Allowable costs to which the government will contribute are the direct costs of the work, such as labor, supervision, materials, supplies, and equipment that are specified in the contract. These costs must be supported by documentary evidence available for audit. Indirect costs, such as general overhead and corporate management expense, are not allowable costs.

Q. Is the cost of work performed before the date of the contract with the government allowable?

A. No.

Q. If the work that is provided for in an actual cost contract cannot be completed within the estimated total cost, what are my obligations?

A. Your obligations under the contract vary to suit the circumstances of your project. Usually, your share of the estimated total allowable cost stated in the contract will be the limit of your obligation; but the contract may require you to complete certain items of the work at your own expense above the maximums fixed for those items by the contract.

Q. Who determines whether a fixed unit cost or actual cost contract is to be used and what factors are considered?

A. The government decides which would be more appropriate in each particular case, depending on estimated costs, size and type of operation, etc. A unit cost contract is a much simpler document than an actual cost contract. It is easier to administer and is generally preferable, both to the applicant and the government, when the unit costs can be reasonably estimated in advance.

Q. In what other way is a fixed unit cost contract simpler than an actual cost contract?

A. Under the fixed unit cost contract, the operator is not required by the government to maintain itemized cost records or to submit documentary invoices of the actual costs that are incurred.

Q. May rehabilitation of old workings be allowed as a part of the cost of exploration?

A. Yes, in certain cases, limited rehabilitation may be provided by the contract when considered necessary to conduct the exploration work.

Q. May permanent installations and improvements or their rehabilitation be allowed as a part of the cost of the exploration work?

A. Only to the extent necessary to conduct the exploration work provided for in the contract.

Q. May the cost of operating equipment be included?

A. Yes, in one of three ways:

(1) Depreciation is allowed on equipment owned by the operator to the extent of one-sixtieth of its fair market value or of its book value, whichever is less, for each month used on the project.

(2) Rental of equipment not owned by the operator is allowed at reasonable rates.

(3) Purchase of equipment is allowed, in which case the government has an equity in the salvage value.

Q. What becomes of equipment that is purchased or installations that are made to which the government has contributed?

A. The government owns them jointly with the operator in proportion to the amounts contributed, although title will be in the name of the operator. The government's interest in any salable or salvageable property must be liquidated and accounted for when the government's obligation to contribute to costs terminates or when the property is no longer needed for the work.

Q. May I retain such equipment and installations for my own use?

A. Yes, but at a price at least as high as could be obtained by sale to others.

Q. Will I be given any allowance for the use or acquisition of the land and any existing building or installations thereon?

A. No. The land and all fixtures are devoted to the project without allowance.

Q. Will the cost of necessary rehabilitation or repair of existing buildings or installations be allowed?

A. The contract may allow these costs for work performed after the contract is executed if such rehabilitation or repair is necessary for the conduct of the exploration work.

Q. If the operator is an individual, a partnership, or a corporation, may the services of the individual, the partners, or a corporate officer be charged as allowable costs of the work?

A. Yes, to the extent that the time and services of each are devoted exclusively and directly to the performance of the work, that they are qualified to perform the services, and that the services are provided for in the contract.

Q. Who is to perform the exploration work?

A. You or your employees may perform the work, or the exploration contract may provide for an independent contractor to perform all or any part of the work.

Q. Will the government give me any technical aid in the conduct of the exploration?

A. Yes. Representatives of the government inspect the work from time to time and will be available to advise on its performance. However, the government assumes no responsibility for the performance of the work, and unless you are qualified to direct the work yourself, you will be required to employ competent technical help.

Q. If it is found after some of the work described in the contract has been done that the work should proceed in another direction or be replaced by another type of work, can these changes be authorized?

A. Yes. Both the government and the operator benefit when the work is performed in the most effective manner, and the contract may be amended accordingly.

Q. If it is found while the work is in progress or upon its completion that additional work is needed to accomplish the original purpose of the contract, can the additional work be authorized?

A. In a suitable case, the contract may be amended to provide for added work and a related increase in the estimate of the total allowable cost. However, the government's contribution cannot exceed $250,000.

Q. If, after some of the work has been done, I decide that there is no advantage in completing the work, can the contract be terminated short of completion?

A. Yes, if the government agrees with you, it will terminate the work by agreement. If it does not agree, you may be required to complete the work. Should you discontinue the project or abandon it altogether without the government's agreement, you would obviously be subject to a claim for damages or other remedy that the law may provide for breach of contract and a demand for refund of the government's contribution.

Q. Who is responsible for damages to persons or property arising out of negligence in the course of the work?

A. You are. The work is entirely under your direction and control.

Q. What reports will I have to make as the work progresses?

A. You are required to submit a two-part monthly report to the government. The first part is a monthly voucher claiming costs for contribution by the government. In the case of actual cost contracts, costs must be supported by certified true copies of invoices, transcripts of payrolls, etc. The second part is a concise narrative description of the work performed, results obtained, and any unusual situations encountered. It may be illustrated and supported by engineering-geological maps or sketches, drill hole logs, assay reports, etc. The voucher form (Form 9-1648) is furnished by the government.

Q. Is any final report required?

A. Yes. Upon the completion or termination of the work, you must furnish the government with an acceptable final report, giving a summary of the work performed under the contract and costs thereof together with pertinent geologic and engineering data and an estimate of any ore reserves resulting from the exploration work.

Q. What records will I have to keep?

A. For either a fixed unit cost contract or an actual cost contract, you should maintain records in accordance with the items to be reported on a voucher for reimbursement. For a fixed unit cost contract, itemized cost records are not required by the government. For an actual cost contract, a simple distribution sheet is usually sufficient for small projects; large projects require more formal cost accounts. For either a small or large project under an actual cost contract, documentary support in the form of invoices, payrolls, tax returns, etc., must be available for audit.

Q. May the government inspect and audit my records and accounts?

A. Yes, either by its own auditors or by a certified public accountant. In virtually all cases, an appointment is arranged at a time convenient to the operator.

Q. How do I obtain the government's contribution?

A. At the end of each month, you submit a voucher. If the voucher is prepared properly, payment usually is made within 2 or 3 weeks after the voucher is submitted.

229.11 *Repayment by the operator*

Q. Must I repay the government the money that it contributed?

A. You are required to pay royalty to the government on any production from the land that is covered by the contract from the date of the contract until the government notifies you, not later than 6 months after receipt of a sufficient final report, of certification or of its intent not to certify. If the government considers that the exploration indicates that production may be possible from the area covered by the contract, it shall so certify. Thereafter, you are required to pay royalty on any subsequent production for the period specified in the contract. The royalty is 5 percent of the "gross proceeds" or "value" of production in the form in which it is sold, held, or used.

Q. If a certification of possible production is issued, how long do I pay royalty?

A. You pay royalty on production for a period of not less than 10 years and not more than 25 years from the date of the contract or until the government's contribution is repaid with interest, whichever occurs first.

Q. Will I be required to pay royalty on production of minerals not eligible for exploration assistance?

A. Yes. All minerals or metals mined or produced are subject to the repayment provisions of the contract.

Q. What security other than my legal obligation to pay does the government have for payment of this royalty?

A. The contract gives the government a lien on your interest in the land and on any production from it. The lien or subordination agreement gives the government additional security.

Q. If the government does not issue a certification of possible production, will the property be free from the government's claim for royalty and the liens to secure its payment?

A. Yes, except with respect to any production that may occur during the time in which the government has a right to certify, but not after the government has given notification of its intent not to certify.

Q. If the government certifies that production may be possible, am I required to start producing?

A. No. The exploration contract does not require you to produce.

Q. If I am a lessee of the property and my leasehold is surrendered to the owners, does this action destroy the government's liens?

A. No. If you are a lessee, the lien agreement that you secured from the owner establishes a lien on the land and any production from it. You also remain liable as surety for royalty on any production.

Q. If my interest in the property is subject to a mortgage and the mortgage is foreclosed, does this foreclosure destroy the government's liens?

A. No. The liens are protected by the subordination agreement you secured from the mortgagee. In this case, your liability as surety also remains in effect.

Q. Who pays the royalties to the government?

A. If you sell your production, the purchaser should be notified of the government's royalty interest and directed to make the payment to the government, supporting the payment with copies of the settlement sheet. If the purchaser does not pay the royalty, you are liable for payment and must supply copies of settlement sheets or sales invoices.

Q. How is the amount of the royalty verified?

A. By audit of production records, sales amounts, and settlement sheets.

229.12 *Interest on amount of government participation*

Q. Is interest charged on the government's contribution?

A. Simple interest is calculated from the first day of the month following the dates of the government's contribution. Interest continues until the period for royalty payment expires as specified in the contract or until the amount of the government's contribution is fully repaid, whichever occurs first.

Q. What is the interest rate?

A. The rate is stated in the contract. It is no less than the rate the Department of the Interior would be required to pay if it borrowed from the Treasury, plus a 2 percent interest charge in lieu of the actual cost to the government of administering the contract.

Q. How is the interest paid?

A. Interest is paid from the royalty on production. You are not obligated to pay royalty unless you produce, and the interest does not increase the royalty rate.

Q. If I surrender, transfer, or convey any of my rights in the land after certification, who becomes liable for the royalty?

A. You remain liable for the royalty on all production until the royalty payment expires or the government's contribution is fully repaid with interest.

Q. What federal laws and regulations must I comply with if a contract is granted?

A. You must comply with the following: Nondiscrimination in Employment, Section 202, Executive Order 11246, September 24, 1965; Work Hours Act of 1962—Overtime Compensation; and Copeland (Anti-Kickback) Act—Nonrebate of Wages.

229.13 *Limitation on the amount of government participation*

229.14 *Government not obligated to buy*

229.15 *Title to and disposition of property*

OFFICE OF MINERALS EXPLORATION FIELD OFFICES

Region I (Includes Idaho, Montana, Oregon, and Washington)	Room 656 West 920 Riverside Ave. Spokane, Washington 99201
Region II (Includes Alaska, California, Nevada, and Hawaii)	Building 2 345 Middlefield Road Menlo Park, California 94025
Region III (Includes Arizona, Colorado, Kansas, Nebraska, New Mexico, North Dakota, Oklahoma, South Dakota, Texas, Utah, and Wyoming)	Room 203, Building 53 Denver Federal Center Denver, Colorado 80225
Region IV (Includes all other states)	Room 11, Post Office Building Knoxville, Tennessee 37902

For further information, write to the Office of Minerals Exploration, U.S. Geological Survey, Washington, D.C. 20242.

The British Columbia Department of Mines has provided financial assistance to worthy prospectors since 1943 and offers some supervision. The Saskatchewan Department of Resources and Industrial Development operates a prospectors' assistance plan, which includes technical advice, transportation, loan of equipment, and a program for training Indians in prospecting. Literature is available from these government agencies.

All the states of the United States and the provinces of Canada maintain one or more agencies dealing with matters of concern to the prospector and miner. They are prepared to furnish information, provide identification of specimens, give advice, and perform other services within reason. The names of these agencies do not always indicate clearly the extent of their activities, which often (as one should suspect) overlap. Nevertheless, they may consider themselves restricted in certain directions, and it is wise to write to only one agency in a single letter and to address it carefully to the agency or its director, not to individuals who may be in the field or no longer employed there.

Keep in mind that these agencies try not to compete with private geologists, engineers, or assayers but will refer you to them when possible. Nevertheless, the extent of the services offered varies from place to place and from time to time; you can properly expect help only within your state or province of residence or the one in which the property is situated. It is better to send an inquiry first than to lose your samples.

The following directory of geologic and mining agencies in the United States and Canada gives addresses to which correspondence and samples can be sent and from which information, literature, and maps can be obtained.

STATE GEOLOGIC AGENCIES

Alabama
Geological Survey of Alabama
P.O. Drawer "O"
University, Alabama 35486

Alaska
Division of Mines & Geology
Box 5-300
College, Alaska 99701

Arizona
Arizona Bureau of Mines
University of Arizona
Tucson, Arizona 85721

Arkansas
Arkansas Geological Commission
State Capitol
Little Rock, Arkansas 72201

California
Division of Mines & Geology
California Dept. of Conservation
Resources Building
1416 9th Street
Sacramento, California 95814

Colorado
Colorado Geological Survey

254 Columbine Building
1845 Sherman Street
Denver, Colorado 80203

Connecticut
Connecticut Geological & Natural History Survey
Box 128, Wesleyan Station
Middletown, Connecticut 06457

Delaware
Delaware Geological Survey
University of Delaware
Newark, Delaware 19711

Florida
Department of Natural Resources
Bureau of Geology
P.O. Drawer 631
Tallahassee, Florida 32302

Georgia
Department of Mines, Mining and Geology
19 Hunter Street, S.W.
Atlanta, Georgia 30334

Hawaii
Division of Water & Land Development
Department of Land & Natural Resources
P.O. Box 373
Honolulu, Hawaii 96809

Idaho
Idaho Bureau of Mines & Geology
Moscow, Idaho 83843

Illinois
Illinois State Geological Survey
121 Natural Resources Building
Urbana, Illinois 61801

Indiana
Department of Natural Resources
Geological Survey
611 North Walnut Grove
Bloomington, Indiana 47401

Iowa
Iowa Geological Survey
Geological Survey Building
16 West Jefferson Street
Iowa City, Iowa 52240

Kansas
State Geological Survey of Kansas
The University of Kansas
Lawrence, Kansas 66044

Kentucky
Kentucky Geological Survey
University of Kentucky
307 Mineral Industries Building
Lexington, Kentucky 40506

Louisiana
Louisiana Geological Survey
Box G
University Station
Baton Rouge, Louisiana 70803

Maine
Maine Geological Survey
Department of Economic Development
Room 211, State Office Building
Augusta, Maine 04330

Maryland
Maryland Geological Survey
214 Latrobe Hall
Johns Hopkins University
Baltimore, Maryland 21218

Massachusetts
Massachusetts Department of Public Works
100 Nashua Street
Boston, Massachusetts 02114

Michigan
Michigan Department of Natural Resources
Geological Survey Division
Stevens T. Mason Building
Lansing, Michigan 48926

Minnesota
Minnesota Geological Survey
University of Minnesota
Minneapolis, Minnesota 55455

Mississippi
Mississippi Geological, Economic & Topographical Survey
2525 North West Street
P.O. Box 4915
Jackson, Mississippi 39216

Missouri
Missouri Division of Geological Survey & Water Resources
P.O. Box 250
Rolla, Missouri 65401

Montana
Montana Bureau of Mines and Geology
Montana College of Mineral Science & Technology
Room 203-A, Main Hall
Butte, Montana 59701

Nebraska
Conservation and Survey Division
University of Nebraska
113 Nebraska Hall
Lincoln, Nebraska 68508

Nevada
Nevada Bureau of Mines
University of Nevada
Reno, Nevada 89507

New Hampshire
Department of Resources & Economic Development
James Hall
University of New Hampshire
Durham, New Hampshire 03824

New Jersey
New Jersey Bureau of Geology & Topography
John Fitch Plaza, Room 709
P.O. Box 1889
Trenton, New Jersey 08625

New Mexico
New Mexico State Bureau of Mines & Mineral Resources
Campus Station
Socorro, New Mexico 87801

New York
Geological Survey
New York State Museum & Science Service
Room 973
New York State Education Building Annex
Albany, New York 12224

North Carolina
Division of Mineral Resources
Department of Conservation & Development
P.O. Box 2719
Raleigh, North Carolina 27602

North Dakota
North Dakota Geological Survey
University Station
Grand Forks, North Dakota 58201

Ohio
Department of Natural Resources
Division of Geological Survey
1207 Grandview Avenue
Columbus, Ohio 43212

Oklahoma
Oklahoma Geological Survey
The University of Oklahoma
830 South Oval, Room 163
Norman, Oklahoma 73069

Oregon
State Department of Geology & Mineral Industries
1069 State Office Building
1400 S.W. Fifth Avenue
Portland, Oregon 97201

Pennsylvania
Bureau of Topographic & Geologic Survey
Pennsylvania State Planning Board
Harrisburg, Pennsylvania

Puerto Rico
Economic Development Administration of Puerto Rico
Program of Geology & Mineral Resources

P.O. Box 38
Roosevelt, Puerto Rico 00929

Rhode Island
Planning Division
Rhode Island Development Council
Roger Williams Building
Providence, Rhode Island 02908

South Carolina
Division of Geology
P.O. Box 927
Columbia, South Carolina 29202

South Dakota
South Dakota State Geological Survey
Science Center
University of South Dakota
Vermillion, South Dakota 57069

Tennessee
Department of Conservation
Division of Geology
G-5 State Office Building
Nashville, Tennessee 37219

Texas
Bureau of Economic Geology
The University of Texas
University Station, Box X
Austin, Texas 78712

Utah
Utah Geological & Mineralogical Survey
103 Utah Geological Survey Building
University of Utah
Salt Lake City, Utah 84112

Vermont
Vermont Geological Survey
Geology Building
University of Vermont
Burlington, Vermont 05401

Virginia
Virginia Division of Mineral Resources
P.O. Box 3667
Charlottesville, Virginia 22903

Washington
Washington Division of Mines & Geology
Department of Natural Resources
P.O. Box 168
Olympia, Washington 98501

West Virginia
West Virginia Geological & Economic Survey
P.O. Box 879
Morgantown, West Virginia 26505

Wisconsin
Geological & Natural History Survey
University of Wisconsin
1815 University Avenue
Madison, Wisconsin 53706

Wyoming
Geological Survey of Wyoming
Box 3008
University Station
University of Wyoming
Laramie, Wyoming 82070

PROVINCIAL DEPARTMENTS OF MINES

Alberta
Department of Mines and Minerals
Edmonton, Alberta

British Columbia
Department of Mines and Petroleum Resources
Victoria, British Columbia

Manitoba
Mines Branch
Department of Mines and Natural Resources
Winnipeg, Manitoba

New Brunswick
Department of Natural Resources
Fredericton, New Brunswick

Newfoundland
Mineral Resources Division
Department of Mines, Agriculture and Resources
St. John's, Newfoundland

Northwest Territories, Yukon
Northern Economic Development Branch
Department of Indian Affairs and Northern Development
Ottawa

Resident Geologist
Federal Building
Whitehorse, Y.T., or Yellowknife, N.W.T.

Nova Scotia
Department of Mines
Halifax, Nova Scotia

Ontario
Department of Mines
Toronto, Ontario

Prince Edward Island
Deputy Provincial Secretary
Provincial Government Offices
Charlottetown, Prince Edward Island

Quebec
Mineral Resources Branch
Department of Natural Resources
Quebec, Quebec

Saskatchewan
Department of Mineral Resources
Regina, Saskatchewan

3

Transportation

The kind of transportation that is used for prospecting depends on the nature of the country, the distance to be covered, and the time and money that are available. Foot travel with a back pack has by no means been abandoned, and the airplane has come into a more general use than might be supposed. In between these extremes is a wide range of transportation, some combination of which will probably be used by most of today's prospectors, who prefer to walk little and cannot afford an airplane. Considering the rapid changes in roads that have occurred in many places, it is well to make inquiry as close to the area to be explored as possible. It can generally be assumed that the more accessible an area, the more likely it is to have been prospected somewhat thoroughly. The procedure, therefore, is usually to use highway, railroad, boat, or air transportation and then go by one of the special methods that are described below.

In Alaska, Canada, and other northern places, conditions may be so different between winter and summer that transportation must be planned with the weather in mind. During the in-between seasons called "break-up" and "freeze-up," usually 4 to 6 weeks in duration, travel may be entirely impossible, and the prospector who is already in the field remains in camp or nearby—hence going and coming by winter travel in order to get maximum time for prospecting. The best prospecting season in Canada usually begins in May (near the border) to July (near the Arctic Circle) and ends in October. Alaskan weather is similar. It is generally futile to prospect, except with geophysical instruments, in the western mountains of North America when snow is on the ground. Only summer prospecting is considered in this book; "anyone engaging in the special kinds of prospecting that can be done in winter will probably be familiar with the methods of travel or know where to obtain advice," says A. H. Lang. Let us hope so. The climate in other parts of the world should be investigated before starting out; the American Geographical Society formerly published a series of small books dealing with the practical problems of exploring in various remote parts of the globe, but this was prior to the postwar revolution in transportation and other equipment.

All the exploratory work in the Rocky Mountains, it was stated in 1938 by George O. Argall, Jr., was done with pack animals: mules, burros, horses. To a large extent, this was true also of other parts of the world where prospecting has been done. Camels have been used mostly in Australia, Central Asia, and Siberia. Llamas are employed in South America, especially in the Andes.

The account of the use of pack animals in the previous editions of this book appears now peculiarly out of date. It does not seem necessary, therefore, to retain Von Bernewitz's instructions on packing, loading, and caring for animals. Instructions on saddling and packing horses are

Pack trains are still used in remote parts of British Columbia. *(Socony Mobil Oil Company)*

given in most books on camping. The burro no doubt still has, and always will have, "many traits that try his owner's patience." Yet, an occasional mule, burro, or horse was used during the uranium rush in the West in the 1950's; burros were even sold through the Sears, Roebuck catalog, though probably mostly as pets. The following quotation from an article on life in Aspen, Colorado, in the 1880's, written for *The Colorado Magazine* by the author's father-in-law, William W. Wardell, may be of interest:

> Burros were plentiful. When not in use packing goods and provisions up to the mines and bringing down sacks of ore, they were turned loose on the city streets and allowed to forage for themselves. Consequently, the women who did not have fences around their houses were likely to lose their laundry from the line, as the burros were exceptionally fond of eating such rare delicacies as negligees. We children always had a burro to ride to school if we could catch one. It was not thought unusual to catch a burro and ride him for his feed. When a burro puncher was ready for his animals, he simply rode around the town and gathered up those which had his brand. No one ever thought of stealing a burro, and the puncher was glad to have someone feed the animal, for it saved him the expense.

The use of a good mule is still recommended for southeastern Alaska, where few pack horses are available.

The Geological Survey of Canada has published instructions for handling pack horses where they are still presumed to be the most suitable means of transportation in the western mountains. The author has, fairly recently, seen camel caravans being loaded for long journeys across the Arabian Peninsula, not for prospecting, of course, but for carrying some mineral products, certainly salt. Until only recently replaced generally (but not entirely) by snowmobiles, sled dogs were still used to carry packs in the arctic regions, especially where trees are too dense for pack horses or where water routes do not exist.

What, then, about the canoe as a means of transportation? Aren't they nostalgic, as old-fashioned as a Lewis and Clark flintlock? By no means! Canoes are widely used today in Ontario, Quebec, Manitoba, and other parts of Canada where ground access is not otherwise possible. The truly extensive network of lakes and streams in the Canadian Shield of eastern Canada, with their long, sluggish stretches, is admirably suited to canoe travel. The rivers of interior Alaska, however, are mostly too swift for canoes.

Usually recommended for an extended trip by two prospectors is a 16- or 17-foot canoe having a square stern, an outboard motor (unless portages are long or frequent), and three paddles. The Indian canoe

These prospectors are shooting the rapids in southeastern Canada. *(Old Town Canoe Company)*

The Acker Packsack prospectors diamond core drill fits into a canoe. *(Acker Drill Company, Inc.)*

made of birchbark has been superseded by the more durable canvas-covered canoe or the plywood canoe, both being more popular than the canoe made of aluminum, which is harder to repair. The "prospector" model canoe is lighter than the "cruiser" model and has a flatter bottom, which is better for shallow streams and for quick handling in rapids. The cruiser, which is a little sturdier and has a rounded bottom, is easier to paddle and carries a heavier load. "Semifreighter" canoes are heavier than these, and "freighter" canoes are still heavier, but their length of 18, 20, or 22 feet is usually too burdensome for portaging by two men. A wood keel protects the bottom of a canoe, although it adds to the difficulty of paddling except in a side wind.

Experience in loading, paddling, and portaging is required for canoe travel. The canoe should be large enough to be stable, the load (including occupants) should be kept low, overloading and lack of balance should be avoided, and risky rapids and storms should be avoided (shore travel is advisable on large lakes). A piece of canvas and a tube of waterproof cement should be taken for making patches. The canoe should be painted yearly in a bright color so that it can be seen readily from the air and from a distance.

Other types of boats can be used. Sectional and collapsible canoes are seldom favored by prospectors. Rowboats are suitable on lakes but not for portaging. Flat-bottomed "poling" boats, 16 to 30 feet long, built of wood and powered by a strong outboard motor, are the best means of travel on the quick rivers of Alaska and western Canada. Rafts are not

A flat-bottomed boat is well adapted to prospecting in Alaska and western Canada. *(Richland Manufacturing Company)*

unknown on some streams, but they present problems of construction and safety; rubber rafts are hard to steer. With the tremendous growth of family boating in recent years, many persons have become familiar with buying, using, and repairing outboard motors, the only kind of motor used by prospectors. They even know enough to carry along an oil-company credit card.

Both airplanes and helicopters have become major modes of transportation in prospecting. Canada, Australia, and New Guinea saw plane travel for this purpose several decades ago, and it has spread widely since then. The many lakes and rivers in eastern Canada are especially suitable for flying boats, and charter services are available in many places to carry prospectors, baggage, and small canoes if desired. Parachute drops are used, especially in the western mountains. Arrangements are suggested to record the plans that have been agreed on, in case of accident.

The helicopter aids prospecting in the American West.

A small airplane carries this prospecting party into the north country. *(Pan American Petroleum Corporation)*

Land-based motor vehicles have supplanted feet and animals as the principal means by which the prospector and mineral collector get around. The character of the terrain matters little. It should be emphasized that unless geophysical devices (of the sort discussed in chapter 10) are in operation during movement of the vehicle, so that a record can be made of the country being traversed, motor transportation tends generally to go too fast to enable the prospector to observe the features that he needs in order to discover mineral deposits effectively. Just as, in the old days in California, the Chinese placer miner was able to rework gravel abandoned by the less patient Caucasian, so too the foot prospector can often find what the hurried motorist has passed by.

Motor vehicles have been put to use by prospectors in driving a small rock crusher, sawmill, or other machine. Going to town for supplies for a more or less permanent camp is excuse enough for having a motorcycle or automobile. Von Bernewitz, in the previous edition of this book, remarked that "'dude' prospectors in motor-cars have [even] hauled a trailer." The number of "dudes" has increased astonishingly since then!

Compared with the means of transportation that were available a quarter of a century ago, today's prospector and outdoorsman have an exciting array of motorized vehicles at their disposal. Inasmuch as no manufacturer has offered to provide the author with a free sample to encourage a testimonial on its behalf, the choice is left freely to the reader. The fundamental motor vehicle is, of course, the automobile. Track of

All-terrain vehicles (above and on following page) are of a wide variety in today's prospecting. *(Allis-Chalmers; ATV Manufacturing Company)*

All-terrain vehicles (above and on preceding page) are of a wide variety in today's prospecting. *(Rokon, Inc.; Heald, Inc.)*

standard width for unsurfaced roads, heavy-tread tires for sandy or muddy roads, and a high center for deep ruts are obvious advantages. Trucks and pickups are often more serviceable than a regular automobile, and these grade into campers and trailers, although a trailer can be hauled by an ordinary car and disconnected when desired. More expensive and likewise more versatile are the four-wheel, Jeep-type vehicles. A whole range of rugged trail vehicles, which have the motorcycle as their ancestor, can be found on the market. Then, there are the motorized vehicles that are designed to travel across sand, snow, marshes, swamps, beaches, and other kinds of uncooperative land and water, under the group name ATV (all-terrain vehicles). New items of this sort are often described in the mining periodicals that are listed on pages 405–406. You may, if you wish, walk on snowshoes in the winter; many prospectors did, in the old days.

4

Prospecting Equipment

Before spending money for the purchase, lease, or loan of equipment, the prospector should decide what his purpose is: a weekend in the hills? specimen collecting? serious mining? uranium only? gold only? agates for polishing? or come what may? Perhaps, of course, he cannot tell in advance, for a spectacular find could change his notions entirely. Yet, there is an advantage in planning ahead. Only a rich and capricious man would start out by buying every conceivable item of prospecting equipment, from an airplane (with hired pilot) to a supply of food designed to keep him in epicurean luxury. Such a person would probably not enjoy prospecting anyway.

This chapter deals with the equipment that is typically used for ordinary prospecting and mineral collecting as well as for preliminary exploration of a prospect. Ordinary prospecting has come to include the use of devices for identifying radioactivity and fluorescence, and so these are mentioned here, although they are discussed in chapters 8 and 9. Travel and transportation equipment is covered in chapter 3; camping equipment, in chapter 5; equipment for geophysical and other special methods of pros-

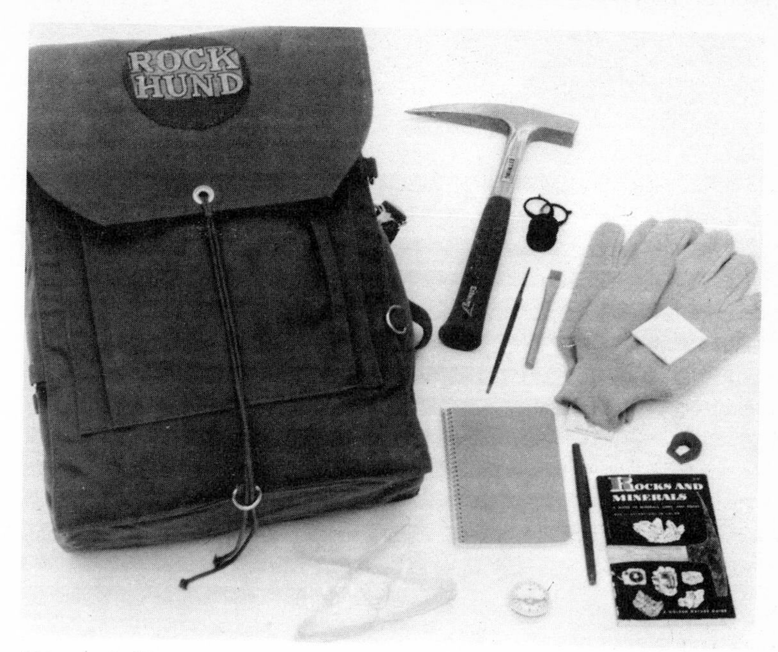

This sample kit serves for both prospecting and mineral collecting. *(C. L. Crump Company)*

pecting, in chapters 10 and 11; and equipment needed for developing a prospect into a producing mine, in chapter 33. Some of the more successful of today's professional prospectors own enough equipment to represent a sizable investment of capital.

A. H. Lang suggests that, whenever possible, you consider buying equipment near the prospecting area. The cost of hauling it may offset any saving in buying it at home or in the large cities. Furthermore, you may obtain valuable advice about local conditions. Mail sources for most prospecting equipment (catalogs are available) are:

Colorado Geological Industries, Inc.
5818 East Colfax Avenue
Denver, Colorado 80220

Exanimo Establishment
Segundo, Colorado 81070

Miners
177 Main Street
Newcastle, California 95658

Van Waters and Rogers, Inc.
4300 Holly Street
Denver, Colorado 80216

Ward's Natural Science Establishment, Inc.
P.O. Box 1712
Rochester, New York 14603

Heavy mineral exploration in South Africa uses modern equipment and ancient energy. *(Acker Drill Company, Inc.)*

Maps The latest and most detailed maps of the area that you intend to work in should be obtained before starting out. Sources of maps that are of greatest value to the prospector are discussed at the end of this chapter. Maps can be folded to fit into a transparent plastic cover, either bought or made at home. Or they can be cut into sections as large as convenient. The easiest way of mounting a map whole is to iron on a piece of commercial cloth backing. The map can also be cut into pieces of the right size for pasting onto the pages of a notebook, or for pasting onto pieces of cardboard (soak the map and cardboard in water, and mount with thin paste while wet). It can also be pasted onto a large enough piece of cloth (bleached muslin sheeting free of starchy filler), which can be folded between the segments; stretch the cloth fully, tack it onto a smooth surface, wet both cloth and map, and mount with thin paste, using either homemade flour-and-water paste or commercial wall-paper paste. Spraying the map will waterproof it but hinders marking it later.

Map case Unless mounted as described above, maps should be protected in a waterproof map case (sold in stationery and art stores, blueprint shops, and engineering-supply houses) or carried in a mailing tube that is made of pasteboard or metal.

Compass An ordinary pocket compass (sold in sporting goods, war surplus, and secondhand stores) is a necessary item, at least in staking a claim. The much more expensive Brunton or Brunton-type compass is prized by professionals. Remember that a compass points to magnetic north, not true north; in some places, the two are in quite different directions. In some country, a compass cannot be relied on.

The Brunton compass registers the dip of a sandstone outcrop in Fremont County, Wyoming. *(Standard Oil Company, New Jersey)*

Taking a Brunton compass reading of the dip of sedimentary strata near Mackay, Idaho. *(Standard Oil Company, New Jersey)*

Measuring tape A metal or cloth tape (sold in hardware, auto, and variety stores), measuring 50 or 100 feet, is needed to lay out a mining claim and to determine other distances. Cloth is cheaper and may be more durable.

Sketching materials A drawing pad (of which various kinds are sold in many places) can also be used for general note-taking; the hard-covered, engineer's type, with square-ruled and tracing pages, is most convenient; a stenographer's notebook is handy and cheap. Some black and colored pencils and one or more transparent protectors and 6-inch scales (preferably divided into tenths or to correspond to the map scale being used) should be taken along.

Magnifying glass Folding pocket lenses can now be bought cheaply (at optical, photographic, and jewelry stores). Usually, the stronger the magnification, the smaller the field of view. A 10-power lens, standard for diamond evaluation, is ample for examining specimens and gold pannings.

Magnet A horseshoe or bar magnet (sold in hardware and variety stores) is useful for cleaning magnetic minerals (mostly magnetite) from

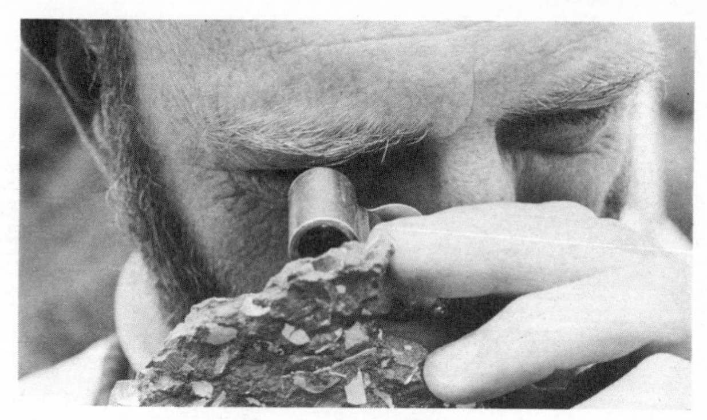

Examining a rock under magnification may reveal important knowledge about its composition. *(Standard Oil Company, California)*

the gold pan and for making certain mineral identifications. Alnico magnets are very powerful for their sizes.

Gold pan Virtually a standard item for the prospector, the gold pan is made of sheet iron and comes in several sizes, the larger size (16 inches is common) being used mainly in placer operations. Even lode, vein, or hard-rock prospecting finds much use for a gold pan in sampling and concentrating the heavy (and usually the more valuable) minerals. Corrugations (riffles) on the lip save fine particles. Panning is explained in chapter 7 of this book.

Of the above items, the only one that is not apt to be found in almost any town is the gold pan; it and the equipment described below are sold in many western American and Canadian hardware stores and by prospectors' supply firms. The following equipment is employed in work that involves removal of pieces of rock and their treatment. A prospector may pay to have this sort of work done for him, or he may do it at a later time for development or assessment purposes.

Sample bags Sample bags should be made of canvas and be used with cloth tags, which can be marked with a soft, nonsmearing pencil. Specimen bags for crystals or other valuable minerals may be of paper or cloth, for it is presumed that you will promptly wrap such finds carefully in newspaper, tissue paper, or pages that are torn from a catalog or telephone book.

Location notices These are sold (in stationery stores in mining areas) as printed and can be put in a baking-powder or tobacco can when placed in your location monument.

R. L. Bennet is prospecting for gold in the Scottish Highlands. *(R. L. Bennet)*

Geiger counter or scintillation counter These are discussed in chapter 8, on radioactive minerals. Metal- or mineral-detecting devices are now prohibited in federal areas that are designated as having natural, historical, or recreational value in the United States.

Ultraviolet lamp This is discussed in chapter 9, on luminescent minerals.

Rock hammer A prospector's pick may turn out to be the piece of equipment most often carried by the mineral hunter in the field. Several kinds are in common use: one has a square hammer on one side, a pick on the other; another kind has a chisel edge (suitable for prying apart layered rocks) instead of a pick. Special varieties are also known, including one with a long handle, helpful for climbing uphill. A bricklayer's or stonemason's hammer may do for most purposes, but not a carpenter's hammer. Rock hammers now come in attractive designs and with several

The prospector needs a rock hammer above all other tools. This view is in the Northwest Territories of Canada. *(Socony Mobil Oil Company, Inc.)*

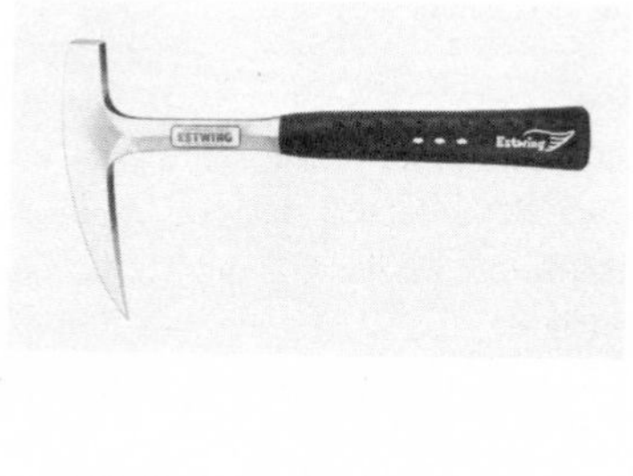

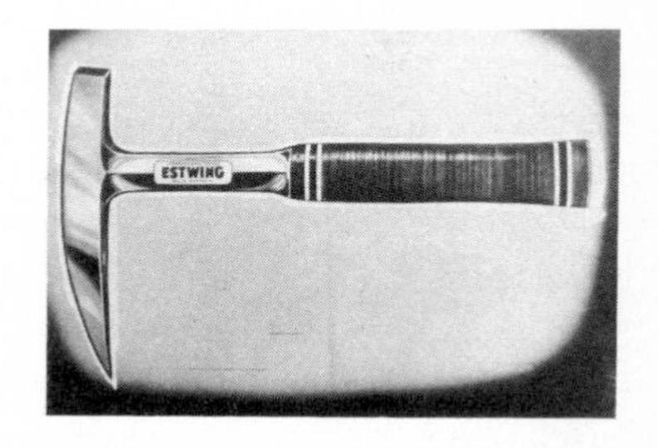

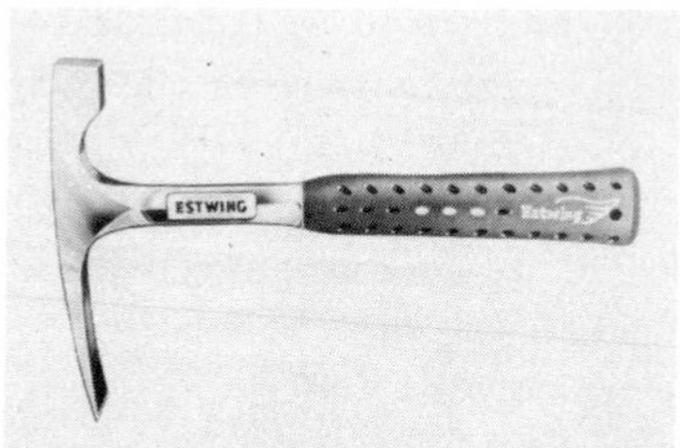

Estwing rock hammers of several types. *(Estwing Manufacturing Company)*

kinds of handles, even in colored, shock-absorbing plastic. A new type of rock-cracking hammer—a smaller version of the sledge that is described below—has become popular.

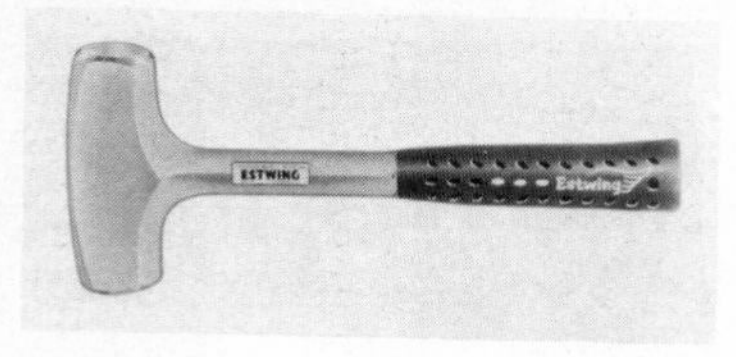

The Estwing crack hammer is also useful for rock work. *(Estwing Manufacturing Company)*

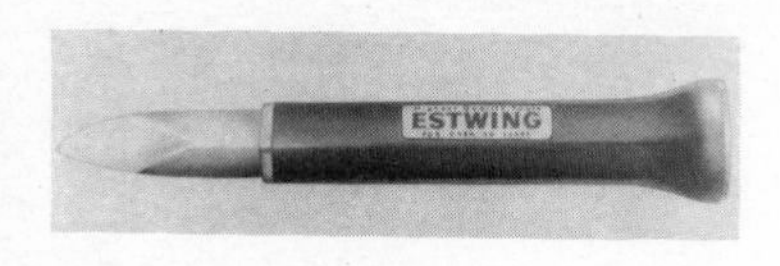

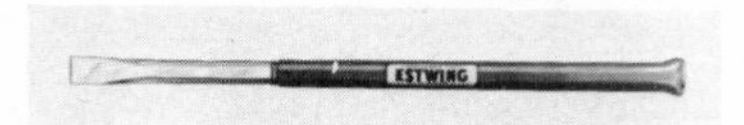

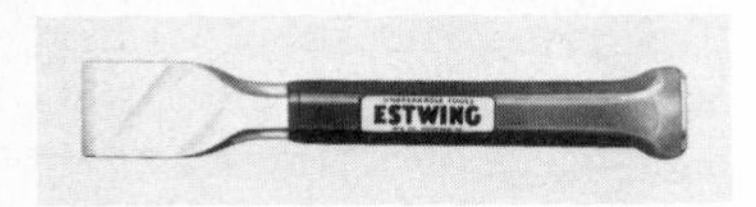

Several varieties of rock chisels may be needed in different kinds of rock. *(Estwing Manufacturing Company)*

Chisels Cold chisels of several sizes are often useful for wedging out rocks in sampling or extracting crystal specimens. The terms "gad" and "moil" are used for various wedges and spikes of different lengths (6 to 18 inches) and shapes of points (some detachable), serving diverse applications.

Sledges A short-handled sledge, or striking hammer, weighing 2 to 4 pounds, is called a single jack, because it is used in one hand by one man, who holds a chisel, gad, moil, or drill in his other hand. A double jack, weighing 8 to 10 pounds, is a sledge swung by one member of a pair of miners, the other manipulating the drill with more or less confidence in his partner's aim.

Drill steel Used mainly for trenching, drills have chisel-shaped bits, which are usually detachable and sharpened by grinding instead of by blacksmithing, as in the old days. The cuttings of rock can be removed with a small scoop, or spoon, made from a metal rod.

Drifting pick A 3- to 4-pound miner's pick, having two pointed ends

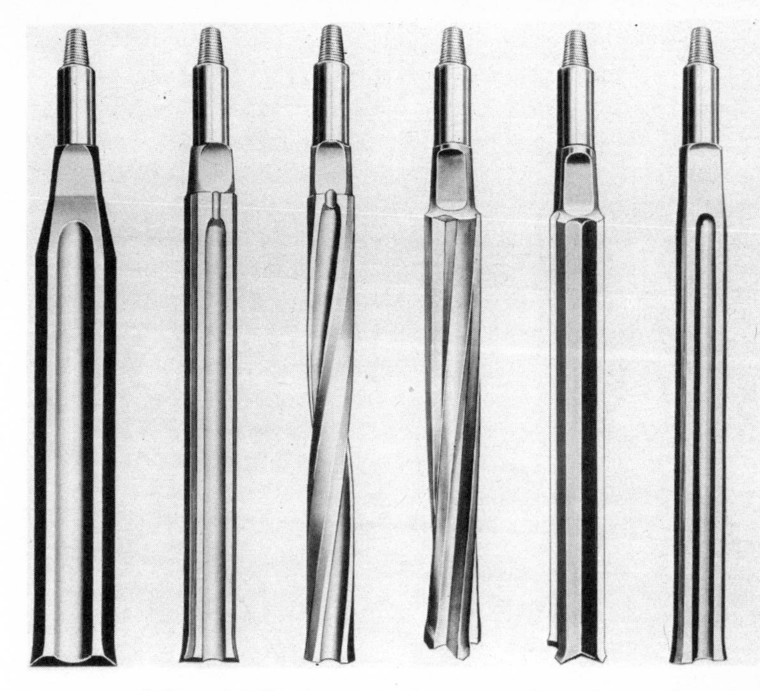

Cable tool drilling bits are made in different designs.

and an axelike handle, is used to extract rock. Several kinds of picks, ore hammers (pick hammers or poll picks), and mattocks are available, but a railroad or clay pick is not suitable, even though it is heavier.

Shovel The use of a shovel is principally in trenching. A folding shovel of army type may be preferred (for lightness) to the usual one that has a no. 2 round point and a long handle, with a hand grip (miner's shovel) or without (dirt shovel).

Other hand tools A two-handed axe (perhaps with a spare handle), a sheath for it, a whetstone and a single-cut flat file for maintaining the edge, a small hatchet kept in a belt sheath, and a prospector's grub hoe (with one hammer face and one hoe face) for clearing the ground surface may be as useful for prospecting and claim staking as for camping. Assorted garden tools and bits of miscellaneous hardware often prove handy to the specimen collector when removing fine material from cracks and cavities in rock; something new turns up on nearly every club trip.

Mortar and pestle Made of iron or steel, these are used for grinding

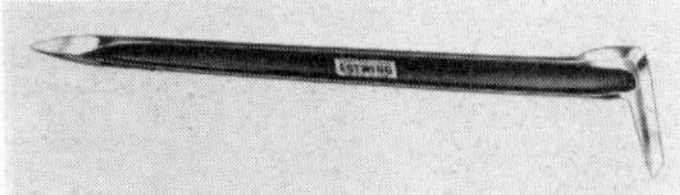

This Estwing tool is a combination of gad and pry bar. *(Estwing Manufacturing Company)*

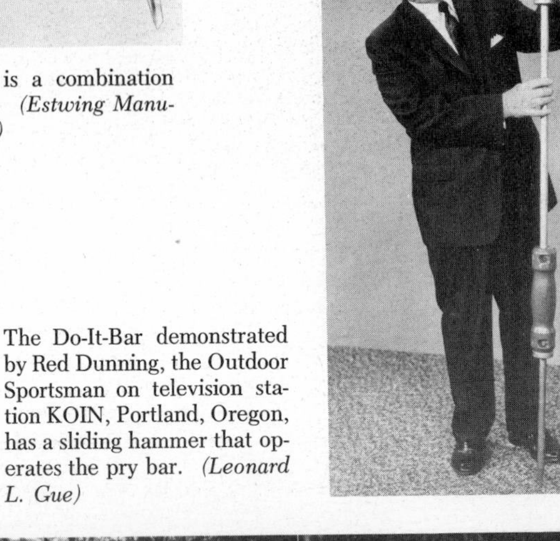

The Do-It-Bar demonstrated by Red Dunning, the Outdoor Sportsman on television station KOIN, Portland, Oregon, has a sliding hammer that operates the pry bar. *(Leonard L. Gue)*

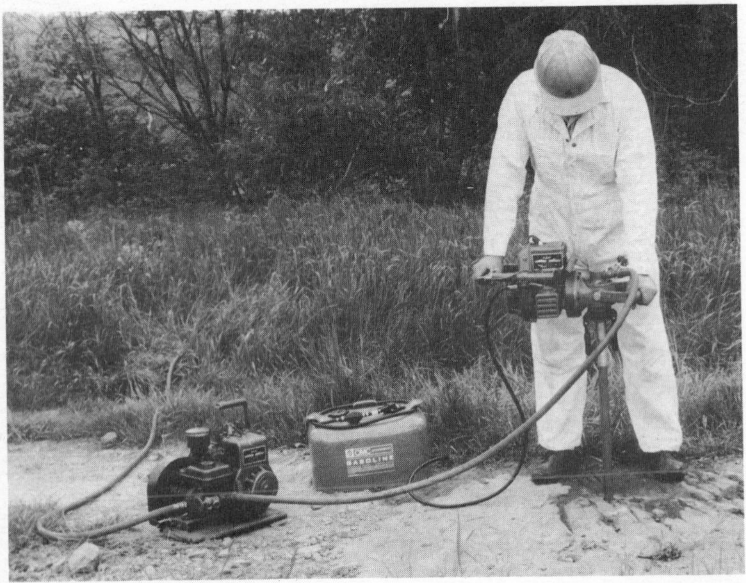

The Packsack diamond core drill is portable enough for remote areas. *(Acker Drill Company, Inc.)*

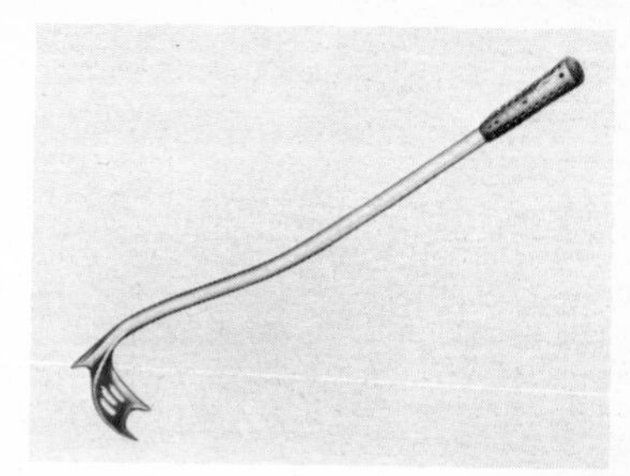

The Estwing gem scoop is handy on loose gravel and the beach. *(Estwing Manufacturing Company)*

samples. Owing to their lighter weight, substitutes (such as a 2-inch pipe cap or piece of pipe and the edge of a prospector's pick) are often favored, and samples can even be crushed by hammering them inside a canvas bag.

Sieve Bought or made with wire screen, a sieve separates loose or crushed rock according to size. The number of holes per inch (usually 60 to 80) depends on the particle size; placer material of value (such as gold or gems) is often very small.

Safety glasses Goggles of safety glass or plastic are advisable to wear while using a rock hammer or drill; they are inexpensive and easy to carry.

Explosives Although required for trenching in solid rock, explosives are discussed later (in chapter 33), where equipment for further development of an early prospect is considered. In much the same category are such items of intermediate usage (between prospecting and mining) as portable power drills and light diamond drills.

Further lists of prospecting equipment and supplies are given in Section 10, Article 13 of the *Mining Engineers' Handbook,* 3d edition, edited by Robert Peele (John Wiley & Sons, Inc., New York, 1941 (2 volumes).

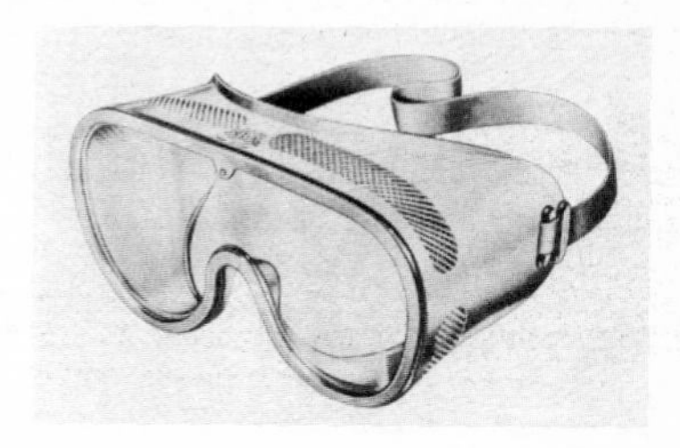

Safety goggles are especially recommended when working in granite and other hard rock. *(Estwing Manufacturing Company)*

MAPS

More people are becoming familiar with the use of maps all the time, but they are generally satisfied with road maps and other standard, geographic maps and atlases. A map must show both direction and distance correctly, that is, to scale; otherwise, it is merely a sketch, or sketch map, although it may be quite useful nevertheless. The prospector will do well to get ac-

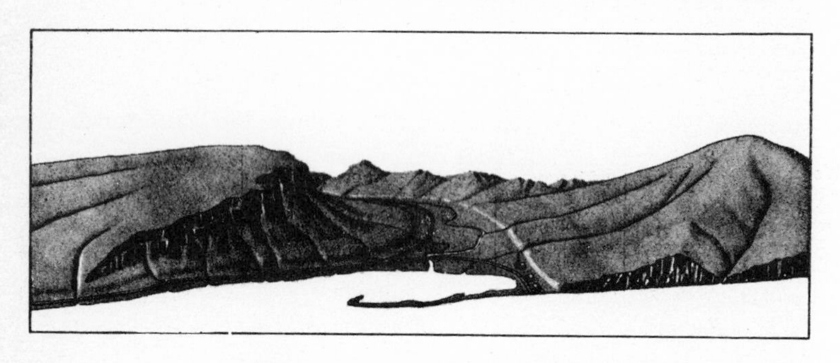

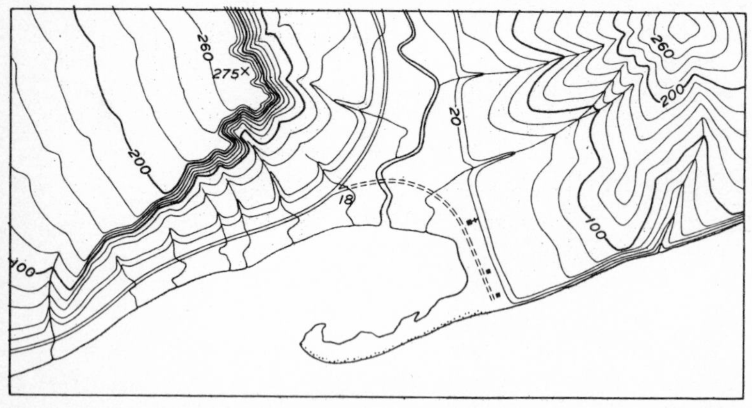

The Use of Symbols in Mapping. These illustrations show how various features are depicted on a topographic map. The upper illustration is a perspective view of a river valley and the adjoining hills. The river flows into a bay which is partly enclosed by a hooked sandbar. On either side of the valley are terraces through which streams have cut gullies. The hill on the right has a smoothly eroded form and gradual slopes, whereas the one on the left rises abruptly in a sharp precipice from which it slopes gently, and forms an inclined tableland traversed by a few shallow gullies. A road provides access to a church and two houses situated across the river from a highway which follows the seacoast and curves up the river valley.

The lower illustration shows the same features represented by symbols on a topographic map. The contour interval (the vertical distance between adjacent contours) is 20 feet. *(U.S. Geological Survey)*

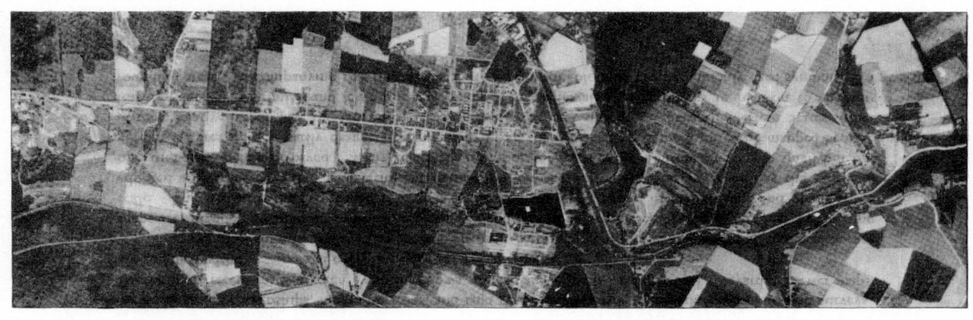

(a) Aerial photograph used in the preparation of map shown below.

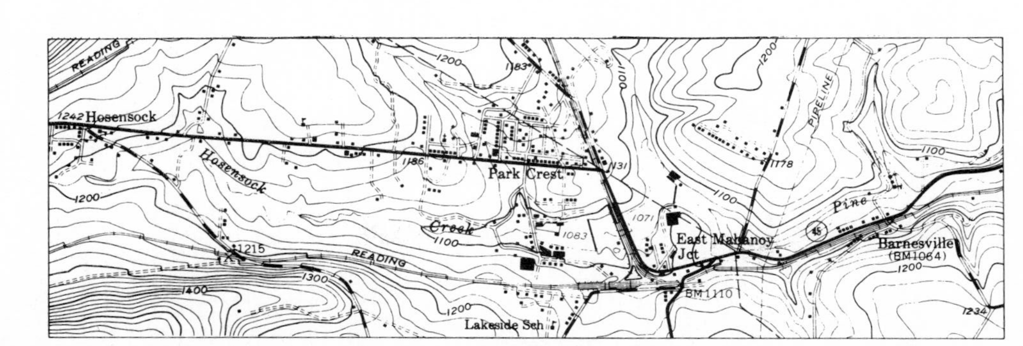

(b) A portion of the Delano, Pa., 7.5′ quadrangle map (Scale 1 : 24,000. Contour interval 20 feet. Mapped in 1946). *(U.S. Geological Survey)*

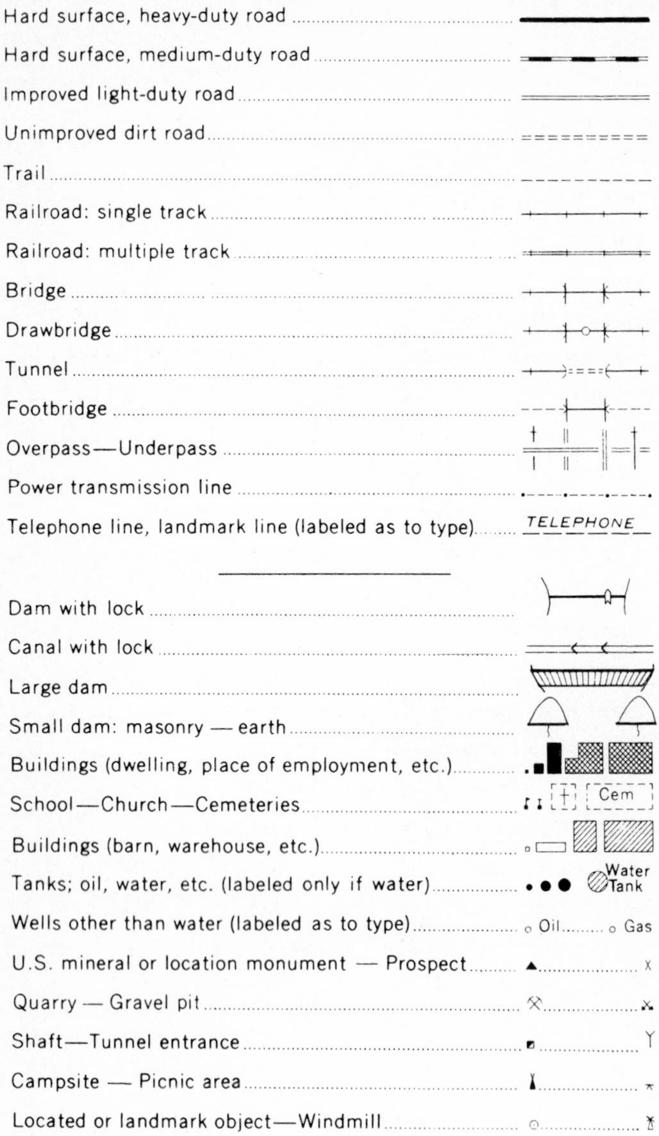

Hard surface, heavy-duty road

Hard surface, medium-duty road

Improved light-duty road

Unimproved dirt road

Trail

Railroad: single track

Railroad: multiple track

Bridge

Drawbridge

Tunnel

Footbridge

Overpass—Underpass

Power transmission line

Telephone line, landmark line (labeled as to type) TELEPHONE

Dam with lock

Canal with lock

Large dam

Small dam: masonry — earth

Buildings (dwelling, place of employment, etc.)

School—Church—Cemeteries Cem

Buildings (barn, warehouse, etc.)

Tanks; oil, water, etc. (labeled only if water) Water Tank

Wells other than water (labeled as to type) Oil Gas

U.S. mineral or location monument — Prospect

Quarry — Gravel pit

Shaft—Tunnel entrance

Campsite — Picnic area

Located or landmark object—Windmill

Topographic Symbols (variations will be found on older maps).

Exposed wreck

Rock or coral reef

Foreshore flat

Rock: bare or awash

Horizontal control station

Vertical control station ×672

Road fork — Section corner with elevation 429 +58

Checked spot elevation × 5970

Unchecked spot elevation × 5970

Boundary: national

State

county, parish, municipio

civil township, precinct, town, barrio

incorporated city, village, town, hamlet

reservation, National or State

small park, cemetery, airport, etc.

land grant

Township or range line, U.S. land survey

Section line, U.S. land survey

Township line, not U.S. land survey

Section line, not U.S. land survey

Fence line or field line

Section corner: found—indicated

Boundary monument: land grant—other

Index contour

Intermediate contour

Supplementary cont.

Depression contours

Cut — Fill

Levee

Topographic Symbols, continued (variations will be found on older maps).

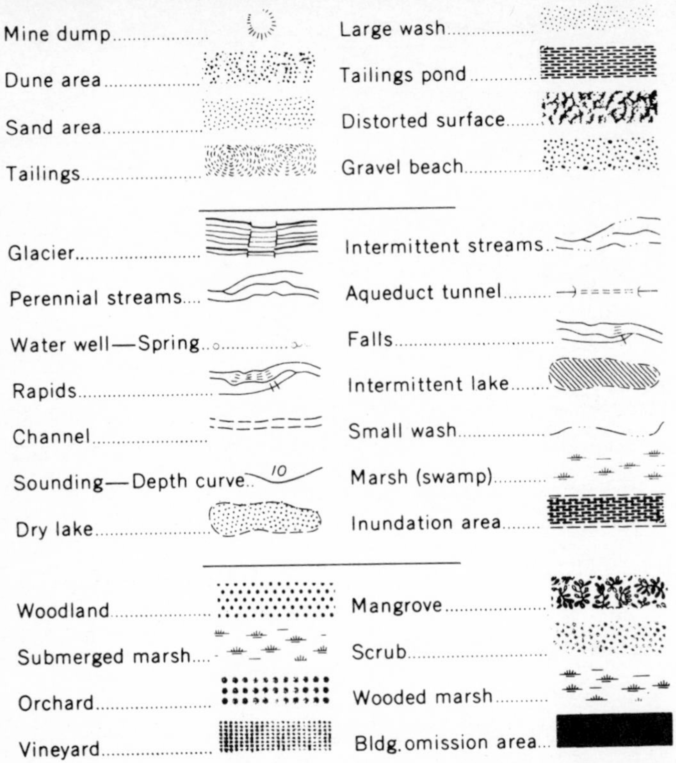

Mine dump
Dune area
Sand area
Tailings
Large wash
Tailings pond
Distorted surface
Gravel beach

Glacier
Perennial streams
Water well—Spring
Rapids
Channel
Sounding—Depth curve 10
Dry lake
Intermittent streams
Aqueduct tunnel
Falls
Intermittent lake
Small wash
Marsh (swamp)
Inundation area

Woodland
Submerged marsh
Orchard
Vineyard
Mangrove
Scrub
Wooded marsh
Bldg. omission area

Topographic Symbols, continued (variations will be found on older maps).

quainted with topographic maps, geologic maps, and claim maps as well; each of these kinds of maps is described briefly below.

1. *Topographic maps.* A geographic map becomes a topographic map when elevations are shown on a base map, usually by contour lines that connect points that have the same altitude above sea level.

Map properties that should be understood include the following:

a. PROJECTION: This enables the curved surface of the earth to be represented on a flat map. The kind of projection (Mercator, Lambert, polyconic, etc.) does not matter much for a map of a small area.

b. SCALE: The ratio between the distance on the map and the corresponding distance on the ground may be indicated by a written scale (such as "1 inch equals 1 mile"), a graphic scale (drawn for comparison), or a fractional scale (such as 1 to 24,000). United States maps

Vein, showing dip

Vein of high-grade ore

Vein of low-grade ore

Stringers or veinlets of ore

Omit dots if veins are shown in red, except where needed to indicate grade of ore

High-grade ore

Low-grade ore

Irregular ore bodies

Altered wall rock
(Showing intensity of alteration)

Symbols for mineral deposits and mine working. *(American Geological Institute)*

Large-scale maps

Vertical shaft

Inclined shaft

Portal of tunnel or adit

Portal and open cut

Trench

Small prospect pit
or open cut

Large open pit,
quarry or glory hole

Dump

Small-scale maps

Vertical shaft

Inclined shaft

Portal of tunnel or adit

Trench

Small prospect pit
or open cut

Mine, quarry,
glory hole, or large open pit

Sand, gravel or
clay pit

Surface openings. *(American Geological Institute)*

Shaft at surface

Shaft going above and below levels

Bottom of shaft

Inclined workings (Chevrons point down)

Spacing of chevrons can be used to indicate steepness of workings

Foot of raise or winze

Head of raise or winze

Raise or winze extending through level

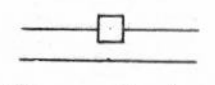

Ore chute

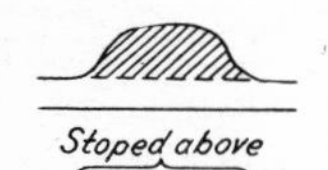

Stope

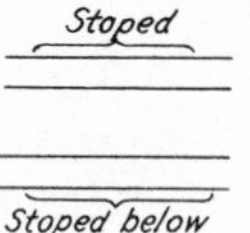

Stopes

400

Elevation of roof

375

Elevation of floor

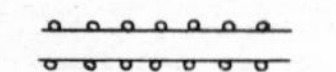

Lagging or cribbing along drift

Caved workings or otherwise inaccessible

Filled workings

Underground workings. *(American Geological Institute)*

of the National Topographic Map series are published in six different scales: 1 to 24,000; 62,500; 63,360; 125,000; 250,000; and 1,000,000.

c. ORIENTATION: Most maps have true north at the top, but magnetic (compass) north in most places is in a different direction. The magnetic declination is the angle between the two kinds of north.

d. CONVENTIONAL SYMBOLS: Water and ice features (hydrography) are shown in blue when color is used; works of man (culture) are shown in black; topography (relief) is shown in brown; main highways, towns not mapped in detail, and public-land survey lines are often shown in red; woodland and other vegetation may be shown in green. Special symbols are used for many particular features, as shown on pages 72–74, and for rocks and geologic structure.

e. TOPOGRAPHY: The following rules of contours are taken from *Geology*, 3d edition, by Richard M. Pearl, Barnes and Noble, Inc., New York, 1963:

(1) Contours separate all points of higher elevation from all points of lower elevation.

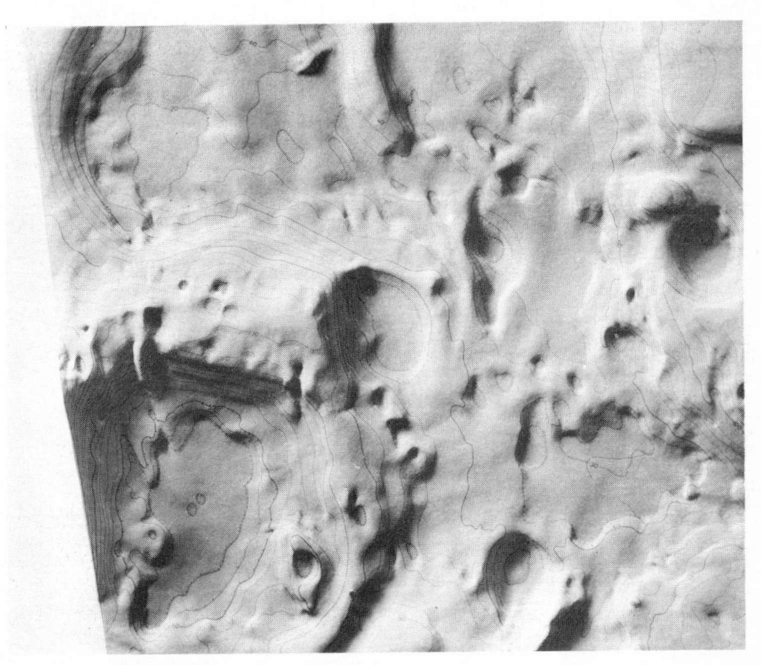

These contours are being drawn on a map of the moon, but the principles of topography are the same everywhere. *(U.S. Geological Survey)*

(2) Contours are always continuous until they close, even though outside the limits of the map.

(3) Contours never intersect or cross one another except at a very steep or overhanging cliff.

(4) Closely spaced contours indicate steep slopes; widely spaced contours indicate flat land or gentle slopes; a change in spacing indicates a change in slope.

(5) Contours are closer at the top of a concave slope and at the bottom of a convex slope.

(6) Contours cross streams at right angles.

(7) Contours bend in valleys and form a V pointing upstream.

(8) Contours loop around spurs and ridges on hilly topography and form a U pointing downhill.

(9) Contours on hills or in valleys are repeated on the opposite side of the hill or valley.

(10) Contours in water bodies may be replaced by a generalized pattern that does not indicate actual depths.

Typically, the contour interval (the difference in elevation between adjacent contour lines) ranges from 5 to 200 feet, smaller intervals being used for flat land, larger intervals for mountainous land; very flat land may even require supplementary contours, which are dashed or dotted. Every fourth or fifth line is an index contour, drawn heavier and marked with its constant elevation. Permanently marked points of known altitude are called bench marks and are individually noted on topographic maps. Hachures (short lines that point in the direction of slope) are used on some maps (and commonly on European maps and those of other countries). Form lines and symbol patterns are used to show certain kinds of landforms. Relief shading is an overprint that is sometimes used to give a three-dimensional effect.

In the United States, topographic maps are sold by the U.S. Geological Survey at the following addresses: 1200 South Eads Street, Arlington, Virginia 22202 (for eastern maps); Federal Center, Denver, Colorado 80225 (for western maps); 520 Illinois Street, Fairbanks, Alaska 99701 (for Alaskan maps). They are also sold over the counter in offices in Washington, D.C.; Anchorage, Fairbanks, and Juneau, Alaska; Menlo Park, Los Angeles, and San Francisco, California; Denver, Colorado; Dallas, Texas; Salt Lake City, Utah; Spokane, Washington. A topographic index (with order blank) is obtainable free for each state and territory, showing all the avail-

A plane table and alidade are used in surface surveying. *(Cities Service Company)*

able quadrangle (four-sided) maps and special maps as well as giving library and ordering information. New maps and indexes are listed in the monthly supplements to the catalog of publications and its annual supplements.

In Canada, topographic maps are issued by the Surveys and Mapping Branch, Department of Energy, Mines and Resources, Ottawa, and many are also sold by branch offices (but not the Ottawa office) of the Geological Survey of Canada. Index maps that show which maps are available (and at what scale) are obtainable free.

2. *Geologic maps.* Adding information about the rocks that are present, and their structures, converts a topographic map into a geologic map. The distribution of the rocks is shown by individual patterns, often colored. Special symbols are used to indicate such features as strike and dip, structural alignment, fold and faults. Economic data, such as the location of mines, quarries, and oil wells, may be added or shown on a separate map.

The U.S. Geological Survey publishes (at various prices) geologic maps of certain of the states, and others are sold by the respective states. A list of these is given in the free catalog of publications of the U.S. Geological Survey. This catalog (and its annual and monthly supplements) also has a list of the state indexes to geologic mapping; these are issued by the U.S. Geological Survey and outline in color the areas in each state for which geologic maps have been published anywhere.

In Canada, the Geological Survey of Canada, Department of Energy, Mines and Resources, 601 Booth Street, Ottawa 4, sells geologic maps, a list of which is available free on request. Complete indexes of publications from the year 1845 are sold or may be consulted in libraries.

3. *Claim maps.* These are produced in Canada, showing the status of

The geologic map is made after the topographic map, using it as a base. *(Standard Oil Company, New Jersey)*

mining claims in an area, usually a township, where active prospecting or mining is going on. They can be obtained from provincial and territorial agencies (see pages 46–47) or from local mining-recording offices.

BOOKS ON MAPS

Map Reading, U.S. Department of the Army, Washington, 1969.

Interpretation of Topographic and Geologic Maps, With Special Reference to Determination of Structure, C. L. Dake and J. S. Brown, McGraw-Hill Book Company, New York, 1925.

Elementary Topography and Map Reading, Samuel L. Greitzer, McGraw-Hill Book Company, New York, 1944.

The Interpretation of Topographic Maps: A Laboratory Manual for Use in Connection with the Topographic Maps of the United States Geological Survey, to Accompany Courses in Physiography, Rollin D. Salisbury and Arthur C. Trowbridge, Holt, Rinehart and Winston, Inc., New York, 1930.

AERIAL PHOTOGRAPHY

Besides maps, air photographs can be of much value to prospectors, for they show aspects of the terrain and geology that are not readily ob-

Aerial photographs are often studied best with a stereoscope. *(Standard Oil Company, New Jersey)*

served in other ways. Large companies use them extensively. Three main types of aerial photographs are used, as follows:

1. VERTICAL PHOTOGRAPHS: Taken directly from above, these are like plans. Overlapping of exposures permits stereoscopic viewing with either a mirror or lens stereoscope, which is especially helpful in rugged country, as it throws the topography into relief. A mosaic can be prepared by assembling adjoining photographs.

2. OBLIQUE PHOTOGRAPHS: Of an older type, these pictures were taken from various angles—usually three: left, center, and right—all directed backward as the plane flies.

3. TRIMETREGON PHOTOGRAPHS: These combine vertical and oblique (left and right side) photography, having both advantages and disadvantages of other methods.

Photographic enlargements and mosaics are often useful in prospecting.

The U.S. Geological Survey issues free a map called "Status of Aerial Photography," which shows all areas known to have been photographed by federal, state, or commercial agencies. Air photographs can be obtained from them. The chief government sources are the Agricultural Stabilization and Conservation Service, Soil Conservation Service, Forest Service, Geological Survey, Air Force, and Coast and Geodetic Survey. All photographic coverage secured on nitrate film from 1936 to 1941 has been transferred to the National Archives and Records Service, Cartographic Branch, General Services Administration, Washington, D.C. 20408. You might inquire locally first. The most useful addresses to write are the following:

Western Aerial Photography Laboratory
ASCS-USDA
2505 Parley's Way
Salt Lake City, Utah 84101

Eastern Aerial Photography Laboratory
ASCS-USDA
45 South French Broad Avenue
Asheville, North Carolina 28801

Director, Division of Engineering
Forest Service—USDA
Washington, D.C. 20250

Director, Cartographic Division
Soil Conservation Service—USDA
Federal Center Building
Hyattsville, Maryland 20781

In Canada, the Department of Energy, Mines and Resources maintains at 615 Booth Street, Ottawa, the National Air Photographic Library, which sends orders for prints, enlargements, and mosaics to the Royal Canadian Air Force, which fills them. The library will distribute free maps that indicate the type of photography that is available for any part of Canada. You may place a tentative order by stating the desired scale, whether matte or glossy prints are wanted, and if stereoscopic coverage is needed. *A Useful Catalogue of Selected Airphotographs,* by H. S. Bostock, was issued as Geological Survey of Canada Paper 68-21(1968).

Certain provincial agencies in Canada also maintain libraries of air photographs, as follows:

Alberta
Technical Director
Department of Lands and Forests
Edmonton

British Columbia
Air Photo Library
Department of Lands and Forests
Victoria

New Brunswick
Director of Photogrammetry
Department of Lands and Mines
Fredericton

Ontario
Surveyor General
Department of Lands and Forests
Toronto

BOOKS ON AERIAL PHOTOGRAPHY

Outstanding Aerial Photographs in North America, American Geological Institute, Washington, 1951.

A Useful Catalogue of Selected Airphotographs, H. S. Bostock, Geological Survey of Canada Paper 68–21, 1968.

Aerial Photographs in Geologic Interpretation and Mapping, U.S. Geological Survey, Washington, 1960.

Photogeologic Procedures in Geologic Interpretation and Mapping, A. G. Ray, U.S. Geological Survey Bulletin 1043-A, 1956.

Short Guide to Geo-botanical Surveying, Sergei Viktorov et al, Pergamon Press, Oxford, 1964.

Photography and Regional Mapping, J.A.E. Allum, Pergamon Press, Oxford, 1967.

Aerial Photographs in Field Geology, Laurence H. Lattman and Richard G. Ray, Holt, Rinehart and Winston, Inc., New York, 1965.
Aerial Discovery Manual, Carl H. Strandberg, John Wiley & Sons, Inc., New York, 1967.
Aerogeology, Horst F. Von Bandat, Gulf Publishing Company, Houston, 1962.

5

Camping, Clothing, Cooking

Camping, which has spread widely to Europe, Australia, and other parts of the world since the Second World War, has become an American phenomenon. At the present time, it is estimated that about one vehicle in every three on the road in the United States is a family camper or trailer that is designed for camping. Except in the use of specialized prospecting tools and equipment, described in chapter 4, prospecting and rockhounding present no general problems of in-the-field accommodations that are essentially different from those of camping.

The subject of camping is therefore confined here to a concise summary of the most up-to-date methods and materials. The information has been furnished by experts, and a list of several of the best books on the subject is given.

The extent of the prospector's outfit will depend largely on his ultimate destination. If he plans to remain within walking distance of an auto camper or trailer, he needs nothing more than what is usual for a day of camping or fishing. If he will occupy a tent, this may become an important concern. His clothing will depend on the climate and how long he expects to be gone. Food is largely a personal matter; water cannot be taken for granted; cooking equipment of many kinds is readily available. Suggestions regarding each of these matters are given below for those who do not intend to make use of a regular overnight cabin, camper, or trailer, where life will probably be conducted pretty much as at home, even down to the ubiquitous, iniquitous television set.

Unless provided for by a superior style of transportation and camping, the main factors to consider in selecting a campsite are availability of water, firewood, and tent poles. Care must be taken in using water from streams and lakes; it may need to be boiled or treated. Drainage (gravel drains readily; level ground is preferred, or dig a trench); sanitation (including burning and burial of waste); protection (of food from animals, of people from insects, of property from other people); fire protection; and survival in case of getting lost—these should be especially emphasized, along with the other advice that is given in books on camping and woodcraft.

Tent Excluding the use of a camper or trailer, and perhaps even with them, most prospectors and collectors will sleep in a tent unless the trip is short and the weather fine. So many kinds of tent are available—differing in size, shape, kind and weight of material, and cost—that suggestions are difficult to make. Most often recommended is a wall tent measuring 8 by 10 feet and having a 4-foot wall and (unless a wood stove is used inside) a canvas floor with doorsill. Next most often recommended

These are two of the types of tents most popular for prospecting purposes. (Coleman Company, Inc.; Camp Trails)

might be a single-pole, umbrella (pyramid, miner's) tent with a few pegs, which is popular with prospectors. The explorer tent is especially good when mosquitoes are thick. An A-shaped (or pup or mountain) tent needs good weather to serve well; a tarp (shelter half) or lean-to is similar. Special arrangements, beyond that of insect protection (mosquito netting or fly bar), may be preferred under arctic or desert conditions. Helpful suggestions for selecting a tent are given in certain catalogs and camping books.

Sleeping bag A blanket and tarpaulin often add to comfort. Catalogs and camping books often give useful advice on choosing and using a sleeping bag.

Air mattress Extra insulation is needed beneath it in cold weather.

Packsack Some sort of back carrier—rucksack or haversack—is usually needed for daytime use in addition to a pack frame or packsack that is used for transporting camp equipment and extra clothing.

Tarpaulins Canvas tarpaulins or plastic sheets will protect equipment, cover firewood, and make shelters. Canvas sacks, treated with paraffin, protect food packages in the field.

Lighting Some sort of portable illumination—whether a lantern, candles, or flashlight—is necessary in camp.

Knife A sheath or pocket knife is needed, and perhaps a small sharpening stone.

Pliers

Saw

Wire and rope A small cord of thin wire and a sash cord are often useful.

Friction tape

Nails and rivets To make repairs and set up camp.

Fishing tackle Simple items without a rod will serve.

Firearms Check permits if these are contemplated.

CLOTHING

The most thoroughly detailed literature on prospecting assures us that special clothing is not needed. Even daily wear (except shoes) will do, if necessary; but some thought should be given to comfort. One's preferences should be established by experimenting a bit. Special items are best bought locally, where the weather and terrain to be encountered are familiar to the dealer; this is especially true in the far north.

Shoes Two pairs of medium-height (6 to 8 inches) laced boots are advised, and a lighter pair of shoes for camp wear. Personal needs as to soles, insoles, hobnails, counters, and other features should be carefully heeded. Waterproof, salt-free grease keeps leather boots pliable.

Socks An ample supply of wool socks is recommended by everyone. If two pairs are worn, the inner one is lighter and may be either wool or cotton.

Underclothing Wool or wool-mixture underwear of sufficient weight may be augmented by a woolen undershirt or woolen sweater (worn under or over a shirt).

Shirts Wool is almost always best.

Trousers These are best made (without cuffs) of khaki, blue denim, or canvas (for heavy rain and thick brush).

Coats A water-repellent and wind-resistant coat, either a light windbreaker or light parka, can be carried or worn as needed; a canvas hunting coat is better when it is raining or cool. A raincoat of light plastic may prove useful.

Hats A waterproof hat need not be heavy; the hood of a parka can be worn over a cap, so either may be preferred.

Gloves Protection when handling rock and tools is given by waterproof workmen's gloves or leather ones.

Sewing kit For long trips.

COOKING

This subject includes two considerations: cooking equipment and the provisions. A means of refrigeration may also be important in the first category. If ample water is not accessible, water becomes the most essential item of food.

Cooking Equipment

The articles of cooking equipment to be taken into account fall into the following categories:

1. Stoves
2. Cooking and eating utensils

What kind of heat you may need will depend on your mode of transportation and the locality in which the prospecting or mineral collecting is done. If you are not going to cook outdoors or in a tent, you need not pay further attention to this matter except when keeping warm, heating water, or drying clothes.

A campfire requires a supply of firewood or pressed-sawdust logs, must be used outside, and takes time to prepare. On the other hand, it provides warmth, may encourage congeniality (or may not!), and has plenty of room for a wide variety of food preparation. A two- or three-burner camp stove may be bulky or heavy, has a limited capacity, and requires special fuel. However, it needs no firewood, is set up readily, and can be moved inside if the weather is bad; it may be useful, in any event, for warming and drying a tent. Consider also any laws prohibiting campfires (as in southern California), and be honest in appraising your skill in handling a fire or stove: it is not at all the same as managing a barbecue at home.

The following are the usual types of camp stove, besides plain candles and the many curious stoves that have been made for temporary or emergency use.

Wood-burning stove, or wood stove Best in a cold climate, it is made of sheet metal; it warms a tent better and more safely than a liquid-fuel stove; it needs a long enough stovepipe, a fireproof collar, and a spark shield; various models are made, and an oven may be attached.

A versatile camp stove warms the prospector both inside and outside. *(King Seeley Thermos Company)*

Canned-fuel stove Lightweight, often folding, it burns gasoline (white or leaded or either), propane, butane, kerosene, benzine, naphtha, alcohol, or solid fuel (such as Sterno); usually needs a folding shield; a container and perhaps a funnel are needed for liquid fuel.

Shelter warmer Used mainly for heating, it serves also as a small cooker; burns kerosene; needs ventilation.

Charcoal grill These simple devices come in numerous models.

Cooking and Eating Utensils

These may be taken from home, kept separately for camping purposes, or bought for the occasion. Camp kits are sold widely in many degrees of completeness. A compromise between lightness and usefulness must often be made. A book on camp cooking may not be a utensil, but it is highly utile.

Dutch oven Made of cast iron or cast aluminum, 10 to 12 inches across, this piece of equipment, though heavy, is generally recommended as the most useful for its ability to cook stew and beans and to fry and boil anything else, as well as to bake bread or biscuits. A simple reflector oven is a most effective device for baking or roasting.

Double boiler Made of aluminum, 2-quart size, it serves as a kettle or two saucepans and even cooks oatmeal right in the ashes of a fire. A pressure cooker may perhaps be worth its cost and weight.

Frying pan Cast iron (10 inch) is best, but heavy; plain black iron or cast aluminum is suitable; steel or stamped aluminum pans will burn their contents over an open fire. A large griddle may come in handy for pancakes or eggs.

Coffeepot Made of light steel rather than enamelware, 2-quart size or larger, with bail and metal handle rather than wood.

Water bucket Galvanized or folding-canvas pail, 10 to 12 quart size.

Kettle Made of steel, 4-quart size, with bail, this is used to heat water.

Dishpan

Washbasin Made of light metal.

Bread pan Or roll pan or both.

Can opener

Spatula

Bowls For mixing and eating; heavy plastic ware is best.

Cups Plastic ware is best.

Plates Plastic ware is best.

Knives A 10-inch butcher knife is needed, as are table knives for eating.

Forks For cooking (with long handle), mixing, and eating.

Spoons For cooking (large, with long handle) and eating (soup and tea sizes).

Soap Bar, flakes, detergent.

Candles

Matches Waterproof or in waterproof containers; or lighter and fluid.

Wrapping paper Or bags or foil for lunches.

Asbestos mitt

Artificial sponge

Scouring pads

Dishcloths

Dish towels Needed in wet weather.

Oven rack Do not overheat.

Provisions

The main factors to keep in mind when packing food are the mode of transportation, the time to be spent in the field, the number of persons, their eating requirements and tastes, and the cooking facilities available.

Sufficient nourishment, a balanced ration, and a pleasing variety are to be considered.

Living off the land, except when fishing, is time lost from prospecting. Firearms, furthermore, may be burdensome, expensive, dangerous, or illegal.

Before giving several lists that have been suggested by organizations that are experienced in prospecting, a few general comments may be helpful.

Concentrated foods that have been developed from military rations are sold in sporting-goods stores and by certain mail-order firms. Prepared foods of a huge variety are familiar to every housewife; refrigeration should be taken into account for the frozen foods.

Prepared foods are standard fare for outdoor activities today. *(Stow-Lite)*

Other than these modern specialties, the staples usually emphasize bacon and beans (the two B's), potatoes, rice, flour, baking powder, oatmeal, canned tomatoes, fruit juice, dried or canned fruit, coffee, syrup (for hotcakes), sugar, canned milk (for coffee and cereal), and salt. Packing with animals may require omitting much of the bulky potatoes and canned items, which are heavy. Only enough perishable food should be taken for a few days unless refrigeration is available.

The Geological Survey of Canada recommended 4 pounds per man-day in List A, printed below, which is for one man per month on a basis of moderate packing weight. The garrison rations of the U.S. Army also allow 4 pounds per man-day. On a two-man basis, the Ontario Department of Mines advised provisions averaging about 3 pounds per man-day; this (List B) is also given below, as applying to one man per month for summer use on a rather austere diet. The same amount per man used to be provided for parties of the U.S. Geological Survey. For prospecting in Alaska, the Cordova Chamber of Commerce suggested 3½ to 4 pounds daily for one man during summer, as in List C, given below. Von Bernewitz prepared a list of necessaries for one man per month, totaling 100 pounds, based on actual field experience in a remote region, which is printed below as List D. Further lists of provisions are given in Section 10, Article 13 of *Mining Engineers' Handbook*, 3d edition, Robert Peele, editor, John Wiley & Sons, Inc., New York, 1941 (2 volumes).

LIST A

Flour	25 pounds
Baking powder	1 "
Rolled oats (quick-cooking variety)	6 "
Beans (dried)	5 "
Rice	5 "
Potatoes (dehydrated)	4 "
Vegetables (dehydrated carrots, turnips, beans, etc.)	4 "
Bacon and ham	20 "
Cheese	3 "
Egg powder	1 "
Sugar	15 "
Tea	1 "
Coffee	3 "
Chocolate (semisweet, for lunches and emergency)	2 "
Milk (powdered whole milk)	3 "
Salt	1 "
Fruit (dehydrated prunes, peaches, apples, apricots, figs, raisins, etc.)	6 "
Butter (canned)	4 "
Jam, syrup, honey	5 "
Pudding powders (prepared mixtures)	2 "
Dehydrated soup mixtures	1 "
Yeast	—
Oranges	1 dozen
Canned fruit	6 tins
Canned sausage or beef	6 tins

Pepper, and other spices if desired	—
Lard (or substitute)	2 pounds

Bacon and ham should be canned or of the gelatin-packed variety if they are to be kept for more than a few weeks. Fresh or dried meat can sometimes be obtained locally, but the possibility of being able to do so cannot be depended on in most instances.

LIST B

Flour or bread equivalent	22 pounds
Salt pork (long clear)	3 "
Beans	3 "
Raisins	1 "
Rolled oats	4 "
Pepper	2 ounces
Corn syrup	5 pounds
Coffee or cocoa	1 "
Dried apples	1½ "
Dried peaches	1½ "
Sugar	15 "
Pot barley	½ "
Dried potatoes	2 "
Bacon	12 "
Butter or oleo (canned)	4 "
Rice	3 "
Cornmeal	2 "
Salt	2 "
Tea	1½ "
Baking powder	½ "
Dried apricots	1½ "
Dried prunes	1½ "
Bouillon cubes	1 "
Split peas	½ "
Dried milk	5 "

The syrup could be dispensed with and a little sugar added. Prepared pancake flour can be substituted for some of the flour.

LIST C

Flour, meal, rice, etc.	46 pounds
Beans, peas, corn	10 "
Potatoes and onions	7 "
Bacon, ham, salt pork	21 "
Canned beef and other meat	3 "
Lard, butter or oleo, cheese	9 "
Sugar	14 "
Evaporated eggs (canned)	1 "

Tomatoes, cabbage, and other vegetables (canned)	4	"
Salt (iodized)	4	"
Honey, syrup	3	"
Milk and cream (cases)	¼	"
Jams (canned)	1	"
Evaporated fruits	12	"
Raisins	2	"
Coffee, tea, chocolate	5	"
Spices, baking powder, sauces	2	"

LIST D

Flour, cornmeal, hardtack, rice, grits, oatmeal, or similar foods, at least two-thirds of which should be flour prepared for self-rising	42	pounds
Clear mess pork, bacon, ham (say half pork)	27	"
Beans and split peas (two-thirds beans)	7	"
Evaporated fruits, mostly apple	4	"
Sugar	5	"
Butter or oleo	3	"
Milk (canned)	2	"
Cheese	2	"
Tea, coffee, and chocolate	2	"
Salt, pepper, mustard (two-thirds salt)	3	"
Baking powder (if self-rising flour is not used)	1	"
Lime juice	2	"

Inasmuch as all the items considered in this chapter cost money, it seems proper to mention that all legitimate expenses for prospecting for minerals (other than oil and gas) are tax deductible. Hobby expenses are, of course, not included. During the uranium boom of the 1950's, M. J. Rogers prepared for sale certain forms to be made out, but the author of this book has no further information at this time.

6

Health and Safety

Chapter 4 of the Von Bernewitz edition of this book was titled "Health and First Aid." As a consultant for this revision, Dr. H. C. Dake, formerly editor of *The Mineralogist*, spoke as a medical man in expressing his opinion that the chapter should be omitted entirely. He did not believe that medical advice should be given in such a form. It can probably be said that no one would begin to read this book in order to learn what to do in case of accident, and that general warnings about taking care of one's health represent gratuitous advice. The extensive technical treatise on prospecting in Canada by A. H. Lang barely mentions carrying a first-aid kit before any possible accident and collecting workmen's compensation afterward!

As a compromise, therefore, this chapter makes a few general recommendations in regard to the prevention of accident or illness, provides a list of first-aid and survival equipment as recommended by several organizations, and suggests a few convenient booklets that might be carried by those who are likely to use them.

1. *Safety first* is advice that speaks for itself. Even the best discovery will do you no good if you are not around to benefit from it.

2. Sanitary facilities at a campsite should be investigated thoroughly. The purity of drinking water (mentioned in chapter 5) is more important

than the taste (iron, sulfur, desert brackishness) or the color (dark from tropical vegetation). The disposal of waste by burning and burial has also been mentioned.

3. Protection from insects, whether dangerous or merely annoying, is important and depends on the place and season. They are best avoided, screened out, repelled, and their bites treated—in this order.

4. Snakes are a menace and do not always give warning when encountered in a pile of rock; the kinds of snakes and treatment for their bites vary around the world.

5. Frostbite, which is intensified by wind, should likewise be guarded against, especially of hands and feet.

6. Sunglasses are a protection against the burning effect of glare on snow, water, or the desert. The U.S. Federal Mine-Safety Act requires that safety goggles, safety shoes, and hard hats must be used, even by visiting collectors, in quarries that have federal contracts.

7. In case of trouble, arrangements should have been made for traveling, working, and camping in company, and for obtaining outside assistance when needed. In some sections of Canada and in certain federal areas of the United States, you must secure permission and leave a record before disappearing into the wilderness.

8. Beware of old and abandoned mine tunnels, shafts, and dumps. Certain of the states (such as Colorado) have laws requiring that such dangerous places must be closed or covered, but the logical (if not the legal) procedure is that, even in those states and certainly elsewhere, you should look out for yourself.

Besides the extreme danger of rotted timbers, shaky ladders, and other deteriorating structures, as well as weakened floors and flooded shafts, one should be aware that some mines contain noxious gases. Cripple Creek, Colorado, is an example of a district in which this insidious condition has been proved to exist. The author's wife's grandfather lost his life in his own mine there around the turn of the century; the accident was blamed on a fall but was most likely due to deoxygenated air, a not unfamiliar hazard in that mining camp. A lighted candle or lamp can be carried to give warning of mine gas.

9. Special precautions must be taken with explosives; a list of dos and don'ts was adopted in 1964 by the Institute of Makers of Explosives. There are also rules for storing blasting supplies in magazines. Never pull loose wires that you may find in rock quarries, for they may be connected to live explosives.

10. In these times when nearly every American and Canadian drives

his own car or cars, it may seem presumptuous to offer safety advice on this subject: careful motorists do not need it, and the others do not heed it. Nevertheless, scarcely a year fails to see a prospector go to his death in the Mojave Desert of California on account of an automobile breakdown, insufficient gasoline, lack of water or food, or other causes. This paragraph is being written only a short time after a noted personage, former Bishop James A. Pike, lost his life in the Judaean wilderness—not while prospecting, of course, but while conducting himself exactly as many a prospector does in a remote land. Spare tires, chains, a shovel, extra gasoline and water, some dehydrated food, a first-aid kit, a fire extinguisher, flares and reflectors—these and other safety matters should be considered before setting out. In emergency, the car itself can be taken apart to provide needed equipment for signaling, cooking, resisting flood, keeping warm—as the Calfornia Highway Patrol has shown.

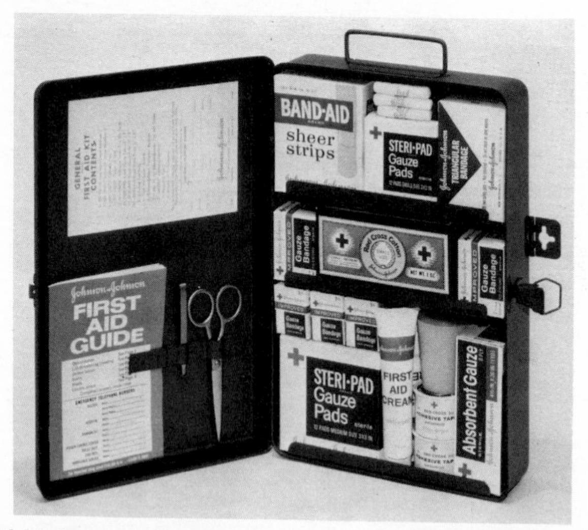

A first-aid kit is essential to all prospecting and mineral collecting. *(Johnson & Johnson)*

The Mineralogical Society of Arizona has prepared the following list for a desert survival kit:

1. 10 to 15 iodine tablets, tetraglycine hydropariodide (1 tablet per quart of water for drinking purposes, 1 tablet per canteen lid of water for antiseptic).
2. Short pencil stub.
3. 10 to 15 strike-anywhere matches, waterproofed with 2 coats of nail polish.

4. Small magnetic compass.
5. Candle stub, about 1/2 by 3 inches.
6. Small magnifying glass.
7. Metal signal mirror, double-faced, with small hole in center.
8. 5 feet of 12-inch heavy-duty aluminum foil.
9. 1 to 3 large, heavy-duty toy balloons; yellow or red.
10. Book of white cigarette papers.
11. Razor-blade, in original package.
12. 2 feet of 1-inch adhesive tape.
13. 4 or 5 assorted-sized fish hooks.
14. Small, sturdy pocket knife, honed razor sharp.
15. Large-eyed needle.
16. 75 to 100 feet of 8-strand cobbler's linen thread.
17. Large size bandage box or tobacco can.
18. 36 by 36 inch square of nylon chiffon; yellow, red, or blue preferred.
19. Tweezers.
20. Fire-Stix firestarter.

The U.S. Naval Institute, Annapolis, Maryland, will provide a checklist of minimum arctic survival equipment, and it publishes a book called *How to Survive on Land and Sea.* The U.S. Air Force, Washington, D.C., publishes the *SAC Land Survival Guide Book.*

Several inexpensive books on first aid are available, among them the following:

First Aid, Carl J. Potthoff, Boy Scouts of America, New Brunswick, N.J., 1957.

First Aid Textbook: Prepared by the American National Red Cross for the Instruction of First Aid Classes, Doubleday & Company, Inc., Garden City, N.Y., (4th edition), 1957. This is augmented by *Basic First Aid,* Books 1–4, 1971.

First Aid for the Mineral and Allied Industries: A Bureau of Mines Instruction Manual, John S. Kelly, U.S. Government Printing Office, Washington, D.C., 1971.

Part 2

7

Basic Prospecting

The fundamental rules of prospecting apply to all types of mineral finding that do not employ special geophysical equipment of the kinds that are described in chapter 10. These include but are broader than specimen collecting, and they also include the use of familiar equipment such as radioactivity detectors and ultraviolet lamps. These last instruments are discussed separately in chapters 8 and 9, but the principles of prospecting with them are the same as without them. In fact, both kinds of instruments can be used, as will be told, for finding valuable minerals that are neither radioactive nor luminescent. The only item that might in any sense be regarded as special equipment is the so-called gold pan; this simple device can be used to as much advantage in looking for other heavy minerals and for gems as for placer gold. The elementary and rather obvious pieces of equipment described in chapter 4 should always be taken along as well: rock hammer, magnifying lens, magnet, and a few of the most necessary tools, together with sample bags, notebook, and pencil. Just what you will need will become fairly apparent in a short while; being overburdened is generally worse than lacking something convenient.

The first decision is what to look for and where to go. No one, surely, can give you foolproof information, or else he would be on the ground himself. Nevertheless, during the uranium boom, for example, more than one millionaire was made by the simple expedient of asking for advice at an office of the U.S. Geological Survey or Atomic Energy Commission. In 1953, the author gave a short talk at a monument dedication near Canon City, Colorado, and was strongly tempted to suggest to his small audience that they look for uranium near Guffey—but he did not take the time, because it was raining hard! Within a few months, the discovery was made. In his prospecting course in Phoenix in 1956, one skeptic asked publicly, "If you know how to find uranium, why are you bothering to give this course?" A good question, with no answer except that some persons prefer to lecture or write rather than dig. Mark Twain observed something to this effect.

Still, rather than starting out haphazardly, these matters—what to look for and where to go—are worth considering carefully. First, look for what you are most apt to recognize when you find it, and concentrate on it. A bookkeeper can examine a large column of figures, looking only for 7s and seeing literally nothing else. You cannot keep your attention on everything at the same time, and yet your chance of success usually increases with having a broader interest than just gold or just jade. This is a personal matter for each one to decide for himself; but it is in general wise for the prospector to do as Leo Mark Anthony, of the University of Alaska, says and "be completely familiar with the minerals for which he plans to search and yet keep an open mind about other minerals which might have economic possibilities."

Certain mineral products—gold or silver or uranium, for example—may be in current demand or seem more exciting than others or be more popular. Or you may have always wanted to be a gem hunter or own a platinum mine. There is no arguing about tastes, says the old proverb.

Secondly: where to go? An old saying says, "If you want to hunt elephants, go to elephant country." The same thing applies to uranium, agates, and gold—even though "gold is where you find it." In other words, you are most likely to make a discovery if you go where your selected mineral or rock has already been found—but preferably an area that has not been so thoroughly exhausted that little of value has come from there in recent years: "proved but not wasted, mature but not senescent." "Follow the ore," said the old timers. Production information is readily available from the agencies named on pages 43–47 or the publications listed on page 407. You can undertake to prospect in untried country after you have acquired enough experience. Do not, however, completely write off an area as worthless because of what others say. The recent Texas Gulf Sulphur discovery at Timmins, Ontario—reckoned at nearly $500 million dollars—was made within 12 miles of a gold camp where mining and intensive exploration had gone on for 55 years. Ironically, a prospector had his cabin just 500 feet from the edge of the ore body, which was exposed at the surface.

Be cautious, nevertheless, of careless statements such as that "the surface has scarcely been scratched." Replying to that remark in regard to Colorado, Charles W. Henderson, of the U.S. Geological Survey, said in 1926, "The surface has been well scratched and even intensively perforated with holes ranging from 10 to 3,000 feet in depth, and with tunnels as much as 5 miles in length." "Much of this 'scratching'," he admitted, "was misdirected." Still, the prospector is well advised to "have the unfavorable localities eliminated by competent geologists" while saving his efforts for more promising places.

In Latin America, Africa, and Asia, the prospector is advised never to pass an old working *(antigua)* without examining it. Even though the ancients were good miners, modern equipment has turned more than one such mine into a substantial producer. "No abandoned gold mine is ever allowed to rest in peace," wrote Herbert Hoover, the world's most accomplished mining engineer, who brought into successful operation some properties of the sort.

Most minerals are found in certain rock types and structures together with others that have the same geologic origin. To be able to recognize these typical rocks and mineral associations is the prospector's main goal. This subject is discussed in detail in chapters 15 to 17 and 18 to 32, which are extremely important.

After acquiring some knowledge of geology and mineralogy—of rocks and minerals—you should begin to relate them to the particular area that you propose to explore. Here, the climate and terrain may be the decisive factors, for they cannot be altered: adjusting to the climate by wearing adequate clothing and conquering the problems of terrain by elaborate means of transportation have nothing at all to do with the matter. Take, for example, the heavily wooded Cascade Mountains, where exposures or outcrops of bedrock are sparse. This situation, located at Puget Sound by "The Song of the Old·Settler," is clearly portrayed in the following verses; the sketches are taken from a place mat at Ivars Acres of Clams, in Seattle.

THE OLD SETTLER

Ivar's says you are welcome to take this as a souvenir.

I've travelled all over this country
Prospecting and digging for gold,
I've tunneled, hydraulicked and cradled
And I have been frequently sold.

So rolling my grub in a blanket
I left all my tools on the ground,
And started one morning to shank it
For a country they called Puget Sound.

Arriving flat broke in mid-winter
I found it enveloped in fog,
And covered all over with timber
Thick as hair on the back of a dog.

I took up a claim in the forest
And set myself down to hard toil,
For two years I chopped and I labored
But I never got down to the soil.

I tried to get out of the country,
But poverty forced me to stay
Until I became an old settler,
Then nothing could drive me away.

And now that I'm used to the climate,
I think that if man ever found
A place to be peaceful and quiet,
That spot is on Puget Sound.

No longer a slave of ambition,
I laugh at the world and its shams,
As I think of my happy condition
Surrounded by ACRES OF CLAMS!

The song of the Old Settler is one of Pacific Northwest's richest pieces of nearly forgotten pioneer lore. The old saying was; "When the tide is out, the table is set," for a man could live in those days by beach combing and digging clams.

The words and illustrations are here revived, for they bring the lusty, natural humor of the old settlers, and it is that spirit which is reflected in the amazing progress of this region.

I have used some of the verses as a radio theme for several years and hope you enjoy them as much as I have.

Ivar Haglund

I've travelled all over this country
Prospecting and digging for gold,
I've tunneled, hydraulicked and cradled
And I have been frequently sold.

So rolling my grub in a blanket
I left all my tools on the ground,
And started one morning to shank it
For a country they called Puget Sound.

Arriving flat broke in mid-winter
I found it enveloped in fog,
And covered all over with timber
Thick as hair on the back of a dog.

I took up a claim in the forest
And set myself down to hard toil,
For two years I chopped and I labored
But I never got down to the soil.

I tried to get out of the country,
But poverty forced me to stay
Until I became an old settler,
Then nothing could drive me away.

And now that I'm used to the climate,
I think that if man ever found
A place to be peaceful and quiet,
That spot is on Puget Sound.

No longer a slave of ambition,
I laugh at the world and its shams,
As I think of my happy condition
Surrounded by acres of clams!

Wooded country cannot be overlooked, however. In much of Gilpin County, Colorado, the surface of the ground was covered with swampy soil that bore aspen trees in the creek bottoms, while on the hillsides grew brush and even luxuriant evergreen forests. Leadville was heavily forested when the silver-lead ores were discovered in 1877.

The advantage of desert country for its sparsity of vegetation may be offset by the discomfort of working there in excessive heat and drastic changes in temperature from day to night. Outcrops in a desert may be covered with so-called desert varnish—a black or brown coating of manganese or iron oxides that is left by evaporation; hence, an orebody cannot be told from barren rock unless it is broken off.

Your ultimate choice of a place to prospect may be influenced chiefly by its suitability for prospecting in general, by its accessibility, by the appeal of new discoveries, or by entirely personal matters: perhaps you have always wanted to strike it rich in Wyoming, or perhaps your car will not travel more than 100 miles before breaking down.

Crystals of topaz shine in the desert sun at the noted collecting locality of Thomas Mountains, Utah. *(H. C. Markman)*

More self-evident problems of import are such things as too many insects to make prospecting feasible, and legal restrictions, as in a national park. Do not, therefore, go mineral hunting wherever collecting and claim-staking are prohibited; for this information, see chapter 13.

Put the odds in your favor by learning as much as you can, selecting the most likely places to search, and then looking for whatever you have most clearly in mind. The wry comment of Tommy Crocombe—a Cornish mine captain in the Little Johnnie mine at Leadville, Colorado—to Harry R. Newitt, reported in 1936, may be typical of the feeling of many an experienced prospector. Asked where he should look to find a mine, Newitt was told: "Damme, my son, I don't know. Where she is, there she is: and where she ain't, there be I." Job the patriarch said it a bit more elegantly: "Surely there is a vein for the silver, and a place for gold where they fine it."

From the standpoint of staking a mining claim, the distinction between a lode deposit and a placer is essential, as defined legally on pages 190–191. It is, if not essential, at least helpful to know the difference when prospecting, because you should realize the unlikeness between minerals that occur in solid rock and those that are found in loose sand and gravel. The formations of sand and gravel may later become tightly cemented by mineral matter until they are as firm as granite, but they still remain placers. We then have to designate whether the rock is a lode—a vein or group of closely spaced veins worked as a unit—or a placer—that is, formerly loose material that was piled up (by water, ice, or wind action) as time passed. A cafe in Boulder, Colorado, used to have a sign over the counter: "Our hash is made, not accumulated." Lodes or veins are "made"; placers are accumulated.

Let us further investigate the matter of outcrops, regardless of whether or not they are former placers that are now cemented into solid rock. (The term *exposure* has exactly the same meaning, often applied especially to igneous rocks, which are those of molten origin.) The surface appearance of a mineral deposit is usually first revealed by the presence of broken rock called float. (Tracing this float is discussed below.) Even if they are not concealed by vegetation, outcrops may be inconspicuous, perhaps barely reaching the surface, as was true of some of the best veins of the San Juan Mountains in southwestern Colorado. The very productive Leadville Blue Limestone, in the same state, was buried under a wash of glacial debris. And billions of dollars' worth of ore must lie hidden beneath barren lava rock in the Western states—but this doubtless is a problem requiring geophysical prospecting. The untapped reserves of gold in California's Tertiary gravels are largely of this kind.

A rise in the price of gold would certainly see a renewed attack on concealed ore deposits. During the Depression, many abandoned mines were reopened in Australia under the combined impetus of a higher price for gold and a patriotic appeal to miners to support the government in this tangible way.

The material that lies above an outcrop is called overburden. Lowlying areas, such as depressions in badly weathered rock, may nevertheless hide valuable minerals beneath an accumulation of overburden. This may consist of soil; loose silt, sand, or gravel (alluvium); slide rock (talus, scree); frozen muck (as in Alaska); or other debris or covering. Frost action is often effective in wedging loose the bedrock. Burrowing animals —ants, gophers, squirrels—may have brought pieces of mineral to the surface; rich gold ore was found in 1933 in Western Australia as a result of testing scrapings from a rabbit burrow. Many superb specimens of amazonstone (green microcline feldspar) and smoky quartz are found entangled in the roots of trees growing in soil of the Pikes Peak granite in central Colorado. Further information on what are known as indicator plants is given in chapter 11; most prospectors, however, are more likely to recognize a mineral in the soil than to observe any peculiar difference in the color or condition of the vegetation.

While you are paying attention to the surface of the ground, every physical feature should be investigated for whatever clue it might suggest. Look at colored, stained, and dendritic rock (having mossy or fernlike patterns), hot springs and cold ones, dry lakes, oily films and tarry ooze. Look closely at the black sand. Get familiar with contacts, faults, and all the structural features that are discussed in chapter 17. Watch out for anything that may appear different: it is, almost by definition, the unusual that is valuable.

Having exhausted the possibilities of the ground surface, you may take the next step of drilling a short distance into the earth wherever your suspicion seems aroused. The result is more or less a matter of chance, however.

Following float to the outcrop comes next. This fundamental aspect of prospecting may be divided into two activities, which are similar in principle but different in the way that they are conducted. One is to trace visible fragments of rock that were broken loose by frost and other processes of weathering and have been carried away from an outcrop by local agents of erosion. In an arid region, float does not travel so far as it does where streams are more closely spaced and more continuously alive. The shape of the land (the topography) is a rather reliable guide to the position of the outcrop and its distance from the float. A little experience will help. Rough gold suggests a nearby source; to the prospector, "coarse gold" means 40 percent or more in nugget form. Jade, which is tough; quartz, which lacks cleavage (see page 220); and corundum (ruby, sapphire), which is hard, can migrate farther than topaz, which cleaves readily, or galena and fluorspar, which are both soft and cleavable.

As a rule, the larger and less rounded the pieces of float, the closer they are to their source.

When float is found, trace it upstream as far as it goes, then up the hillside on both sides of the valley or gulley. Where the train of float ends must be past the outcrop. If the outcrop is not then visible, dig a trench (but don't overdo this hard work) across the line of movement of the float. It should be emphasized that float is not always a value in itself but may serve as a key to commercial mineralization; quartz may be worthless, but its durability preserves it as a guide to veins, pegmatite, and other rock material. Edwin W. Over, Jr., told the author of having found more than 1,000 pegmatite bodies on St. Peters Dome, Colorado, almost all by means of white quartz float.

The nature of float must be considered in connection with the local geology. It is useless, for example, to follow quartz float in an area of schist or slate (see page 247), where the veins of quartz are too small to be of value.

The other way of using float to advantage involves panning, for the desirable minerals may be too small to see otherwise. These are the placer minerals, occurring "in place," but the place is not where they originated. It may later lead to the source—the more or less mythical "mother lode"—or it may be enormously productive in its own right. An ideal placer mineral is heavy, physically durable, and chemically resistant. Its heaviness causes it to settle out of moving water or wind: usually in a stream, often on a beach, sometimes in a sand dune. Physical durability means hardness, toughness, or malleability (see page 220). Chemical resistance protects against weathering. The typical placer minerals include native gold and native platinum, native copper, diamond and many other gemstones, magnetite (the chief mineral in most black sand), chromite and ilmenite, cassiterite (tinstone), zircon, rutile, and monazite.

Many peculiarities exist among these minerals, so that all placer deposits are not alike in any respect, and the skilled prospector for placer minerals is one who has learned (usually by experience) how to judge them. A vast literature has been printed, mostly about gold placers, and you may read further about placers in California, Alaska, or wherever else you intend to prospect in order to understand better the conditions in those areas. Basic information on alluvial deposits in general is given in chapter 17, on elementary geology.

Heavy-mineral methods are especially good for locating gold, chromium, tin, tungsten, and uranium and less so for finding iron, nickel, thorium, and titanium.

Günter Zeschke has listed, according to specific gravity, the following heavy minerals that have commercial importance. The minerals that are most important economically are italicized.

Mineral	Specific gravity	Mineral	Specific gravity
Iridosmium	20–21.2	Hematite	4.9–5.2
Gold	15.6–19.3	*Magnetite*	4.9–5.2
Platinum	14–19	*Uraninite*	4.8–9.7
Silver	10–12	Molybdenite	4.7–4.8
Bismuth	9.7–9.8	Pentlandite	4.6–5.1
Cinnabar	8.0–8.2	*Pyrrhotite*	4.5–4.6
Ferberite	7.5	Gadolinite	4.5
Galena	7.3–7.6	Thorite	4.4–5.4
Wolframite	7.1–7.5	*Zircon*	4.4–4.8
Cassiterite	6.8–7.0	Xenotime	4.4–4.6
Vanadinite	6.7–7.2	*Ilmenite*	4.3–5.5
Antimony	6.6–6.7	*Chromite*	4.3–4.6
Cerussite	6.5–6.6	Stannite	4.3–4.5
Scheelite	5.9–6.2	Rutile	4.2–4.3
Arsenopyrite	5.9–6.2	Chalcopyrite	4.1–4.3
Fergusonite	5.8–5.9	Willemite	3.9–4.3
Pyrargyrite	5.8–5.9	Corundum	3.9–4.1
Cuprite	5.7–6.1	*Garnet*	3.8–4.2
Columbite	5.4–6.4	Brookite	3.8–4.1
Zincite	5.4–5.7	Spinel	3.6–4.4
Franklinite	5.0–5.2		
Monazite	4.9–5.3		

Geologic conditions are responsible for the difference in yield between placer deposits and lode deposits (say, of gold) in a given region. A few examples may suffice: in the Congo, Guyana (British Guiana), and Siberia, the gold is mostly of alluvial origin; in Indonesia, Rhodesia, and Transvaal, it has been nearly all lode gold; in Ghana (the Gold Coast), Korea, Queensland, Tasmania, and Western Australia, lode gold far exceeds placer gold; and in the United States, Canada, New Zealand, and New Guinea, large amounts of both kinds have been produced.

Lodes may not have placers associated with them because of conditions that failed to preserve the placers; and, in turn, the veins that were the original source of placer gold may have been eroded away. In spite of these exceptions and the diversity indicated previously, it is reasonable to expect placers where lodes have been found, and vice versa.

The logical procedure, therefore, is to follow the placer material upstream and then upslope, just as we followed the readily visible float as explained before. Both placer material and float may appear together. Placering, however, requires the proper use of a gold pan; C. McK. Laizure once listed a couple of hundred other machines and processes that have been presented to the public in California, "none of which has been tested or approved by the State Division of Mines, although some have been observed in operation by representatives of the state." These devices serve more for recovering placer minerals than for identifying their presence, and so will be discussed in chapter 33. Like a gold pan, however, are such things as the batea (a wood dish widely used in South America), the Asiatic ladle (a gold pan having a handle and used in the Soviet Union), and the dulong (used in Indonesia). Miner's pan, California pan, and riffled pan are other names. George A. Jackson opened the Pike's Peak or Bust gold rush in 1859 by panning dirt in an iron "treaty cup," one of those that was given to the Indians by the government.

The gold pan is an indispensable prospecting tool: cheap, portable, versatile, and efficient. The amateur can easily recognize a number of minerals that are valuable or are good clues to valuable deposits farther along. These are either heavy minerals (dark or light) or those that tend to float (such as mica). In addition, John B. Mertie, Jr., of the U.S. Geological Survey, showed that an experienced panner can recover from any sample 80 percent of the minerals that are heavier than quartz, and nearly anyone can separate practically all the coarse gold and platinum (these are very heavy) and a high percentage of the other heavy minerals. These minerals include barite, cinnabar, epidote, galena, garnet, hematite, hornblende, ilmenite, magnetite, monazite, pyrite, scheelite, sphene, zircon, and others. The gold pan can also be used for quantitative purposes, but skill is required.

A gold pan is easier to use than to describe in use, as follows:

1. Load it about half full with sand and gravel, dug as deep as possible; a pound or two of crushed ore taken from a lode is enough.

2. Add water from a pool of slack water or a washtub, break up the clumps of dirt, and stir up the gravel.

3. Fill the pan with water, take out the larger pebbles by hand, and toss them aside.

4. Shake the pan and give it a rotary motion, dropping first one wrist and then the other. In this way, the lighter material spills gently over the far side, which is slightly tilted.

The work is repeated, becoming slower toward the end, until nothing remains in the pan but black sand and gold (in the typical operation). A good technique may be rewarded by having the "heavies" strung out in a crescent, the gold gathering at the tail end.

This pan of gold nuggets from Quartz Creek, Yukon Territory, was worth $6,000 at the old price.

A skilled panner can quite literally find a needle in a haystack—in such hands, the process is almost incredibly fast and accurate. The author saw Robert W. Roots, Jr., at about 75 years of age, win the world's panning championship at Sacramento by recovering all 20 flakes of gold, the size of pinheads, from 1 ton of crushed rock in less than 20 minutes. Many women have the dexterity to give much the same performance, and some have won the international championship.

In countries where boulders are covered with moss, the moss itself should be scraped off and panned to release particles of heavy minerals. In northern regions, the ground may have to be thawed before it can be sampled.

If you intend to prospect mainly for gold, you will want to understand its known occurrence in the particular region where you expect to work. The published literature is enormous, but some suggested items are listed on page 292 of this book. The other placer minerals may represent occurrences that cover the whole range of geologic conditions, so that it is impossible to do more than to try to learn what is known about platinum, gems, or whatever else you are interested in.

Always try to recover more than one material at a time. V. E. McKelvey, of the U.S. Geological Survey, said, "If a metal can be recovered as a byproduct of another constituent, its concentration need be only one-tenth as much as it would be otherwise." A little gold can make it possible to work placer deposits for other, less valuable metals. Among the metals that are now valuable but that were often overlooked in older placer operations, tin, platinum, and tungsten are most likely to be found.

The next step in tracing a lode is to move upstream, testing each tributary and dry gulch, following the present or former watercourse that turns up the best paying sample. The trend of the placer may not follow the present drainage pattern at all but rather an ancient drainage that is now on top of a mountain range, perhaps even at right angles to the direction of today's streams. California presents this sort of situation to a high degree. Alternating good and barren stretches of stream suggest that tapping an ancient drainage may be responsible for the favorable showings. For this and related reasons, the geologic history of a region may need to be known in order to decipher the situation adequately—hence, learn some basic geology (see chapter 17) and find out as much as possible about the area in which you are prospecting.

Geochemical prospecting, explained in chapter 11, may be combined to advantage with stream panning.

Having found something of possible value is not enough if you forget where it was acquired. So keep legible notes and mark positions on a map. This is part of the need to be "systematic, consistent, curious about everything . . . sampling everything about which the prospector is in doubt," as expressed by Leo Mark Anthony, of the University of Alaska.

Outcrops may be of any size and appearance. Some stand boldly above the surface; this is especially true of quartz outcrops, which, incidentally, are often the favored material for gold and silver. The Mother Lode, in California, is conspicuous for its outcrops of quartz, but these are largely barren, whereas the large mines had no large quartz outcrops, although their gold veins contained quartz. Silicious (usually quartz) ores in soluble rocks (such as limestone) are especially apt to outcrop strongly; those in granite (and related rocks) show a variable outcrop effect; and those in quartzite weather at about the same rate as the rest of the rock.

Outcrops may not be readily visible at the surface at all, perhaps weathering more thoroughly than the surrounding rock and being less prominent rather than more so. Some outcrops, especially those that contain ores with pyrite, occupy depressions in the surface. Certain extremely rich districts—such as Cripple Creek, Colorado, and Kalgoorlie, Western Australia—showed practically no outcrops. Hence, the first impression as to the size of an outcrop is not trustworthy, for a small and obscure one may be rich. Before sampling it, however, a more detailed study is in order.

The physical aspects of the rock should be observed closely. The spacing of the fractures may suggest the extent of mineralization that is possible. The degree of silicification (the amount of quartz and the spacing of quartz veinlets), kaolinization (the decomposition to clay), and sericitization (a change to mica) are often indicative.

Where different outcrops are related, they may influence one another. Thus, the intersection of two veins is often the site of an ore shoot—the bonanza every prospector dreams of—and the complex crossing of veins or fractures may mean the most productive part of an ore shoot. These conditions are discussed further in chapter 17, on basic geology.

Outcrops often tell us much about lodes. The prospector tries to gauge what he will find at depth before going to the expense of digging, and he can sometimes come to a fairly close estimate. Strong veins, being wide and long in the direction of their trend, are apt to continue in depth. If a lode is more resistant to weathering than the surrounding ("country") rock, it is likely to get narrower as it deepens. On the contrary, if the lode is less resistant than the country rock, it will probably become wider as it goes down. In both of the last two instances, the result stems from the fact that the more resistant material—whether lode or otherwise—tends to dominate the outcrop surface.

Besides the silicification of an outcrop, the color should be especially examined. This requires a broader knowledge of minerals, but a good deal of useful information can be given here, and you can look up the description of any of these minerals in chapters 18 to 32.

As the insoluble minerals in a lode remain intact—either in place or as float or as placer minerals—the soluble minerals are dissolved by groundwater and are carried farther down or farther away. The outcrop becomes stained with various mineral compounds, which may extend to shallow depths. The colors of these secondary minerals serve to indicate the presence of a primary deposit of either precious or base metal lower down or in the vicinity, although its ultimate value is not guaranteed. They may be the only clue to hidden wealth.

Some prospectors have trained themselves to look chiefly for gossan. The insoluble material that is blotched is called gossan, or iron hat. It is a porous, jagged capping on the lode. The color is usually rusty yellow or chocolate brown or red; it is due to iron oxide and hydroxide minerals that are derived from iron sulfides (pyrite, chalcopyrite). Copper minerals yield blue and green colors. Certain nickel ores turn pale green ("nickel bloom"). Cobalt may give a pink or red color ("cobalt bloom"). Molyb-

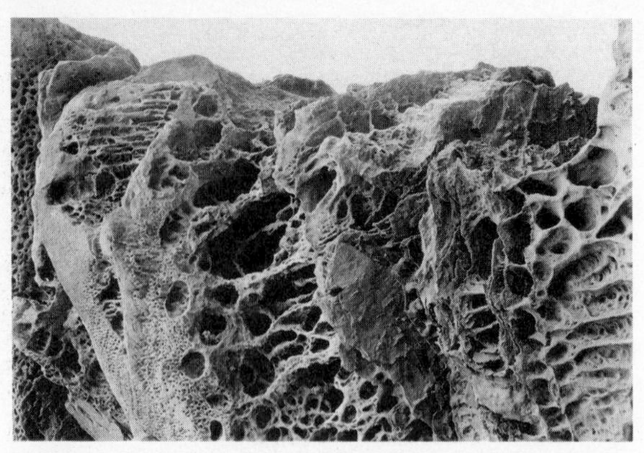

Cellular rock may, if color stained, be a good clue to buried mineral deposits. *(U.S. Geological Survey)*

denum becomes pale yellow; manganese, sooty black; uranium, a range of bright orange, yellow, and green hues.

The color of the soil in deeply weathered areas is generally influenced mostly by the iron content. Be wary of the blue and green colors of copper, the red of mercury, the black of manganese; these metals, as well as iron, can stain a lot of rock and yet not mean anything of value. Beware also of lichens, which resemble mineral stains. Gossan may be thin or feeble in cold regions, where chemical weathering on recent erosion surfaces has been slight. Southeastern Alaska fits this description.

Apart from gossan, a few other indications of color might be watched for. Igneous rock that is white and chalky looking—often granite or porphyry—is a promising sign. Quartz veins of a cloudy or milky appearance are more auspicious than not; likewise, as previously mentioned, is a quartz vein colored yellowish brown by hydrous iron oxides. The brassy look of pyrite in a vein (not in beds of clay or coal) is heartening. The rock next to many mineral deposits has been altered chemically, producing pyrite, calcite, white mica, chlorite (dark green), or quartz. Each district has its special marks.

A random perusal of an outcrop should be followed by a systematic traverse. This is accomplished by walking slowly a series of parallel lines spaced perhaps 100 to 500 feet apart and then walking a similar series at right angles to the other set. These paths should be aligned first along the geologic structure and then opposite to it. A more complete traverse would be rather like making a geologic map. Some prospectors go from one outcrop to another, some move along ridges, others have different preferences. The amount of time that seems worth spending at any given outcrop cannot be specified; in general, few discoveries will prove profitable, and so one's time must be allotted in accordance with one's judgment. But, as the Ontario Department of Mines says, "If the rocks are of an encouraging type, it is better to prospect patiently there than to keep traveling, forever looking for a better country."

Some general advice about ore deposits may be helpful in conclusion. An old prospector's foolish tale has it that veins get richer with depth—does not the ore come from below? Once beneath the so-called enriched zone, which is always more or less shallow, veins do not gain in value (called tenor) except where they become ore shoots.

Likewise untrue is the belief that outcrops of veins always contain valuable minerals; and that their minerals, if once present, are always subject to being dissolved (leached) out by the processes of weathering, perhaps to be enriched below.

Neither is it to be relied on that ore deposits are similar in nearby areas; nature is too arbitrary.

Nor may one assume that a happy combination of conditions, no matter how favorable they seem, is sure to yield ore. Plant growth of some types may indicate the presence of ore (see chapter 11), but it usually does not.

The Ontario experts believe that "the secret of success is curiosity, close observation, and steady hard work." This is no secret!

There are numerous miscellaneous methods of prospecting that seem to have less validity than these standard ones that are based on generally accepted scientific principles. "Creekology" was commonly used in the oil fields of Pennsylvania, and throwing one's hat in the air to see where it landed was a popular method in the same area and elsewhere. Seances with the spirits were held to decide the location of mines in Ute Pass, Colorado. At about the same time, in 1910, Arthur J. Hoskin, of the Colorado School of Mines, wrote, in regard to prospecting, "The time of mystics, in any line, is past." He was a poor prognosticator, for the growing mania today for astrology and similar practices suggests that prospectors can scarcely escape their influence. Doubtless, the ouija board and other devices have been employed frequently in many places. Methods of prospecting that involve alleged psychic powers are not dealt with here further. These would include dowsing, witching, and a wide array of similar techniques that are not in good scientific standing. On the other hand, recent experiments with dogs that have been trained to sniff out

The divining rod used by Medieval miners.

sulfide-bearing rocks, using methods that were tested first in Finland and then Sweden and Russia, are proving successful in Canada, under the direction of Harry V. Warren, of the University of British Columbia.

Richard Gump, San Francisco merchant, quotes Sung Ying-hising in his *T'ien kung k' ai wu,* dated 1637, as reporting that jade was prospected for in the rivers of China by young maidens walking unclothed at night. Being female (*yin*), they would attract the jade, which is male *(yang),* and could pick it up as it touched their feet. This beats the hell out of Ping-Pong. Prospecting has not been the same since.

8

Prospecting for Radioactive Minerals

Even if they are not absolutely required, special instruments so much aid in radioactive-mineral prospecting that the subject is scarcely worth discussing without them. In this respect, prospecting for radioactive materials is a major phase of geophysical prospecting, which is explained at some length in chapter 10. However, the use of the Geiger counter and the scintillation counter has become, in recent years, more widespread even than the use of the gold pan, described in chapter 7; and such popularity has put these familiar instruments into the category of everyday prospecting equipment. Nevertheless, the special techniques that are used for uranium and thorium prospecting justify our considering it in this separate chapter.

Besides the general advice given at the end of this discussion, three basic principles may well be stated early, as follows:

1. Although it is probably true that any child can operate a Geiger counter, it takes a certain amount of experience and judgment to use it to advantage.

2. Although the Geiger counter (and its more sensitive cousin, the

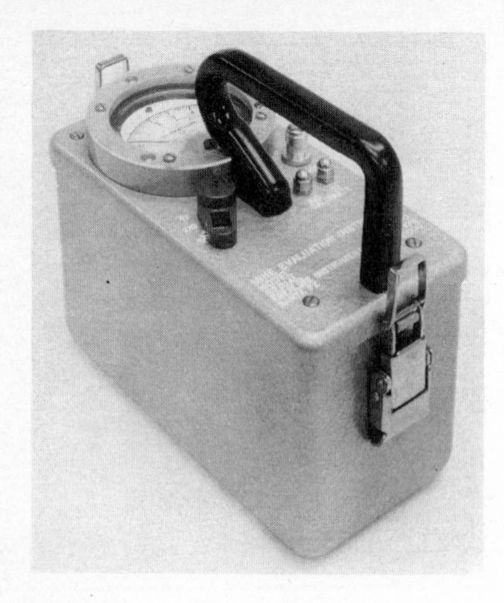

The portable Geiger counter is the basic piece of uranium prospecting equipment. *(Eberline Instrument Corporation)*

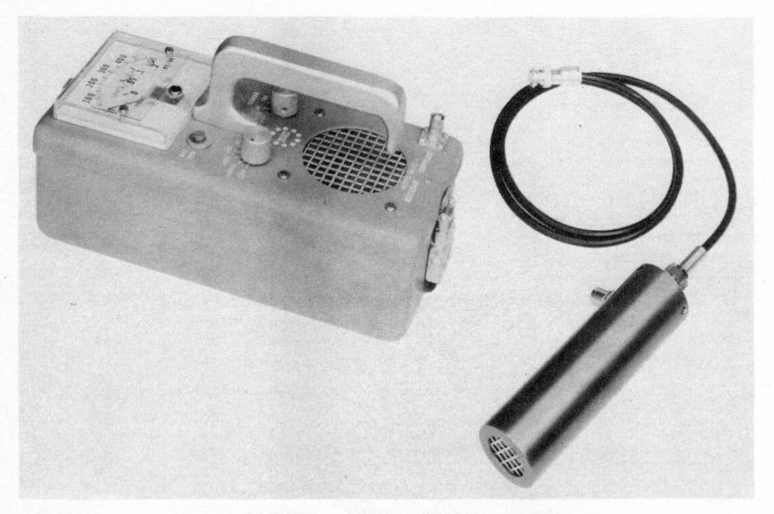

This Geiger counter is called a survey meter. *(Nuclear Supplies)*

scintillation counter) is an electronic device, albeit a fairly simple one, the commercial value of radioactive ores is determined by chemical analysis. The counters, therefore, merely indicate the presence of a reaction but they do not evaluate it, even though they may seem to do so as they give more or less precise readings on their meters.

3. Although these instruments are detectors of radioactivity in general,

This portable scintillation counter includes exploration probes. *(Urinco)*

the government (through 1970) or other buyer purchases only uranium or—much less urgently—thorium. There are, you see, many radioactive elements besides uranium and thorium—including cosmic radiation, which cannot be caught and bottled for sale at all.

Lest this information discourage you, please remember that uranium prospecting can be highly profitable, for it is necessary to today's economy and therefore pays well, and it can be much easier than any other kind of prospecting. The instruments will do most of the work for you, provided you keep control of the thinking! No kind of prospecting is profitable, let alone safe, for the nitwit.

Perhaps the next consideration should be whether uranium prospecting is in demand at present or is apt to be in the future. The demand for uranium declined after large amounts were discovered during the spectacular boom of 1948–55. Exploration fell behind development in 1954. It was thus only a matter of time until the known supply (reserves) would decrease in the face of the slowly but steadily growing need for more raw material as the fuel for nuclear energy. Exploration and development both increased sharply in 1966. A new surge in prospecting for uranium—

a second boom—began in 1968, has halted a bit, but seems likely to be resumed after 1972. Thorium also has a good (but at present a faint) potential as a future nuclear fuel. With the growing maturity of this industry, entirely for peace rather than war, it seems probable that a re-sonable economic balance can be maintained between demand and supply and that uranium prospecting can be conducted with as much assurance of being rewarding as any other kind. It need be neither melodramatic nor obsolete; it may instead become a rather routine activity that is participated in by a number of prospectors who happen to like it.

The descriptions of uranium and thorium minerals appear in chapter 22. It is not so necessary to be able to identify them as other minerals, inasmuch as the Geiger counter is of help. Whereas radioactivity can be recorded by a counter without having any commercial significance, the reverse is not true: any commercial value will give a good reading on the counter. The contact-print test (see page 332) works solely on uranium, not thorium or the byproducts of radioactivity. The fluorescent-bead test (see page 228) is even more specific, for it responds only to what used to be the Atomic Energy Commission's definition of source material; that is, 0.05 percent U_3O_8 or more. Mining and milling (the latter accounts for more than half of the total cost) present other problems, of course.

The particular mineral associations, rock types, and geologic structures that are most favorable for the occurrence of radioactive minerals are described in the appropriate places in chapter 22. If you intend to concentrate your prospecting efforts on uranium (or thorium), you will want to learn the known occurrence in the particular region where you expect to work. This implies a familiarity with some of the vast published literature of the sort referred to on pages 127–128 of this book.

In addition to uranium and thorium, radioactivity methods are particularly effective in detecting occurrences of phosphates, manganese, and petroleum, and less so for beryllium, cobalt, copper, lead, lithium, nickel, silver, titanium, zinc, fluorite, mica, sulfur, coal, and evaporite "salts."

USING A COUNTER

To use a Geiger counter or scintillation counter effectively, it is desirable to understand something about radioactivity. This is the spontaneous automatic breakdown (disintegration, decay) of chemical elements as they change into other, lighter elements. Every natural element has at least one species (called an isotope) that is radioactive, and all the manmade elements are radioactive. For the purpose of dating rocks by their radioactivity, certain pairs of elements (such as potassium-argon) are used; but for prospecting purposes, only the uranium and thorium series are of concern, because only these two elements have commercial value. Both of them change finally to lead, which is the stable end of each series. During the transformations that take place, heat is generated, helium gas is given off, and a number of intermediate "daughter elements" are formed, which in turn break down radioactively.

The radiation that is produced by this natural process cannot be hastened or delayed by pressure, heat, or chemical means and is constant during the passage of time. It consists of so-called alpha, beta, and gamma rays. Alpha rays are positively charged particles of weak penetrating power; beta rays are negatively charged particles of somewhat greater penetrating power; and gamma rays are uncharged radiation, being true rays like powerful X-rays. The Geiger counter records chiefly gamma rays, and some prospecting devices measure beta radiation.

You should be aware that, although some important uranium minerals are luminescent and will glow in ultraviolet light, as described in chapter 9, this effect is not caused by radioactivity, which is invisible. Neither are there any magnetic properties that might influence a watch or compass. The radiation is also soundless until it is directed toward the mechanism of a Geiger counter and produces an electric impulse, which can be heard in a set of earphones. This is the result of radiation colliding with molecules of gas, yielding charged particles. The radiation may also, or instead, be recorded on a meter or as a flash of a neon light. Many instruments do all three things at the same time. For field use, of course, a portable Geiger counter operates on a battery. The cost of the instrument varies; a moderately priced one is probably best for general use, neither too cheap to be reliable nor too expensive and specialized.

If one is going to invest in a costly model (having many tubes), he might do better to buy a scintillation counter instead, for it is more sensitive, though also more delicate. It records on a meter the amplified flashes of light that are produced in a crystal (known as a phosphor) by gamma rays. Scintillation counters are available commercially in small, hand-portable models, airborne-carborne models, and well-logging models; the last two belong with the geophysical equipment discussed in chapter 10.

After they are turned on, the counters are read for the so-called background count. This is due mostly to cosmic rays but also to other factors, and it must be subtracted from the reading obtained in the field. If, for

The radiometric equipment that will be used in flying a survey. *(Cryogenic Research Company, Inc.)*

instance, the total count is a good 40 per minute (on the particular scale being used), but 30 of it comes from the background, you are not likely to have discovered a fortune in uranium. Inasmuch as the background count varies continuously, it must be checked at frequent intervals. Corrections should be made for special conditions (altitude, sunshine, rock type, time of day) as well as for the instrument itself.

Because the counters record the total amount of radiation, not the percentage of uranium or thorium in a specimen, a large specimen seems richer than a small one. Likewise, if the reading is taken within an enclosed space, such as a mine tunnel, it will give a greatly exaggerated result. This is known as the mass effect and should be guarded against. A certain thickness of rock, as well as being in a hollow, may, however, shield off the cosmic radiation and give a lower-than-average reading.

Similarly, without special radiometric tests, radioactivity that is caused by one element cannot be distinguished from that which is due to others. The Garden of the Gods, in Colorado, for example, suggests with the Geiger counter a fairly productive uranium deposit, but the red rocks there contain much feldspar, which contains much potassium—radioactive but commercially worthless. To tell uranium from other radioactive elements, there are several generally inconvenient methods that are known, such as gamma-ray spectrometry.

Owing also to the so-called equilibrium effect, a chemical analysis (upon which sales of radioactive materials are made) may indicate a value that is considerably higher or lower than does the reading of a counter. A deposit needs about 1 million years to reach a state of equilibrium, free from the dissolving action of groundwater. Many mining operations have floundered on a failure to take this situation into account.

Once the above principles are understood, you may undertake the actual prospecting. A slow, systematic traverse, conducted like that described for walking an outcrop—in a grid pattern on flat beds, in straight lines on an inclined structure—will give you data for a map or sketch.

If the counter registers three or four times the background count at any place, you should stop and examine the ground more closely. A consistent reading higher than that justifies taking samples for assay. Remember, too, that 2 to 6 feet of overburden, soil, snow, or a water body may absorb the radiation, no matter how strong, and so this method of prospecting is largely a near-surface one. Do not rush the readings, either of the rock or of the background, or you may miss narrow veins of rich ore. Take care of your instrument: keep it away from luminous dials, overheating (as by a campfire), and cold weather; keep it (and the probe) as dry and clean as possible—radioactive dust will not help matters; employ the correct scale setting so that it will neither be too sluggish nor too agitated (which will damage it); and turn it off when not in use. It should be calibrated for accuracy often. An electronic-instrument repair shop is qualified to check batteries and connections. An 80-year-old woman who took the author's prospecting class in Pueblo, Colorado—"learning keeps me young"—told him that if a Geiger counter is like a radio, she would repair hers herself, for "fixing radios is my hobby."

Both the Geiger counter and the scintillation counter are adaptable to preliminary prospecting from a motor vehicle, making possible a larger coverage in a shorter time. From a practical standpoint, however, this method seems feasible only in an area that has not previously been similarly explored from the same roads.

Let us, then, do as the oldtime gold prospector was heard to say, as he stood at the tavern bar: "I'm a' goin' to get me one of them there jigger counters and go counterfittin' for geranium."

An indirect application of the radiation counter can make it serve to find placer deposits of gold when this metal occurs with a heavy radioactive mineral, such as monazite.

Radiometric measurements are also being used in prospecting for deposits of beryllium. An instrument that is related to the scintillation counter—one brand is called the Berylometer—detects this element, which has entered the nuclear-energy program on an important basis. However,

The radon gas detector is used in surface uranium exploration. *(Cryogenic Research Company, Inc.)*

you should understand that beryllium is not a radioactive fuel, and so the search for uranium or thorium is not directly involved in the search for beryllium. Instruments similar to the Berylometer are now becoming available for the detection of gold, silver, and possibly other metals. The prompt-neutron-capture-gamma-ray method was announced as successful for detecting nickel and may be adapted to about 30 other elements in the future (reported by the U.S. Geological Survey in 1971).

Some progress is being made in prospecting for uranium by measuring the content of radon gas in soil and rock. This rather new form of geochemical-geophysical prospecting is called emanometry. It is concerned with dispersion halos of radioactive elements. Halos can also be determined by relatively cheap fluorescence or X-ray fluorescence analyses.

LITERATURE ON RADIOACTIVE MATERIALS

The literature on the deposits of radioactive minerals is enormous, so that only a professional bibliographer would care to try becoming familiar with it. Almost all the literature and maps that you are apt to want can be had by writing to the U.S. Atomic Energy Commission, Oak Ridge, Tennessee 37204, or Grand Junction, Colorado 81501. They will send you a list of official AEC depository libraries. The books that are named below are classified in several groups, from which you may select reading of interest to you. They will serve as a further guide to this vast subject.

Minerals for Atomic Energy; A Guide to Exploration for Uranium, Thorium, and Beryllium, 2d edition, Robert D. Nininger, Van Nostrand-Reinhold, New York, 1956. (The most comprehensive coverage of prospecting for uranium, thorium, and beryllium minerals.)

The following seven books are on an intermediate to elementary level:

The Uranium Prospector's Guide, Thomas J. Ballard and Quentin E. Conklin, Harper and Row, Publishers, New York, 1955.

Prospecting for Atomic Minerals, Alvin W. Knoerr and George P. Lutjen, McGraw-Hill Book Company, New York, 1955.

Uranium: Where It Is and How to Find it, Paul Dean Proctor, Edmond P. Hyatt, and Kenneth C. Bullock, Eagle Rock Publications, Salt Lake City, 1954.

How to Prospect for Uranium, Harry Kursh, Fawcett Publications, Inc., New York, 1955.

You Can Find Uranium, Joseph L. Weiss and William R. Orlandi, J. B. Weiland and Company, San Francisco, 1949.

Uranium Prospecting, Hubert Lloyd Barnes, Dover Publications, Inc., New York, 1956.

Uranium Prospectors' Handbook, Repro-Tech, Inc., Denver, 1954.

The following three books are small but worthwhile publications that are sold by the Government Printing Office, Washington, D.C. 20402:

Prospecting for Uranium, U.S. Atomic Energy Commission and U.S. Geological Survey, 1957.

Prospecting with a Counter, Robert J. Wright, U.S. Atomic Energy Commission, 1954.

Facts Concerning Uranium Exploration and Production, John E. Crawford and James Paone, U.S. Bureau of Mines, 1956.

The updated and expanded supplement (1968) is:

Prospecting and Exploring for Radioactive Minerals, U.S. Bureau of Mines Information Circular 8396.

Canadian conditions for prospecting are dealt with in the following four publications (see also page 202):

Prospecting for Uranium in Canada, Officers of the Radioactive Resources Division, Geological Survey of Canada, 1953.

Brief Information on Prospecting for Uranium in Canada, Geological Survey of Canada Pamphlet Miscellaneous G. 100–7, 1955.

Information on Services for Testing Radioactive Samples, Department of Mines and Technical Surveys Circular, 1954.

Prospectors' Guide for Uranium and Thorium Minerals in Canada, Canadian Bureau of Mines, 1948.

Several good papers on uranium exploration are given in:

Survey of Raw Material Resources, United Nations Conference on the Peaceful Uses of Atomic Energy, 2d Proceedings, vol. 2, New York, 1958.

Uranium Exploration Geology, Proceedings of a Panel on Uranium Exploration Geology held in Vienna, April 13–17, 1970, International Atomic Energy Agency, Vienna, 1970.

The following few publications will, in the absence of a trained librarian, guide you to the chief literature and maps that deal with prospecting for uranium in specific parts of the United States and Canada and in the individual states and provinces:

Facts Concerning Uranium Exploration and Production, John E. Crawford and James Paone, U.S. Bureau of Mines Handbook, 1956.

U.S. Geological Survey Bulletin 1195 (bibliography 1950–59).

Bibliography and Index of Literature on Uranium and Thorium and Radioactive Occurrences in the United States, Margaret Cooper, Geological Society of America Bulletin, vol. 64, pp. 197–234, 1103–1172 (1953), vol. 65, pp. 467–590 (1954), vol. 66, pp. 257–326 (1955).

Unexplored Uranium and Thorium Resources of Canada, S. M. Roscoe, Geological Survey of Canada Paper 66–12, 1966.

Canadian Deposits of Uranium and Thorium, 2d edition, A. H. Lang et al., Geological Survey of Canada Economic Geology Series no. 16, 1962.

Uranium in Canada, G. C. Garbutt, Eldorado Mining and Refining Limited, Toronto, 1964.

9

Prospecting for Luminescent Minerals

The prospector has two distinct uses for luminescent minerals—those that glow in ultraviolet, or "black," light. A few of these minerals—especially scheelite (an ore of tungsten) and several uranium minerals—are valuable in themselves; and some other luminescent minerals serve as a helpful clue to the existence of nonluminescent minerals of commercial use. The other use for minerals of this kind is as collectors' specimens, this aspect of the mineral hobby having seen a tremendous growth in recent years.

Luminescence is the ability of a mineral (or other substance, of course) to absorb radiation and then emit it in a different wavelength. For our purpose, however, it is concerned only with changing ultraviolet light, which is invisible, into colors of the visible spectrum: red, orange, yellow, green, blue, violet. Purple or violet is probably due to visible light, which cannot be cut out by a piece of window glass held between the lamp and the specimen, whereas true fluorescence is eliminated. While the mineral is exposed to the ultraviolet radiation, the effect is called fluorescence. When the effect persists after the ultraviolet radiation has ceased, it is

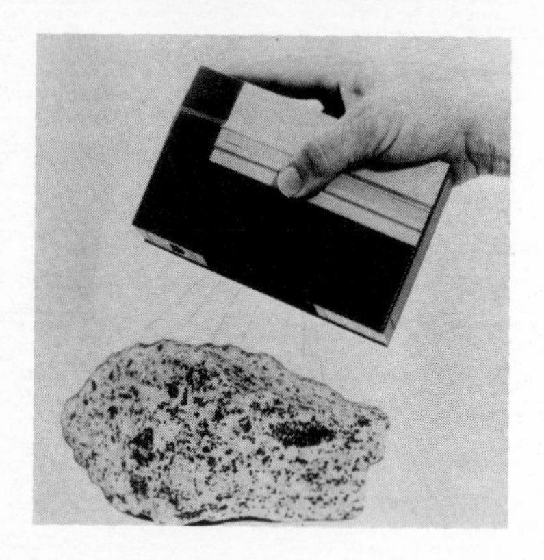

Luminescent minerals glow in ultraviolet light. *(Ultra-Violet Products, Inc.)*

known as phosphorescence, but not the same phosphorescence (bioluminescence) as is given off by organic matter. Other types of luminescence are known—produced by heat (thermoluminescence), friction (triboluminescence), and chemical reactions (chemiluminescence), although these are not of much interest here.

Luminescence can be produced in a number of ways, including lightning, X-rays, cathode rays, and other means. Even sunshine can bring out the fluorescence in ruby, diamond, and certain other gems. The prospector and collector utilize one of several varieties of ultraviolet lamps, and of course the prospector needs one that is portable, operating on a battery. Because luminescence covers a fairly wide range of wavelengths, the lamps are manufactured in models that are termed "long" or "short." The most convenient models combine both long and short wavelengths (concentrated at 3,650 and 2,537 Angstrom units, respectively), so that either radiation can be used, as the operator prefers. If only one variety is available, the shortwave lamp will prove most useful; this radiation does not come from the sun and must be created artificially.

Not only is scheelite the most valuable mineral that always fluoresces, but it is especially interesting because of the fact that the grade of the ore can be told by the vivid bluish-white luminescent color. This is because scheelite often contains molybdenum—as a transition toward the mineral called powellite—and the color of the fluorescence then tends to be white or yellowish white. The more molybdenum, the yellower the color, and powellite is golden yellow. A special color screen that was devised by the U.S. Bureau of Mines can be bought, showing how much tungsten or molybdenum is present.

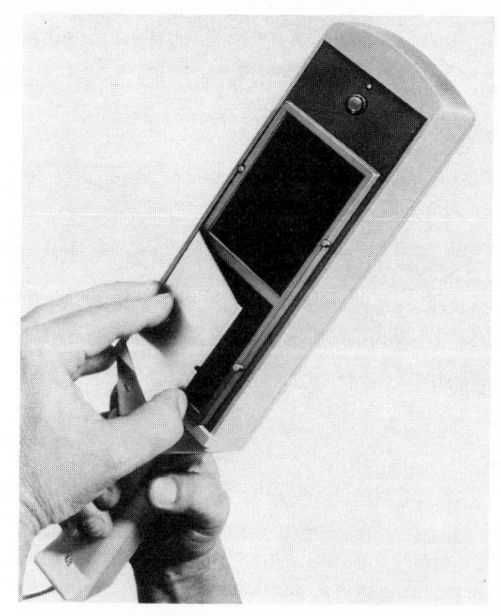

Combining long and short wavelengths, this ultraviolet lamp connects with house current. *(Ultra-Violet Products, Inc.)*

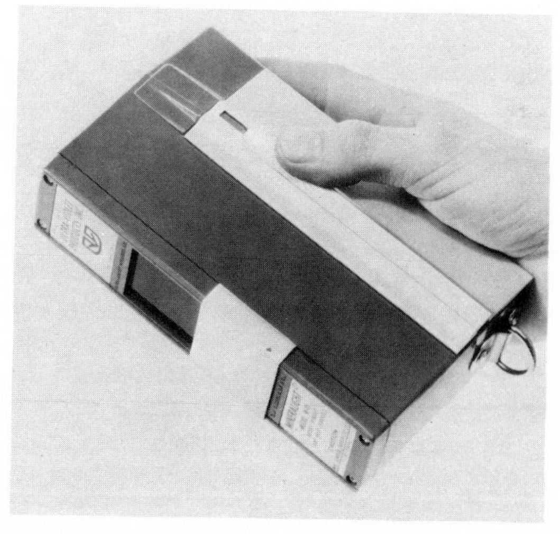

This pocket-sized ultraviolet lamp is the ultimate in portability. *(Ultra-Violet Products, Inc.)*

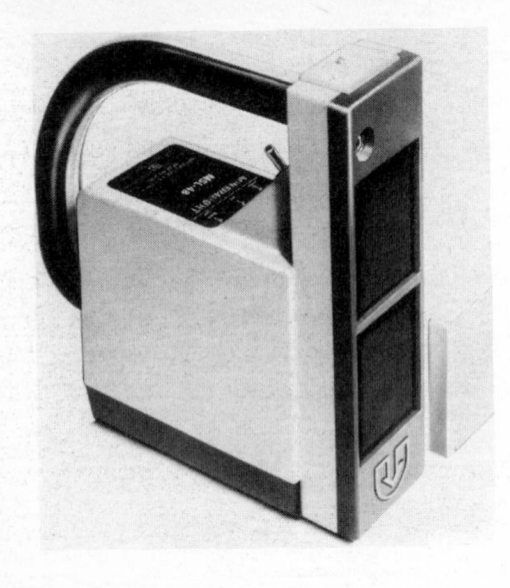

Most ultraviolet lamps, like this one, operate on batteries for field use. *(Ultra-Violet Products, Inc.)*

Primary uranium minerals do not fluoresce at all. Autunite is the most likely of the secondary uranium minerals to fluoresce, the color being a bright yellowish green. Torbernite, which is otherwise very similar and occurs with it, does not glow. Uranium and thorium minerals do not phosphoresce; thorium and rare-earth minerals do not fluoresce.

As can be realized from the above statements, scheelite and autunite may be helpful in finding other tungsten and uranium minerals, which may actually be much more abundant. The search can go on only after dark. During the hectic days of the first uranium boom, some impatient prospectors went around with a Geiger counter in the daytime and an ultraviolet lamp at night. A hood for observing even in the sunshine can be bought or made from black cloth.

Calcite, barite, and fluorite are perhaps the most important of the luminescent minerals that may indicate the presence of metallic ores without being of any other value except as specimens. The colors are not especially diagnostic, for, as with most minerals, the luminescence is due to miscellaneous foreign atoms, typically manganese, copper, silver, lead, or uranium. Willemite, a zinc mineral coming mostly from New Jersey, is beautifully fluorescent, but only in that locality. Fluorite, after which fluorescence was named, often does not show it.

Tests for fluorescence identify uranium (by a bead, as explained on page 228), mercury (using a special screen that is made for the purpose), and beryllium (which is not suitable for ordinary prospecting).

Everything that glows is not mineral! Lichen and fungi may do so, but these can be scraped off, or a fresh surface of the specimen can be exposed. Rattlesnakes also fluoresce.

Present-day equipment, having an adequate filter to eliminate as much visible light as possible (shortwave filters must be replaced occasionally), is close to being ideal. The illumination is intense but uniform and properly balanced, while excessive heat and unpleasant fumes are avoided. A mercury-vapor lamp is best. Although prospecting with an ultraviolet lamp is, in a sense, a form of geophysical prospecting (see chapter 10), it is even simpler than prospecting for radioactive minerals (explained in chapter 8) and so is appropriately discussed here.

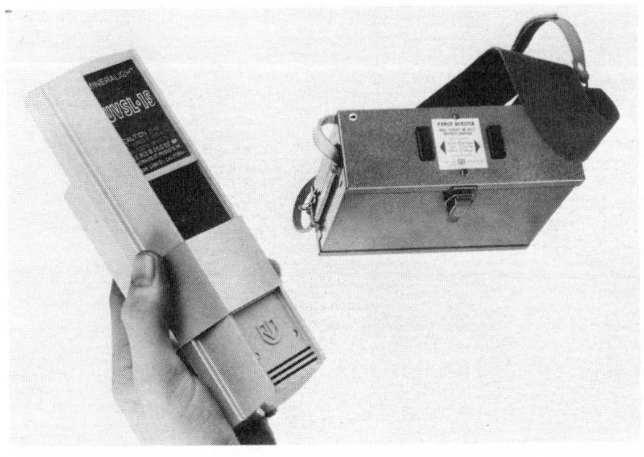

A booster pack is one of several accessories that extends the usefulness of an ultraviolet lamp. *(Ultra-Violet Products, Inc.)*

The following list of luminescent minerals includes only those minerals that have economic value as treated in this book or that are of substantial interest to the collector-hobbyist but whose value is as specimens only. They are arranged in general according to the colors of the spectrum (rainbow), concluding with white; the typical range of hues is suggested. *P* indicates a phosphorescent mineral; *U* means an important uranium mineral. No distinction is made between shortwave and longwave radiation; for most minerals, the color is the same; moreover, the serious prospector and collector will want to have a lamp that gives both ranges of radiation. A further list of minerals that fluoresce only occasionally could be extended indefinitely, for new ones are being observed from time to time; additional names will be seen in the books that are recommended in the references at the end of this section.

RED

Aragonite . . . . red, rose
Calcite . . . . red, rose, orangish red; *P:* red, yellowish red

Mineral	Color
Calomel	reddish
Corundum (ruby)	red
Dolomite	red, rose, orangish red
Eucryptite	rose
Halite	red to rose
Nepheline	orangish red
Phosgenite	*P:* brownish red
Rhodonite	deep red
Spinel	red to rose
Spodumene	*P:* red, yellowish red
Tremolite	rose, orangish red
Wollastonite	rose, orangish red

Orange

Mineral	Color
Albite feldspar	*P:* orangish brown
Anglesite	yellowish orange
Aragonite	orange
Calcite	orange; *P:* orange
Hackmannite	orange
Manganapatite	yellowish orange
Nepheline	reddish orange
Pectolite	orange, yellowish orange; *P:* orange
Sodalite	orange, yellowish orange
Sphalerite	*P:* orange
Thenardite	yellowish orange
Wernerite	yellowish orange
Wollastonite	reddish orange
Zircon	orange, yellowish orange

Yellow

Mineral	Color
Albite feldspar	*P:* orangish yellow
Anglesite	yellow, whitish yellow, orangish yellow
Autunite *U*	greenish yellow
Calcite	*P:* golden yellow
Cerussite	yellow, yellowish white
Chlorapatite	yellow
Cuproscheelite	whitish yellow to greenish yellow
Fluorite	*P:* whitish yellow
Gypsum	yellow, whitish yellow; *P:* golden yellow
Hemimorphite	yellow, yellowish
Howlite	yellow, brownish yellow
Leadhillite	pale yellow
Manganapatite	orangish yellow
Meta-autunite *U*	lemon yellow
Opal	yellowish, greenish yellow
Pectolite	*P:* yellow

Phosgenite . . . . yellow, orangish yellow
Powellite . . . . yellow, golden yellow
Quartz. . . . . yellowish yellow
Serpentine . . . . yellow, whitish yellow
Sodalite . . . . yellow, orangish yellow, yellowish orange
Terlinguaite. . . . . yellow, yellowish
Thenardite. . . . . orangish yellow, greenish yellow
Wollastonite. . . . . yellow, yellowish
Zircon . . . . golden yellow, orangish yellow

GREEN

Autunite *U* . . . . yellowish green
Barite . . . . *P:* bluish green
Beta-uranophane . . . . dull green to pale yellowish green
Chabazite . . . . greenish, deep green
Chalcedony quartz . . . . green, yellowish green
Fluorite . . . . greenish, yellowish green; *P:* bluish green
Hydrozincite . . . . whitish green
Opal . . . . green, yellowish green
Pectolite . . . . *P:* greenish
Quartz. . . . . green
Stolzite . . . . whitish green
Strontianite . . . . *P:* bluish green
Uranophane *U* . . . . pale lemon yellow
Willemite. . . . . green; *P:* green

BLUE

Aragonite . . . . *P:* bluish
Benitoite . . . . sky blue
Borax . . . . *P:* greenish blue
Brucite . . . . blue, whitish blue
Calcite . . . . *P:* blue
Celestite . . . . *P:* blue
Colemanite . . . . *P:* whitish blue
Cuproscheelite . . . . greenish blue, blue
Diamond . . . . blue
Epsomite . . . . *P:* bluish
Fluorite . . . . blue, whitish blue; *P:* whitish blue
Glauberite . . . . *P:* grayish blue
Gypsum. . . . . *P:* bluish, greenish blue
Hanksite . . . . *P:* whitish blue
Hydrozincite . . . . whitish blue
Scheelite . . . . whitish blue, blue
Wavellite . . . . *P:* bluish, blue
Witherite . . . . blue

WHITE

Aluminite	*P:* whitish
Anglesite	yellowish white
Aragonite	yellowish white, white; *P:* whitish
Barite	*P:* yellowish white
Bauxite	*P:* whitish
Calcite	white, yellowish white; *P:* yellowish white
Celestite	*P:* bluish white
Colemanite	white, yellowish white; *P:* bluish white
Cryolite	*P:* white
Dolomite	*P:* yellowish white
Fluorite	white, bluish white
Hydrozincite	whitish, bluish white
Meyerhofferite	*P:* yellowish white
Powellite	yellowish white
Scheelite	yellowish white, white
Sepiolite	yellowish white, bluish white
Stolzite	greenish white, grayish white
Thaumasite	*P:* whitish
Tremolite	whitish, yellowish white, bluish white
Trona	white
Ulexite	white
Witherite	yellowish white, white
Wollastonite	yellowish white

BOOKS ON LUMINESCENCE

Ultraviolet Guide to Minerals, Sterling Gleason, Van Nostrand-Reinhold, New York, 1960.

Minerals that Fluoresce with Mineralight Lamps, Ultra-Violet Products Inc., Los Angeles, 1969.

Prospektion und feldmässige Beurteilung von Lagerstätten, Günter Zeschke, Springer-Verlag, Vienna, 1964.

The Uranium and Fluorescent Minerals, H. C. Dake, Gemac, Mentone, California, 1954.

10

Geophysical Prospecting

The use of Geiger and scintillation counters for detecting radioactivity (as explained in chapter 8) and of ultraviolet lamps for observing luminescence (as covered in chapter 9) are surely examples of geophysical prospecting. They have, however, been given separate chapters here because of the ease with which they can be applied by the everyday prospector and because of their relatively low cost and commonness in the field.

The detection of beryllium by the use of an instrument that is related to the Geiger counter was mentioned on page 126. The probability that about 30 elements—of which nickel was the first—can be detected by the prompt-neutron-capture—gamma-ray method was reported in 1971 by the U.S. Geological Survey.

Some of the equipment discussed in this chapter (certainly the compass) is not expensive either, but, in general, much of it requires specialized training; even when this is not so, the apparatus is less likely to be familiar in routine prospecting, even though experienced operators and crews are accustomed to it. "Remote sensing" is the term now used for all the methods—many very complicated—of reconnaissance from a dis-

A mobile unit makes it possible to record the geophysical nature of a drill hole. *(Cryogenic Research Company, Inc.)*

tance. These include aerial photography, infrared mapping, radar surveys, and other techniques, extending to the information that can be obtained from earth satellites and interplanetary vehicles.

First of all, the concept of an anomaly should be understood. The term simply refers to anything out of the ordinary and may have no economic meaning at all. It may refer to either more or less than normal or average. In most instances, a knowledge of the geologic conditions is helpful, or even necessary, to understand the significance of a given anomaly. For example, an olivine diabase dike that cuts across folded sediments will give a strong magnetic anomaly but may not contain valuable minerals. Many other sorts of anomaly are known, and as many exceptions to the rules also. A geologic map, if available, should be studied in relation to an anomaly, both to find out what it may indicate and to determine whether other geologic structures of a similar nature may exist in a linear or other pattern.

Rather than trying to classify geophysical equipment by its complexity, the grouping here is according to the four standard methods besides radiometric. These methods are magnetic, gravimetric, electrical, and seismic. They all combine geology and physics in the search for mineral deposits. In spite of the reliance on geophysical techniques by large companies, criticism is now being directed at programs of "saturation exploration," at least for radioactive minerals, as not being economic—all of which points more decisively toward the opportunities that seem to await the individual prospector using ordinary methods.

In addition, seismic methods are fundamental to prospecting (for petroleum and natural gas), but their chief use is to solve structural problems rather than to determine the presence of mineralization. Even this knowledge is useful, nevertheless, as folds and faults are structural features of great significance in the search for ores and nonmetallic (especially layered) deposits. Reflection and refraction (bending) seismic surveys can be made.

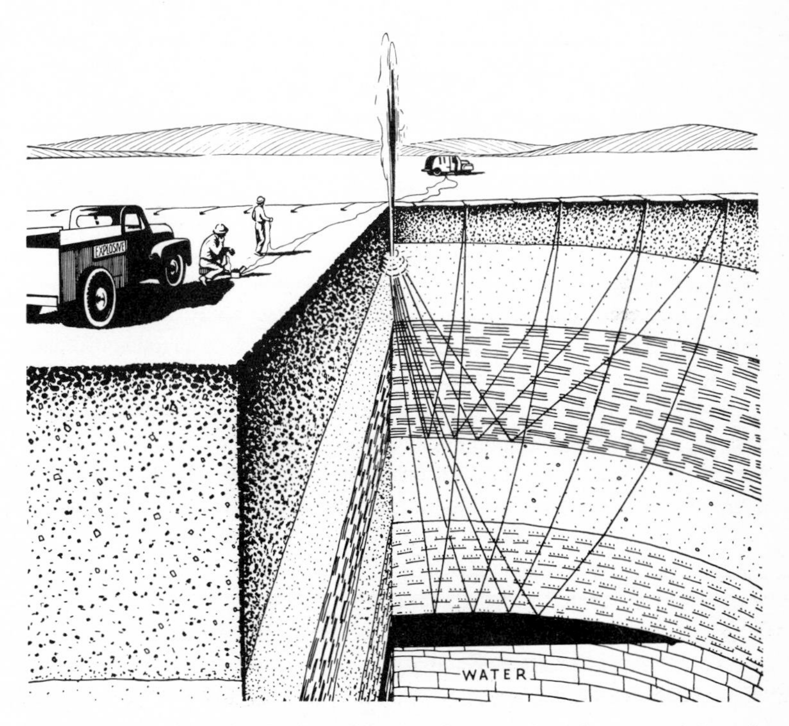

Seismic waves reflected and refracted (bent) at formation boundaries reveal the rock structure. *(Socony Mobil Oil Company, Inc.)*

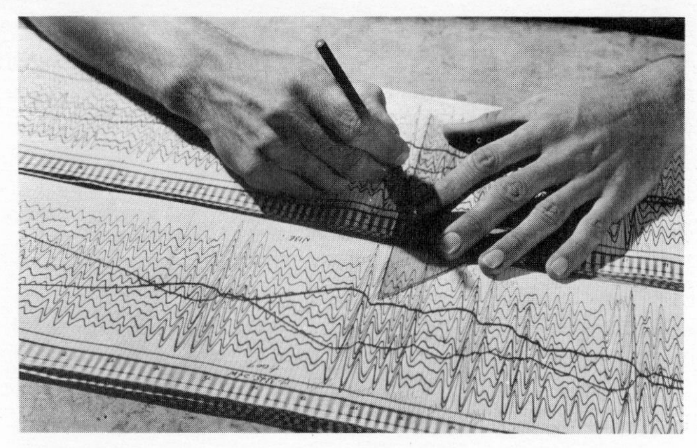

The interpretation of seismic data indicates the underground structure and possible mineralization. *(Standard Oil Company, New Jersey)*

MAGNETIC METHODS

Numerous minerals are magnetic to the extent of responding to sensitive magnetic instruments. A few minerals can be attracted by simple magnets, some of which are fairly strong. The variety of magnetite that is known as lodestone is itself a natural magnet, which attracts iron objects and, furthermore, has polarity, aligning with the magnetic poles of the earth. Native platinum also often has polarity.

The presence of iron, nickel, or cobalt makes minerals that contain them magnetic. Iron, however, is the metal whose existence is revealed by the use of a compass, a dip needle, or a magnetometer. Sensitive but otherwise ordinary compasses have been used to discover deposits of iron ore, as was done during the middle of the 19th century in the Lake Superior country of Minnesota, Wisconsin, and Michigan. Magnetite is the usual mineral that causes the compass to be deflected most strongly. Certain other minerals do so fairly often but to a weaker degree; these include hematite (as at Lake Superior), limonite or goethite, chromite, ilmenite, franklinite, and pyrrhotite (the only common mineral besides magnetite that can be picked up with a cheap magnet). These minerals are described in chapters 19 and 20.

Geomagnetic techniques are especially useful in detecting deposits of copper, iron, lead, nickel, zinc, and sulfur; they are less applicable to antimony, chromium, cobalt, manganese, titanium, and fluorite.

The source of the magnetism that causes the compass needle to point away from magnetic north may be found by making a compass traverse along a row of stakes that are placed in a straight line, perhaps 100 or 200 feet apart. Deflections will then be apparent. The prospector should first acquaint himself with the magnetic nature of the area. He should know what the normal deviation of the magnetic north line is from true, or geographic, north (toward the North Star); this is called the magnetic declination of variation of the compass. Aerial flight maps and the *United States Coastal Pilot,* issued by the U.S. Coast and Geodetic Survey, give magnetic data for the United States. The strength of the earth's magnetic field in the United States ranges from about 50,000 gammas in Texas to 60,000 gammas in Minnesota. Iron and steel objects should, of course, be removed from the vicinity.

Dip Needle

A dip needle is more sensitive than the compass, especially for distinguishing between two magnetic zones. It measures the vertical component of the magnetic field (the angle of dip, called the inclination) and can distinguish between surface rock and buried rock. Thus, it can be used to trace concealed rock that has different magnetic properties from the adjacent rock. The association of certain minerals or metals with certain kinds of rock—such as copper with quartz monzonite—may suggest whether to expect a higher or lower reading than average. The stronger the magnetism, the steeper will be the angle of dip of the needle. The background count should be subtracted, as is done with a Geiger counter (see page 123). The needle must be oriented to the magnetic field—in the vertical plane of the earth's field as you face magnetic west—either by the rest or the swing method. Both mechanical and operational errors are likely to occur. Although iron-bearing rocks are the logical target of dip-needle prospecting, shallow sands that contain magnetite—and, it may be hoped, gold—can also be found with care at depths less than 50 feet.

Magnetometer

More sensitive than either the compass or the dip needle, the magnetometer actually measures the magnitude of a magnetic field in gamma units (not related to gamma rays), in which it is calibrated. The Schmidt-type or magnetic balance, thus described, is not easy to use without instruction or experience in mounting, leveling, and handling it. Varia-

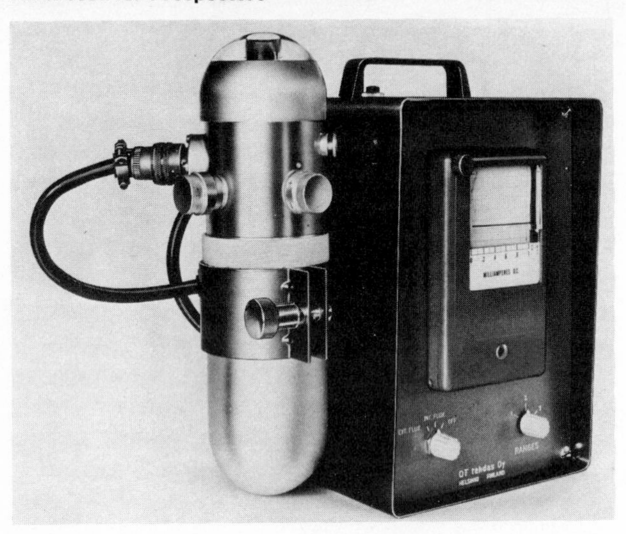

A Jalander electronic magnetometer is here connected to a lightweight recorder. *(Jordan International Company)*

A Siemens torsion magnetometer in operation in Finland. *(Siemens Aktiengesellschaft)*

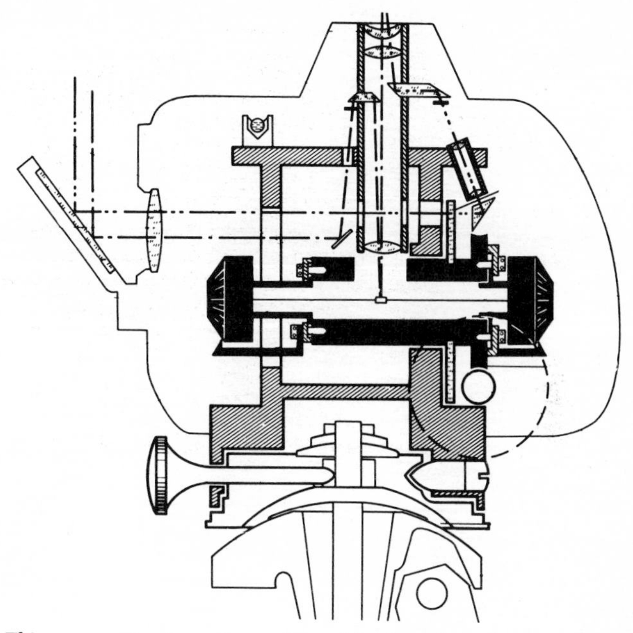

This cross-section shows the working of the Askania torsion magnetometer. *(Siemens Aktiengesellschaft)*

tions in the earth's magnetic field, especially during magnetic storms, must be allowed for. As with a dip needle, the vertical component of magnetism ("vertical force") is usually determined, except near the equator where the horizontal component (northward pull) is measured. Fluxgate and nuclear-(proton-) precession models are also made, utilizing transistors and being more versatile and quicker and easier to use than the magnetic-balance type. A grid traverse may be surveyed, taking care to avoid large metallic objects.

Airborne magnetometers generally measure the total field, typically from an elevation of 300 to 1,000 feet, instead of the difference in magnetic field. The four kinds of instruments that are used in aeromagnetic surveys are the fluxgate, electron-beam, proton-precession, and optical-absorption magnetometers. Aeromagnetic maps are available in several countries of the world, but the use of airborne techniques requires a good deal of geologic knowledge to be successful.

The bigger the magnetic body, the deeper it can be detected. Even if an ore body is not itself magnetic, a common association with one that is magnetic may be a helpful clue in prospecting. Asbestos deposits

A gravity meter is here being used near Crane, in flat land in Texas. *(Standard Oil Company, New Jersey)*

in serpentine and gold veins in quartz diorite have been discovered in this manner in Canada. Finding the western extension, across a large fault, of the Rand gold deposits in the Republic of South Africa was a major triumph of magnetic prospecting.

GRAVIMETRIC METHODS

The well-known force of gravitation exerts a pull that varies throughout the earth. Dense rocks can thus be located with a gravity meter. The

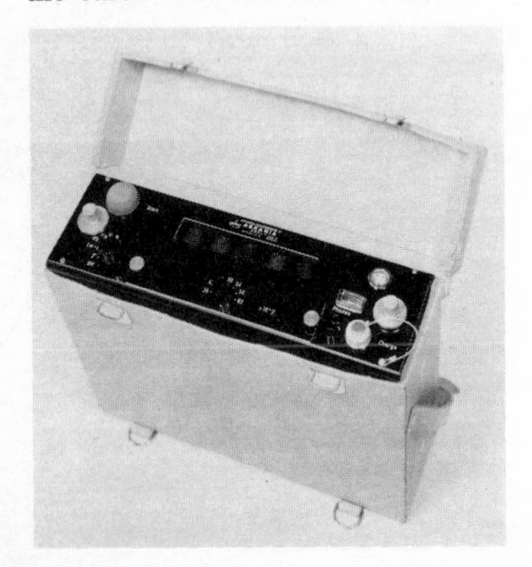

The proton magnetometer is often used in aeromagnetic surveys. *(Siemens Aktiengesellschaft)*

instrument itself, however sensitive, is rugged enough for field use. One type, newer than the torsion balance, works like a spring balance, the values being measured in milligals. The readings require skill to interpret, for many complications are involved. The proper corrections must be made for altitude, latitude, terrain, daily changes, and instrumental peculiarities. A grid survey made from the final figures may tell much about underground conditions if the geology is taken into account.

ELECTRICAL METHODS

Self-potential

Certain mineral bodies generate natural electrical currents, which can be measured in voltage as the self-potential. The currents usually flow downward and into the surrounding rock and return to the surface, completing a circuit. Readings of this "spontaneous polarization" are taken along parallel lines across the outcrop by moving a chemical pot while another pot is moved at intervals of 50 feet or more. The usual cause of the current is the presence of a metal-sulfide deposit that is oxidizing; richness rather than mere size is indicated. Some manganese oxides, anthracite coal, and graphite also produce currents, as do anomalous conditions of no significance. Precautions must be taken to understand what is happening, avoiding buried pipelines and correlating the results with known geologic data.

The Mineral Industries Research Laboratory of the University of Alaska developed in 1969 a portable meter to measure the electrical potential that is generated by an underground orebody; it can be made for about $100.

Among the familiar minerals, the following are especially conductive: native copper, chalcocite, chalcopyrite, magnetite, pyrrhotite, galena, pyrite, graphite, and molybdenite. Moderate conductivity is associated with chromite, hematite, barite, scheelite, antimonite, sphalerite, cuprite, and siderite. Almost all silicates and highly siliceous rocks (such as granite and syenite) are poor conductors, as well as quartz, magnesite, dolomite, and calcite.

Resistivity

Harder to interpret than self-potential, resistivity measures how much electrical voltage is retarded in its flow through a substance. Metallic minerals—many metallic sulfides, some metallic oxides, and graphite—as well as loosely cemented (porous) and water-bearing rocks show lower resistivity than do ordinary, tightly compacted and dry rocks. A direct-current or low-frequency alternating-current generator should be connected to two fixed metal stakes that are set about 7,000 to 15,000 feet apart across the general trend of the rock. Or the stakes and measuring points can be movable if the soil is not overly conductive. The differences in voltage between the surface and underground readings indicate by their apparent resistivity the nature of the zones within the rock. The interpretation of anomalies is not easy, but the method is otherwise simple, inexpensive, and rapid.

Electromagnetic Method

As an alternating current (say, 1,000 cycles per second) flows along a wire loop that is suspended above the ground, a primary magnetic field is created, which causes (induces) currents underground. These, in

This electromagnetic system is called a vertical loop and is held this way. *(Scintrex Mineral Surveys, Inc.)*

turn, create a second magnetic field, which can be measured at the surface by means of a dial or earphone signal. A good subsurface conductor of electricity is needed, and an understanding of geology is required for proper interpretation. The primary and secondary fields may interfere with each other (the amount being measurable as "dip angles"), but topography is seldom a factor. Metal-sulfide bodies and magnetite have been found by this method.

Here is a horizontal-loop electromagnetic system. *(Scintrex Mineral Surveys, Inc.)*

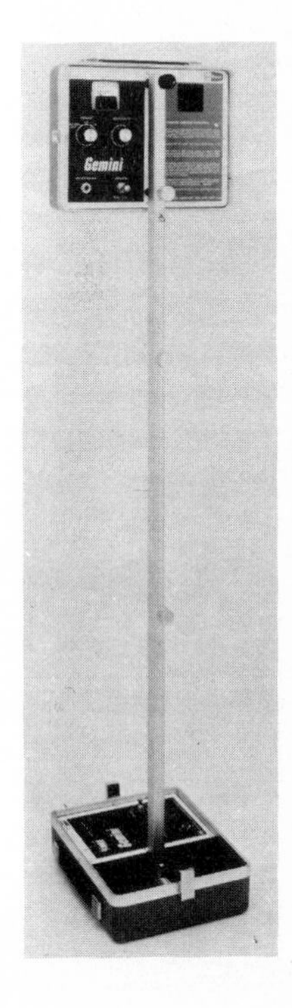

The Fisher M-Scope is principally a metal detector. *(Fisher Research Laboratory, Inc.)*

This portable metal detector is held vertically while in use. *(Goldak)*

BOOKS ON GEOPHYSICAL PROSPECTING

Much information on this subject is given in section 10-A of *Mining Engineers' Handbook,* 3d edition, Robert Peele (ed.), John Wiley & Sons, Inc., New York, 1941 (2 volumes).

Geophysical Prospecting, A. A. B. Edge and T. H. Laby (eds.), Cambridge University Press, London, 1931.

Geophysical Prospecting, G. A. Heiland, Prentice-Hall, Inc., Englewood Cliffs, N.J., 1940.

Exploration Geophysics, J. J. Jakosky, Trija Publishing Company, Los Angeles, 1950.

Manual on Geophysical Prospecting with a Magnetometer, J. W. Joyce, American-Askania Corporation, Houston, 1937.

Geophysical Exploration for Ores, Max Mason, American Institute of Mining, Metallurgical, and Petroleum Engineers, New York, 1929.

Introduction to Geophysical Prospecting, 2d edition, Milton B. Dobrin, McGraw-Hill Book Company, New York, 1960.

Mining Geophysics, D. S. Parasnis, American Elsevier Publishing Company, New York, 1970.

Nuclear Techniques for Mineral Exploration and Exploitation, International Atomic Energy Agency, New York, 1971.

11

Geochemical Prospecting

Using chemistry as a means of finding mineral deposits requires a knowledge of the techniques (unless somebody else does the actual testing) and, often, expensive equipment. There is no reason except expense why a laboratory cannot make the analysis, just as an assayer was expected to do in the days of the oldtime gold prospector. The cost, however, mounts up, and so most prospectors prefer to make a reasonable number of simple tests, such as those employing blowpipe methods (described in chapter 15), and to hire out the more specialized testing.

An explanation of what geochemical prospecting entails is given here first, followed by a brief description of certain of the most frequently used instruments, but their high price and complexity preclude discussing them in detail. Because peculiarities in the growth of vegetation are a matter of local chemistry, what is called geobotanic prospecting—by the use of indicator plants—is included in this chapter; it is perhaps more interesting than important, but it must not be overlooked by today's seeker after mineral wealth.

Geochemical methods are most practicable for finding cobalt, copper, lead, silver, and zinc and less effective for determining antimony, chromium, gold, manganese, tin, and uranium.

Most geochemical prospecting involves tracing a key element in order to find a place where the soil or bedrock contains a suitable concentration of it. Such an element is called an indicator, trace element, or pathfinder element. The total amount of an uncommon element (say, silver) might seem to be ridiculously small (perhaps 1 part per million), yet if it is concentrated in a given spot to a much greater degree than in the surrounding area, a sufficient anomaly might be indicated. Chemical methods are especially important where physical differences cannot be detected. Decomposed bedrock can be sampled and studied wherever outcrops are scarce or lacking. Prospecting of this sort must be done systematically, preferably by making a map of the results as the work proceeds.

Chemical testing in the Canadian mountains begins with a simple dropping-bottle. *(Pan American Petroleum Corporation)*

The discovery of ore deposits in Canada by the use of pathfinder elements, as they are best known there, has attracted much attention. Mercury seems the most effective of such elements, especially for fairly detailed surveys. Arsenic, silver, molybdenum, antimony, sulfur, manganese, and zinc have been successfully used in other countries.

A primary anomaly results from the outward movement of elements from mineral-forming solutions. This causes a zoning, or dispersion halo

Preliminary testing for heavy metals utilizes a larger battery of chemicals. *(U.S. Geological Survey)*

or pattern, around the original source. It may be easy to recognize, perhaps pointing toward the source as a dispersion fan.

As rocks weather, the soluble elements are dissolved and carried away to form secondary anomalies. They may be held in nearby soil—permafrost is not favorable for testing—or absorbed by living trees and other plants, or they may be taken long distances. Inasmuch as streams are the main agents of transportation, of both dissolved matter and suspended sediment, sampling the water and sediment reveals the presence of valuable elements. These are then traced upstream, following each likely tributary in the same way that float is tracked and gold panning is carried on (see chapter 7). Care should be taken to avoid contamination from mines and smelters, industrial waste and common litter, motor-vehicle fumes and insecticide sprays.

If you are especially interested in a particular metal, you ought to be able to set up a portable laboratory that will enable you to make frequent tests for copper, lead, silver, or whatever else at a low cost. Certain colorimetric tests are very convenient. A constant nuisance is contamination, as mentioned above, for the procedures are generally delicate and need to be conducted carefully.

Of the specialized instruments that are used to identify chemical ele-

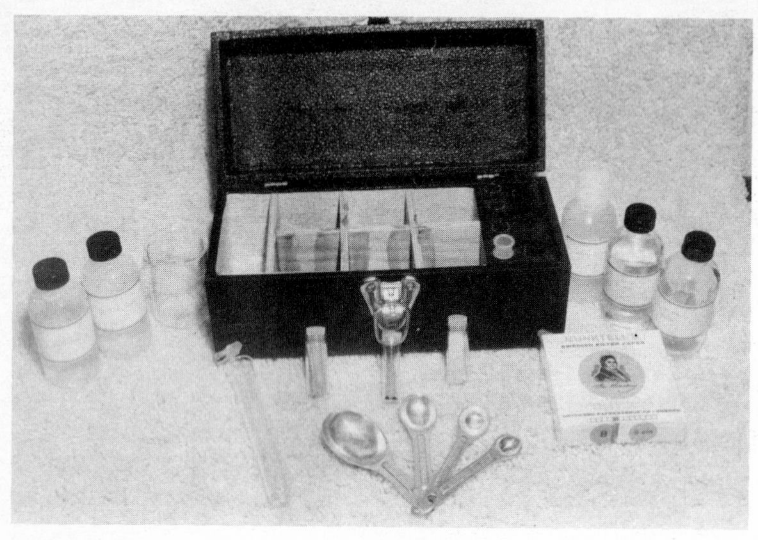

Standard kits are sold for geochemical testing in the field. *(Warnock Hersey International Limited)*

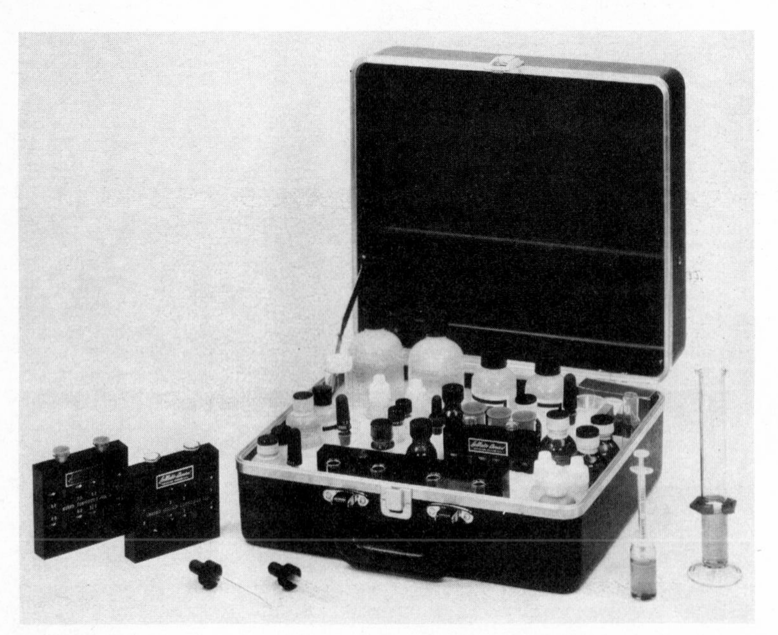

Soil-testing kits have uses in geochemical prospecting. *(La Motte Chemical Products Corporation)*

ments, the spectroscope and spectrograph are the most likely ones in a laboratory or assay office. The emission spectrograph records the configuration of bright lines for as many as 20 to 50 different elements in each sample that is vaporized to incandescence. Each kind of atom gives off a different pattern, the intensity of the lines being proportional to the amounts of the elements. This commercial service costs about $10 to $20 in the United States and Canada; a motorized laboratory operated by a large company may include the instrument at a cost of about $10,000.

When cool, the vapors of atoms absorb light in a similarly distinctive arrangement of lines, which are shown in an absorption spectrograph.

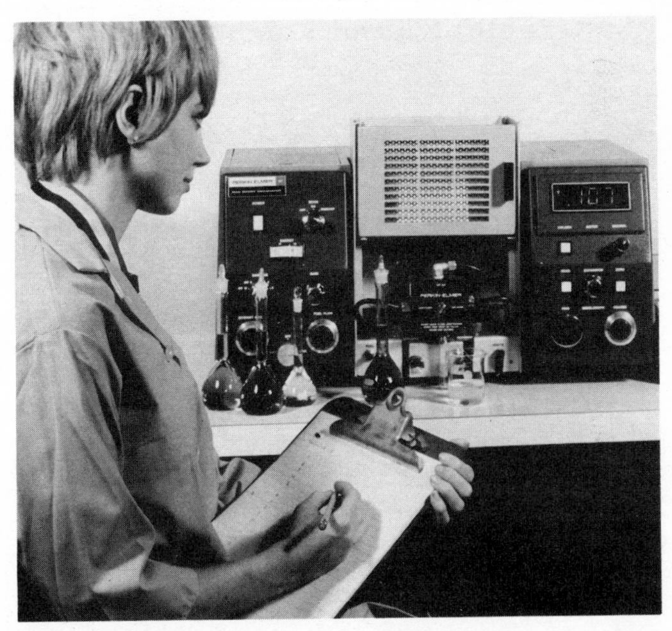

Both column and paper chromatography are done with this equipment. *(Mallinkrodt Chemical Works)*

Combining atomic-absorption spectroscopy and solvent-extraction techniques has made it possible to detect as little as 10 parts per billion of gold and 5 parts per billion of silver, thereby greatly aiding exploration.

An X-ray spectrograph is an expensive instrument, possibly $50,000 or more. By hitting solid samples with X-rays, it brings out a pattern of radiation that is characteristic of each kind of atom. This service, also called X-ray fluorescence spectrography, can be bought commercially for $2 or more per element.

When certain elements are hit by high-energy gamma rays, they set neutrons free. These can be trapped and counted in the same way that a Geiger counter records other radiation. Beryllium is the best-known element being prospected for with radio-activation devices, and portable models are being made to search for gold, silver (the Silver Snooper), and other metals. The prompt-neutron-capture–gamma-ray method, introduced for detecting nickel in 1971, seems certain to be followed for the detection of other elements, probably including copper, titanium, aluminum, mercury, and cobalt.

Other techniques for geochemical determinations include optical methods—colorimetry, paper chromatography, spot tests, fluorescence, and turbidimetry; radiation methods—flame spectroscopy, radiometry, and thermoluminescence; and electrical methods—specific ion activity and polarography.

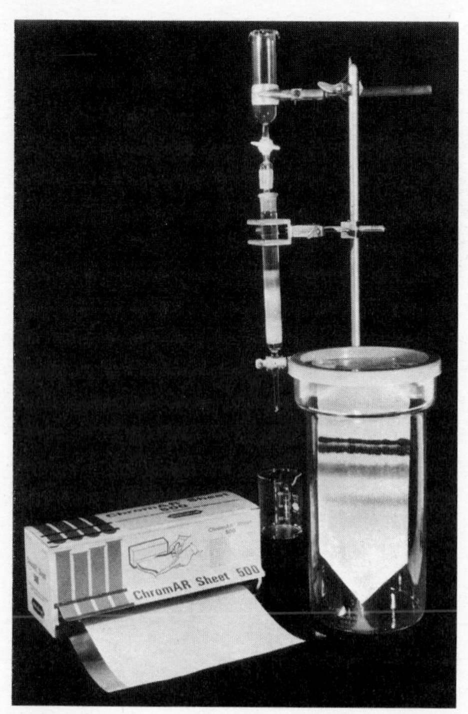

An atomic absorption spectrophotometer reads directly the concentration of a wide range of metallic elements. *(Perkin-Elmer Corporation)*

Plants have been used in prospecting for metals since the Middle Ages. Their utility to the prospector is that their roots can absorb chemical elements from the rock in which they grow, transferring them aboveground to branches, stems, leaves, and fruit, which can be analyzed chemically. Certain plants favor soil in which particular elements are more abundant than average—such as copper, in the so-called copper mosses and in basil, a member of the mint family (in Zambia); zinc in the calamine violet (in Silesia); and uranium and selenium in the poison vetch (in the western United States). On the Colorado Plateau, for example, this method was shown to be about twice as successful as random drilling in locating uranium-bearing material. Plants of the mustard family excel in absorbing uranium, but coniferous trees are most useful in prospecting. Some vegetation reflects an excess or deficiency of nutrients by its color or other peculiarities. Examples include stunted growth and yellow-orange coloration from molybdenum poisoning; white and dead patches on leaves from nickel and cobalt poisoning; stunting and scorching from boron.

Apart from the knowledge of botany that is needed, geobotanical prospecting by means of indicator plants is complicated by a number of factors, so that it is not entirely simple. Furthermore, the chemical analyses are delicate to perform, the material (usually leaves or buds or second-year stems) being burned to ash for testing. Grid surveys and maps can be prepared, as with other geophysical and geochemical methods, but geochemical grids may be distorted by creep or landslide effects and should conform somewhat to the topography. A spacing of 100 feet is usually recommended for soil and plants, increasing to 500 feet on glacial soil and perhaps much more in a broad stream system. The recent Soviet discovery that some mold fungi effectively extract gold from solution may extend geochemical prospecting into a means of actual recovery.

Field testing ranges from a number of simple procedures that can be carried out almost anywhere, to those that are adaptable to a simple campsite laboratory, and on to those that require a well-equipped laboratory. The last tests are generally conducted in a fixed place, but some mobile laboratories (public or private) have come into use.

Copper, lead, zinc, and certain other heavy metals can be analyzed by means of special solvents, employing commercial kits that contain the necessary materials and equipment. The kits are sold by firms that advertise in the mining magazines that are listed on pages 405–406.

Tests to be made at a campsite involve heating the samples and dissolving (leaching) out the elements for analysis. About 24 elements can be identified this way, and the kits are sold commercially.

Training is required to make the recently discovered field tests for gold and silver (and indirectly for copper, lead, zinc, and bismuth), as described in U.S. Geological Survey literature. Both training and more elaborate equipment are needed to make the usual run of assays and chemical analyses.

Blowpipe tests, defined rather broadly to include wet chemical methods, are described in chapter 15. These, like the tests referred to above, determine the elements that are present in a sample of rock, mineral, soil, plant, or water. Some indication is often given as to their amounts. Most minerals require observation of their physical properties (such as color and hardness), as well as a chemical test or two, in order to be identified, and so mineral recognition is not exactly the same as geochemical analysis and is therefore discussed separately in chapter 15. Chapter 12 deals with sampling, testing, and assaying on a broader basis.

BOOKS ON GEOCHEMICAL PROSPECTING

Principles of Geochemical Prospecting, H. E. Hawkes, U.S. Geological Survey Bulletin 1000-F, 1957.

Principles of Geochemical Prospecting: Techniques of Prospecting for Non-Ferrous Ores and Rare Metals, I. I. Ginzburg, V. P. Sokoloff (trans.), Pergamon Press, New York, 1960.

Analytical Methods Used in Geochemical Exploration by the U. S. Geological Survey, F. N. Ward, H. W. Lakin, F. C. Canney et al., U.S. Geological Survey Bulletin 1152, 1963.

Geochemistry in Mineral Exploration, H. E. Hawkes and J. S. Webb, Harper & Row, Publishers, Incorporated, New York, 1962.

Geochemical Prospecting. General Reconnaissance Methods, Nalin R. Mukherjee and Leo Mark Anthony, University of Alaska School of Mines Publication Bulletin 3, 1957.

Elemental Associations in Mineral Deposits and Indicator Elements of Interest in Geochemical Prospecting, R. W. Boyle, Geological Survey of Canada Paper 68-58, 1970.

Geobotany and Biogeochemistry of Mineral Exploration, R. R. Brooks, Harper & Row, Publishers, Incorporated, New York, 1971.

12

Sampling, Testing, Assaying

If you own property, you may operate it as a mine regardless of loss. But if you want to stake a mining claim on public land, you must have reasonable evidence that you have found a mineral deposit that will, in the language of the courts, convince "a prudent man" that it is worth developing. This requires proper sampling of the rock and adequate testing and assaying to verify the discovery. Hence, this chapter belongs between the previous one on finding minerals and the next one on staking a claim.

Sampling is more than a hit-or-miss proposition. It is, in fact, one of the prospector's most important skills. Incorrect sampling leads one either to overlook a good deposit or to expect too much and be disappointed. It has caused much unjustified mistrust of assayers and smelters (although instances of dishonesty are not unknown), many lawsuits, and many mine failures. The typical prospector, being an incurable optimist, selects attractive ore and tries to pass it off as average for the whole deposit. Visible gold, for example, is seldom seen even in rather rich ore; and yet, the specimens likely to be saved to show the assayer usually abound in flecks of the yellow metal. The phantom origin of what has been termed phan-

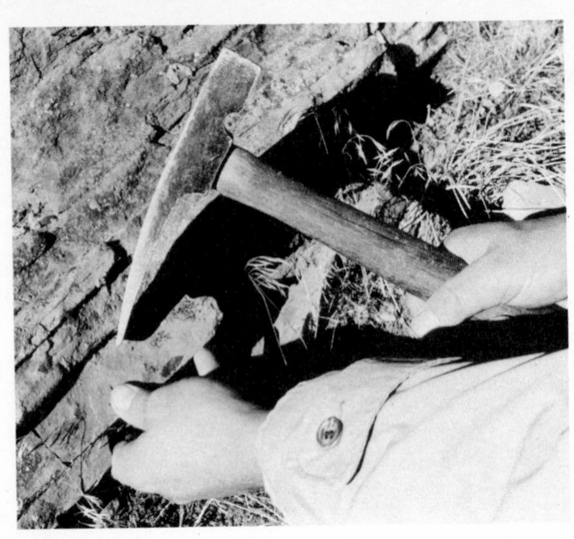

Selected sampling of sedimentary rock is for purposes of examination and identification. *(Standard Oil Company, New Jersey)*

tom gold is a subject of much talk in mining camps, but it can be disregarded, if for no other reason than that mills and smelters will not pay for what they cannot find themselves. Good ore, furthermore, is apt to be softer than lean rock and is therefore easier to break off, and it may crumble to powder. A sample, to be useful, must represent the average tonnage of the entire deposit. You can be fairly sure that when you reach the point of trying to sell your property to a mining company, the professional engineer who examines it will sample correctly, and so there is little point in having deceived yourself.

The only occasion when it might be desirable to select a high-grade sample would be in connection with uranium. Owing to the complications of the mass effect and the radioactivity of other elements besides uranium (as discussed in chapter 8), a specimen of exaggerated grade would serve to indicate the maximum value that could be expected. Surely, if laboratory tests eventually show such a sample to be of low quality, the rest of the deposit is certain to be worth even less, probably much less. Furthermore, minerals in good specimens are easier to identify.

Sampling is possible on a straight statistical basis, using mathematical formulas like those employed in industrial work. The prospector does not need to know these procedures provided that he cuts his samples in a systematic way, as stated below. The necessary tools are a hammer head or pick, a gad or moil for removing rock, a mortar and pestle for crushing and grinding it, a gold pan, perhaps nested sieves, and clean cloth or a box to catch the samples.

The following types of samples should be familiar to the prospector:

1. Channel samples: Take small samples (2 to 4 inches wide and ½ to 1 inch deep) at regular intervals across the entire width of the deposit or vein, at right angles to the trend (or strike). The interval can be 5 or 10 feet underground, 10 or 20 feet on outcrops, or twice the width of the deposit. The sample should not be more than 4 feet long and should contain 1 pound or more of rock per foot. The larger the total sample, the smaller the size of the particles. Include impurities and waste rock in the same proportion as they are present. Veins, bands, or zones that can be mined separately, however, can be sampled as separate bodies. Also sample the wallrock, in case it is mineralized. Each sample should be numbered (on waterproof tags) and recorded as to its position in the solid rock and its width relative to the width of the vein or deposit. Keep good records, label adequately, avoid surface examples, avoid contamination, and remove weathered or soft material to expose fresh rock beneath insofar as possible.

2. Chip samples: Similar to channel samples but obtained more quickly and cheaply, these are knocked off at regular intervals of one or several inches in a regular pattern. Large pieces should be split to give a piece of average size.

3. Panel samples: Similar to channel and chip samples, these are taken from an area one or several feet square rather than from a strip. The panel can represent a square between grid lines that you have marked with crayon.

4. Grab samples: These are taken at random from a deposit or dump or mine car. They should be taken carefully, nevertheless, or they will not be representative. Skipping large, barren rocks results in an excessive proportion of fine material, a common enough mistake. Grab sampling is most justified when the mineralization or the structure is believed to be unsatisfactory.

5. Selected samples: These are useful for mineral identification but not for purposes of assaying.

6. Churn-drill samples: Churn drilling (generally an older method) yields loose cuttings (sludge) by the repeated pounding of a weighted bit.

7. Auger-drill samples: Loose cuttings from an auger drill, either hand or (nowadays) air-power operated, constitute useful samples.

8. Percussion-drill samples: A portable blast-hole drilling machine,

typically powered by a small gasoline motor mounted on the drill, produces loose cuttings, but it is rather heavy and expensive. A light diamond drill is even more so.

9. Cores: Rotary drilling yields solid cores and loose cuttings. A geologist or engineer may be needed to interpret the results. In uranium exploration, discussed at greater length in chapter 22, the diamond-core drill has largely been replaced by the rotary shot-hole drill that was originally developed for geophysical surveying. A rotary drill can be made to "take core" by changing to coring bits. Drilling for uranium veins in hard rock is usually done with a standard diamond-core drill that can drill holes at an angle.

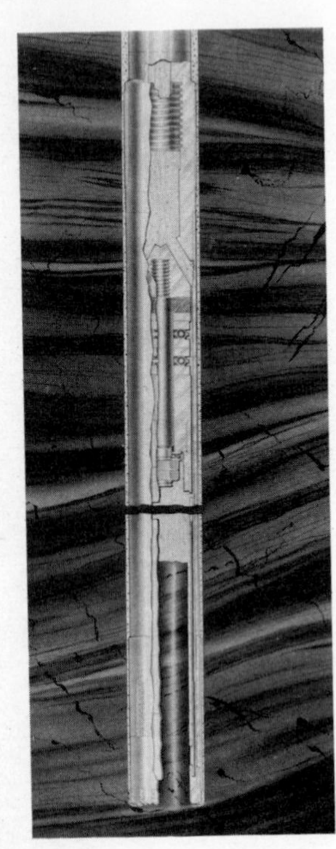

This cross-section shows a diamond core drill penetrating fissured rock. The fine cuttings are worked out by circulating water. *(E. J. Longyear Company)*

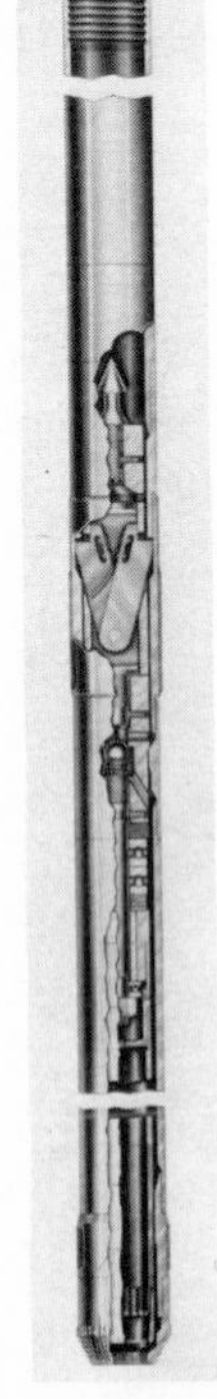

The wire-line core barrel has a retractable inner tube, making unnecessary the removal of the entire drill rod when the core is taken out of the ground. *(E. J. Longyear Company)*

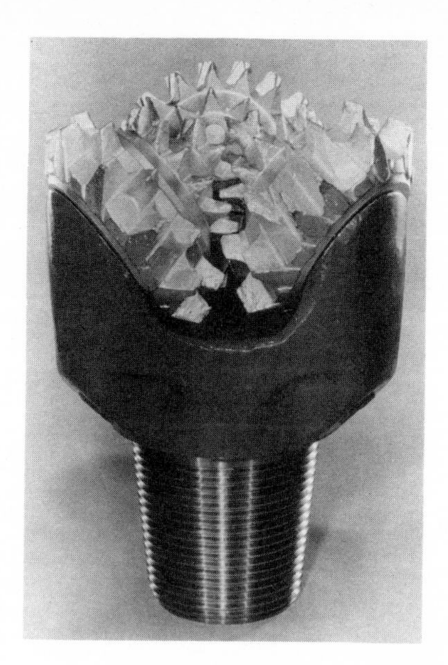

What the drilling bit shows is the final proof of mineral occurrence, regardless of rock type or structure. *(E. J. Longyear Company)*

Cores from a metal mine are laid out in order of recovery so as to show the original structure. *(E. J. Longyear Company)*

Sampling a muck pile or a dump can be done by either the grab method (picking up pieces of small rock, preferably fresh and in some regular order) or the test-pit method (digging round holes as deep as possible). Sampling fine material, as in a tailing, is done by the drive-pipe method, forcing a hollow pipe down into the pile and then knocking the filling out of the pipe.

Even though the tests made by the amateur prospector tell "what" instead of "how much," there is no reason why the same material cannot be used for both testing and assaying. Examining the physical properties of mineral specimens in order to identify them, however (see chapter 15), requires that solid pieces be used.

Preparing the samples for evaluation follows time-honored means. First, crush the specimen to ½-inch size with a standard or homemade mortar and pestle. Then, put the ½-inch pieces on a clean canvas, oilcloth, sheet of steel, or even on the floor. Mix the ore thoroughly, heap it into a cone, flatten it, and divide it into quarters. Rejecting two opposite quarters, crush the rest to about ⅛-inch size. Mix this sample, make another cone, flatten it, and quarter it. Finally, crush to fine sand the remaining sample, weighing about 1 pound, half of which ought to go to an assayer. With experience, a prospector may be able to estimate the gold, mercury, and tin content of his sample by panning it.

The individual widths of the samples should be divided by the value per ton of each, and the total value should then be divided by the total of the widths to get the value per ton of the ore. Exceptionally high assay values should preferably not be included in calculations.

Sacking the samples is easy. Any clean, tight container will do. (The author from time to time gets from different people what are alleged to be placer diamonds, and they always arrive in coffee tins; this mystery has not yet been solved.) Handy envelopes with metal fasteners are sold in several sizes for mailing samples, and so are cotton sampling bags for larger samples.

The tests that the ordinary prospector is often equipped and qualified to do are made with a blowpipe and a variety of chemicals. These are explained, in simple form, in chapter 15.

Public agencies of the sort that are listed in chapter 2 perform a certain range of free tests that may be of great help to the prospector. These services change on occasion, and so you should inquire in your own state or province, giving as much information as you can about the source of the specimen and what you want to know. In this connection, please understand that chemical tests, like medical tests, can be done only within a limited range according to what seems advisable under the circumstances. No chemist can determine "all about" any sample; if you suspect gold or are looking for silver and copper, you will have to say so. Even the knowledge that the specimen was found in a stream bed or in a black ridge may guide the chemist in knowing what to test for. (Anthills in New Mexico, for example, have indicated the course of manganese veins.) Furthermore, public agencies almost universally will not attempt an assay or other quantitative test except for uranium, but they will recommend a commercial firm instead.

The U.S. Bureau of Mines maintains a number of research stations, which will identify, free of charge, mineral samples as to the minerals that are present (but they will not give a detailed chemical analysis). Each sample should be securely packaged and have firmly attached a label that indicates the location where the sample was collected and the sender's name and address. A letter of request should accompany the sample. The directory of these so-called field offices of metallurgy follows:

Tuscaloosa Metallurgy Research Laboratory
P.O. Box L
University, Alabama 35486

College Park Metallurgy Research Center
College Park, Maryland 20740

Twin Cities Metallurgy Research Center
P.O. Box 1660
Twin City Airport, Minnesota 55111

Rolla Metallurgy Research Center
P.O. Box 280
Rolla, Missouri 65401

Boulder City Metallurgy Research Laboratory
500 Dale Street
Boulder City, Nevada 89005

Reno Metallurgy Research Center
1605 Evans Avenue
Reno, Nevada 89505

Albany Metallurgy Research Center
Box 70
Albany, Oregon 97321

Salt Lake City Metallurgy Research Center
1600 East 1st South Street
Salt Lake City, Utah 84112

Custom plants—which are those mills, concentrators, or smelters that work on contract for independent miners—may do free assaying for potential customers.

Commercial assayers perform a wide range of chemical tests—one of the best known did a good deal of protein analysis for the U.S. Department of Agriculture during the Depression—but chemists are rarely equipped to do fire assaying, a rather specialized form of analysis. Regular assayers, moreover, are familiar with mining and are often able to give valuable advice to the prospector. They are situated in the large cities and in many smaller centers in mining country, and they are listed in the mining periodicals (see page 405) and in the yellow pages of the United States and Canadian telephone books.

Several of the assayers issue interesting literature. Firms should be queried in advance as to the amount of sample needed, prices, and other information.

The 1971 assay charges were approximately as given below, from the list of Charles O. Parker & Company, Denver, Colorado, who recommend a 1-pound sample. Other assayers and chemists give quotations on additional tests; but, just as many restaurants will prepare dishes that may not appear on today's printed menu, nearly all testing offices offer much the same services on request. For example, you may want an engineer, or duplicate, assay; a control assay; an umpire assay; or an identification report that names the minerals that are present in a sample. Gold and platinum may be priced on a single sample, as quoted at $5 by the Colorado Assaying Company, Denver. Rare earths and thorium may be combined ($5). A 50-cent sample-preparation charge is often added to the total price. The best advice is to be as open and specific as possible in dealing with assayers and chemists. They generally know so much more about mineral occurrences than the average prospector that their suggestions are profitably taken.

Gold	$1.50
Silver	1.50
Gold and silver (same sample)	2.50
Copper, lead, zinc, each	1.50
Aluminum	5.00
Antimony	5.00
Arsenic	5.00
Barium	5.00
Beryllium	7.50
Bismuth	7.50

Chromium	7.50
Cobalt	7.50
Columbium-tantalum (combined)	10.00
Fluorspar	4.50
Iron	2.50
Insoluble	2.50
Lime	2.50
Magnesium	3.00
Manganese	3.00
Mercury	5.00
Molybdenum	7.50
Nickel	5.00
Phosphorus	5.00
Platinum	7.50
Silica	3.00
Sulfur	2.50
Thorium	10.00
Tin	7.50
Titanium	7.50
Tungsten	7.50
Uranium	7.50
Vanadium	3.00
Rare earths	10.00
Fusion, silica, lime, iron, each	3.00
Complete qualitative analysis	15.00
Complete X-ray spectrographic analysis	10.00

U.S. Bureau of Mines Information Circular 7695 is a free directory (dated 1954) of *Laboratories That Make Fire Assays, Analyses, and Tests of Ores, Minerals, Metals, and Other Inorganic Substances.* A brief explanation of the different kinds of testing techniques is also given. The U.S. Bureau of Mines also will send you free a *Partial List of Commercial Testing Laboratories;* the current one (1970) is given below:

1. Academy Testing Laboratories, Inc., 352 West 31st Street, New York, New York 10001
2. American Standard Testing Bureau, Inc., 44 Trinity Place, New York, New York 10006
3. Arizona Testing Laboratories, 817 W. Madison Street, Phoenix, Arizona 85007
4. Chemical and Geological Laboratories, P.O. Box 279, Casper, Wyoming 82601
5. Colorado Assaying Co., 2244 Broadway, Denver, Colorado 80201
6. Commercial Testing and Engineering Co., 228 N. LaSalle Street, Chicago, Illinois 60601

7. Grand Junction Laboratories, 435 North Ave., Grand Junction, Colorado 81501
8. Industrial Testing Laboratories, 218-222 East 23rd St., New York, New York 10010
9. International Testing Laboratories, Inc., 578 Market St., Newark, New Jersey 07105
10. Jacobs Assay Office, 1435 S. 10th Avenue, P.O. Box 1889, Tucson, Arizona 85702
11. Jones Smith and Geiger Inc., 309 South 9th St., Louisville, Kentucky 40203
12. Ledoux & Co., 361 Alfred Ave., Teaneck, New Jersey 07666
13. Parker, Charles O. & Co., 2114 Curtis St., Denver, Colorado 80295
14. Petterson, A. W., Assayer, 618 A Sanchez, San Francisco, California 94114
15. Phoenix Chemical Laboratory Inc., 3951 West Shakespeare, Chicago, Illinois 60647
16. Reed Engineering, 620-F South Inglewood Ave., Inglewood, California 90301
17. Southern Testing Laboratories, 2227 First Ave., South Birmingham, Alabama 35233
18. United States Testing Co., Inc., 1941 Park Ave., Hoboken, New Jersey 07030
19. Union Assay Office, Inc., Post Office Box 1528, Salt Lake City, Utah 84110
20. Bruce Williams Laboratories, 618-624 Joplin St., Box 557, Joplin, Missouri 64801

A *Directory of Testing Laboratories* is published at $3 by the American Society for Testing and Materials, 1916 Race Street, Philadelphia, Pennsylvania 19103.

The American Council of Independent Laboratories, Inc., 1026 17th Street, N.W., Washington D.C. 20036, issues a free membership list.

Coors Spectro-Chemical Laboratory and Rocky Mountain Technology, Inc., both in Golden, Colorado 80401, provide a wide range of testing services that utilize the most modern techniques. So does the recently established C.D.C. Associates, Boulder, Colorado 80303. In Canada, assaying and geochemical testing are done by, among a few other firms, Warnock Hersey International Laboratory Limited, 125 East 4th Avenue, Vancouver 10, British Columbia.

13

Staking a Mining Claim

Without overemphasizing the opinion of Don Marquis that "an optimist is a guy that has never had much experience," the few warnings and pieces of advice that are given below may prove worthwhile to the prospector.

Test your samples Not to do so, within reasonable limits of time and expense, is to rely on guesses rather than facts. The example of Thomas Walsh in southwestern Colorado is especially instructive. Although he mistook the mineral that he found in an abandoned tunnel, thinking that it was fluorite (as at Cripple Creek) instead of rhodonite, he had an assay made. The results marked the beginning of the Camp Bird mine, overlooked by Walsh's predecessor, William Weston, who had a better technical education but who failed to have an assay made for gold. Weston, in fact, had been a student in England at the Royal School of Mines, but he was looking for (and hence thinking only of) silver.

Try not to exaggerate Overconfidence seems to be part of the typical prospector's nature and is perhaps essential to his success. It should not be carried too far, however. T. A. Rickard said that the overvaluation of the rich mines of the world has caused more financial loss than the fraudulent schemes of irresponsible tricksters. He used Cobalt, Ontario, and Western Australia as bad examples, but such have occurred everywhere.

Watch out for "salting" This old and dishonorable method of making a prospect or mine seem better than it is has unfortunately not vanished. Many are the suspicious events that suggest salting, whereby rich ore is mixed with the sample before (or even after) it has reached the assayer. A shotgun blast can implant considerable metal in the face of a barren mine. Salting can also be done by hiding metallic gold or gold chemicals in the material that is to be sampled, putting them in the sample itself or in the container, placing them in the mining or washing equipment, or dropping them in the gold pan during the last stages.

Carelessness can result in salting as definitely as (and much more easily than) intentional dishonesty. To guard against both kinds of salting, the mine operator should clean and inspect thoroughly all mine surfaces before sampling, as well as all equipment that is to be used. To protect oneself against being deceived, the place where sampling is to be done should be kept secret until the work starts, and the job should be done without stopping for the night. Examining samples under magnification will probably reveal the presence of gold shavings, a common source of fraud. Check samples, a systematic nature, and a knowledge of the history of a given mining district will often help to arouse or allay suspicion as to possible salting.

An early (1884) rush to Cripple Creek took 4,000 men to Mount Pisgah. The cause of the excitement seems to have been the salting of several prospect holes by a chap whose accomplice was caught with a bottle of yellow chloride of gold. When the actual perpetrator of the fraud could not be found, "the affair ended in a big picnic and a general drunk," reported T. A. Rickard.

Another Colorado episode ended very differently. Chicken Bill sold H. A. W. Tabor the Chrysolite claim at Leadville for $40,000. He then boasted of his success at having salted the mine shaft, leaving Tabor in possession of a worthless embarrassment, which Tabor then proceeded to deepen. Within a few feet, his workmen struck a rich vein, from which Tabor took $1½ million before selling the mine for an equal amount.

Bill Nye, Wyoming newspaperman on the *Laramie Boomerang,* told all about "How They Salt a Claim":

> "I wish you would explain to me all about this salting of claims that I hear so much about," said a meek-eyed tenderfoot to a grizzly old miner who was pan-

ning about six ounces of pulverized quartz. "I don't see what they want to salt a claim for, and I don't understand how they do it."

"Well, you see, a hot season like this they have to salt a claim lots of times to keep it. A fresh claim is good enough for a fresh tenderfoot, but the old timers won't look at anything but a pickled claim.

"You know what quartz is, probably?"

"No."

"Well, every claim has quartz. Some more and some less. You find out how many quartz there are, and then put in so many pounds of salt to the quart. Wild cat claims require more salt, because the wild cat spoils quicker than anything else.

"Sometimes you catch a sucker too, and you have to put him in brine or you lose him. That's one reason why they salt a claim.

"Then again, you often grubstake a man . . ."

"But what's a grubstake?"

"Well, a grubstake is a stake that the boys hang their grub on so they can carry it. Lots of mining men have been knocked cold by a blow from a grubstake.

"What I wanted to say, though, was this: you will, probably, at first, strike free milling poverty, with indications of something else. Then you will, no doubt, sink till you strike bedrock, or a true fissure gopher hole with traces of disappointment.

"That's the time to put in your salt. You can shoot it into the shaft with a double-barreled shotgun, or wet it and apply it with a whitewash brush. If people turn up their noses at your claim then, and say it is a snide, and that they think there is something rotten in Denmark, you can tell them that they are clear off, and that you have salted your claim, and that you know it is all right."

The last seen of the tenderfoot, he was buying a double-barreled shotgun and ten pounds of rock salt.

There's no doubt but a mining camp is the place to send a young man who wants to acquire knowledge and fill his system full of information that will be useful to him as long as he lives.

In this connection, it is advisable for the prospector to take his own samples, as recommended frequently by Jim Barrows, of Las Vegas, Nevada.

Accidental salting, however, occurs fairly often during grinding and through the use of old crucibles and dusty equipment and reagents.

A curious example of salting is given by George Gibbons Hayes. A prospector returned to Delaware from California with a souvenir buckskin poke of gold dust and small nuggets. His son took to raising chickens, but, meeting with competition, he recalled his father's stories about the salting of prospects and mines out West, and decided to feed a small nugget to an occasional chicken. As a housewife found it, and her husband had a stickpin made to hold it, the excitement spread, and the man's business flourished, until he "got religion" and confessed his sins.

For some strange reason, gold prospecting began to decline in Delaware at that time.

Look out for promoters No organized market exists whereby a middleman will undertake to handle the sale or lease of mining property. You will have to do this yourself, by advertising in a mining or financial publication or by making inquiry at various mining companies, many of which are listed in the directories that are given on page 407 of this book. If the property is good, no commission will be required.

Little Cottonwood Canyon, in Utah, was the site of the notorious Emma swindle. Allegedly located in 1868, the Emma mine seemed to contain "one of the most remarkable deposits of argentiferous (silver) ore ever opened." This "discovery," widely promoted in England, helped to create in Utah a speculative boom, which collapsed with the revelation of the fraud in 1873.

Among the most conspicuous examples of failure to keep control of a mining claim is the Comstock case in Nevada. Patrick McLaughlin and Peter O'Riley located two placer claims on promising land in 1859, but they were quickly bluffed out of their holding by Henry Comstock, who invented the notion that he owned a ranch that covered the entire site. Further threats and manipulations ended with the two discoverers receiving about $56,200 for property that, as the Ophir mine, yielded $17,655,000. Even before McLaughlin and O'Riley came along, ground in the same area had been staked by James Fennimore, who neglected to record his claim and lost it.

Obey the laws Many court cases have been decided favorably for the prospector who failed to follow the rules but whose intentions were good. Dr. John D. Ridge, an adviser to the publisher of this book, told, in 1968, about just such a Utah case, which the Anaconda Company won against a plaintiff who was technically correct but whose agent had not acted in good faith. It is best not to rely on such accidents of fortune. Do everything that is required, file the claims and affidavits on time (preferably as soon as possible), take corroborating photographs of the property, and secure dated statements from witnesses. As Charles Steen, the uranium king, discovered to his surprise, "multimillion-dollar lawsuits are actually filed." And so they may be, if you too "strike it rich."

No license is needed to prospect on federal land in the United States (except certain lands where minerals are leased only—see page 188), but some of the states require permits to prospect on state land, and these rules should be determined for each state in which you plan to explore.

On private land, agreement with the owner should be reached in advance, for, even if mineral rights are entirely separate from surface rights (as they often are), any interference with farming, ranching, recreational, or residential use might be regarded as trespassing.

In Canada, annual prospecting licenses are issued by the provinces and territories and are variously called a prospecting license, miner's license, miner's certificate, miner's permit, or free miner's certificate. The costs (about $5 to $10) and rights are somewhat different from place to place, but they allow any person at least 18 years old, Canadian or foreigner, to prospect and stake claims on crown land and occupied land where the crown has reserved the mineral rights (unless these are otherwise withdrawn). Company and individual licenses are differentiated in some provinces.

In Australia, prospecting licenses are issued by the states and territories, not the federal government.

In Ireland—as an example of a country that is encouraging prospecting (with good results)—an exclusive license is awarded by the Minister of Industry and Commerce for a definite area (often 10 to 13 square miles) for 1 year, renewable. The fee averages about £30 ($72).

An outline of mining laws, at least of the United States, should be preceded by a simple statement of the meaning of the terms "lode" and "placer." A vein is a more or less tabular body that is enclosed within hard rock and dips at an angle. A lode is properly a group of closely spaced veins that are worked as a unit. A liberal interpretation is generally applied, to indicate "a continuous body of mineralized rock lying within any other well-defined boundaries on the earth's surface and under it," according to the Nevada ruling of 1881, which made clear that the law was drawn for the protection of miners, not the pleasure of geologists. Nor the profit of lawyers, may it be hoped.

A placer is a deposit of valuable minerals that occur loose in sand and gravel, commonly along streams or on beaches. The distinction from a lode seems clear enough, but many a court case has been fought over the difference, for the laws pertaining to lode and placer claims are different. In some instances, the classification has been an arbitrary one, glossing over the technical definitions.

An example of somewhat recent date was the Powderhorn case in 1957 in Colorado, where lode claims had been located on top of earlier placer claims. When the Du Pont company bought and began to develop the property, a court decision of far-reaching importance had to be reached.

The finding was for a placer deposit, it lacking the continuity that a lode should have.

On occasion, both lode and placer classifications have been replaced by one that specified "accumulation by concentration," as is the case with water-soluble minerals. These can only be leased, subject to payment of royalties. The Kramer borate deposits in California were put in this category in 1929. Since then, both the formerly accepted mode of origin and the restriction itself have been abandoned as has been brought out by Robert L. Bates of Ohio State University.

Staking a mining claim requires an understanding of the classification of land and the layout of the public-land systems. Real estate people and attorneys have a similar concern with these subjects. A seemingly baffling diversity of land classification exists. There is public land and private land, of course, but many varieties of each, and the prospector needs to be sure of the status of the land on which he intends to search. Mining claims cannot be staked on private land, obviously, but neither can they be on a large part of the public domain. On the other hand, private ownership of land may not include the mineral rights. Furthermore, the mining laws are different in each country, some of the difference depending on whether there is a background of English, French, Spanish, or other law. All these are involved in American mining law. An excellent and readable summary of the historical development of the world's mining laws is given in chapter 7 of *Mineral Resources: Geology, Engineering, Economics, Politics, Laws,* by Peter T. Flawn (Rand McNally & Company, Chicago, 1966).

Dr. Flawn outlined, from *American Law of Mining,* a chronology of the major United States mineral legislation, which is expanded as follows. It does not include the various public-land acts that authorized reservations or withdrawals.

1. Mining laws of 1866 and 1870.
2. Mineral Location Law of 1872, which replaced previous laws.
3. Coal Act of 1873, which was superseded in 1920.
4. Building Stone Act of 1892, which was partly superseded by the Surface Resources Act of 1947 and superseded by the Common Varieties Act of 1955.
5. Oil Placer Act of 1897, which was superseded in 1920.
6. Saline Placer Act of 1901, which was superseded in 1920.
7. Mineral Leasing Act of 1920, which superseded the Coal Act of 1873, the Oil Placer Act of 1897, and the Saline Placer Act of 1901.
8. Atomic Energy Act of 1946.
9. Mineral Leasing Act for Acquired Lands, 1947.
10. Surface Resources Act of 1947, which partly superseded the Building Stone Act of 1892.

11. Public Law 250, 1953, which permitted multiple use and was superseded by the Multiple Mineral Development Act of 1954.

12. Submerged Lands Act and Outer Continental Shelf Lands Act of 1953.

13. Atomic Energy Act of 1954.

14. Multiple Mineral Development Act of 1954 (Public Law 585), which superseded Public Law 250.

15. Common Varieties Act of 1955, which amended (as the Multiple Surface Use Act) the Surface Resources Act of 1947 and superseded the Building Stone Act of 1892.

The rectangular subdivision of public land—a triumph of systematic engineering practice that was championed by George Washington and Thomas Jefferson, among others—is much the same in the United States and western Canada. The unlikenesses are mentioned below. Entirely different systems are in effect elsewhere in the world, but the prospector in those countries will be presumed to be familiar with them. A simple summary of the North American system is now given here.

Understanding the legal description of land is not at all difficult if you take it up one step at a time, as follows:

1. The rectangular system of land classification in the United States applied to all the states except the original 13, the states made from them or their colonial territory (Maine, Vermont, West Virginia, Kentucky, Tennessee), Texas, and Hawaii.

2. The public-land states, in which mining claims can be staked, are these: Alaska, Arizona, Arkansas, California, Colorado, Florida, Idaho, Louisiana, Mississippi, Montana, Nebraska, Nevada, New Mexico, North Dakota, Oregon, South Dakota, Utah, Washington, Wyoming.

3. The starting point (initial point) of land measurement begins at the intersection of each north-south principal meridian (along true meridians of longitude) and each east-west base line (along parallels of latitude). The principal meridians are identified by name, such as Sixth Principal Meridian and Willamette Meridian. (Every fourth line is a guide meridian or a standard parallel, but these are not important except to surveyors.)

4. North-south strips 6 miles wide are called ranges. The ranges (of townships) are numbered on both sides of the principal meridians, such as R 14 W, which is read "range 14 west (of the principal meridian)."

5. East-west strips 6 miles wide are called tiers. The tiers (of townships) are numbered on both sides of the base lines, such as T 12 5, which is read "township 12 south (of the base line)."

6. The two sets of strips (range and tier) cross to block out an area of 36 square miles (6 by 6). This square is called a township.

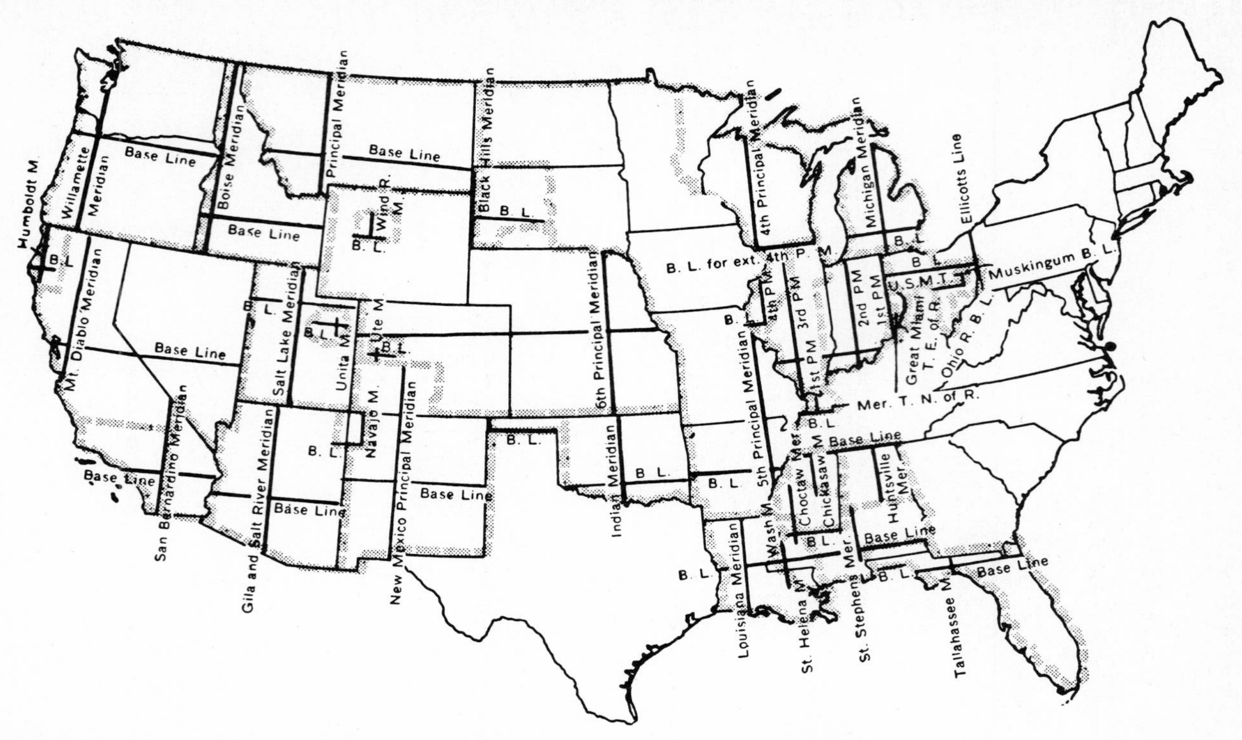

Location of the several prime meridians and their base lines. *(U.S. Department of Agriculture)*

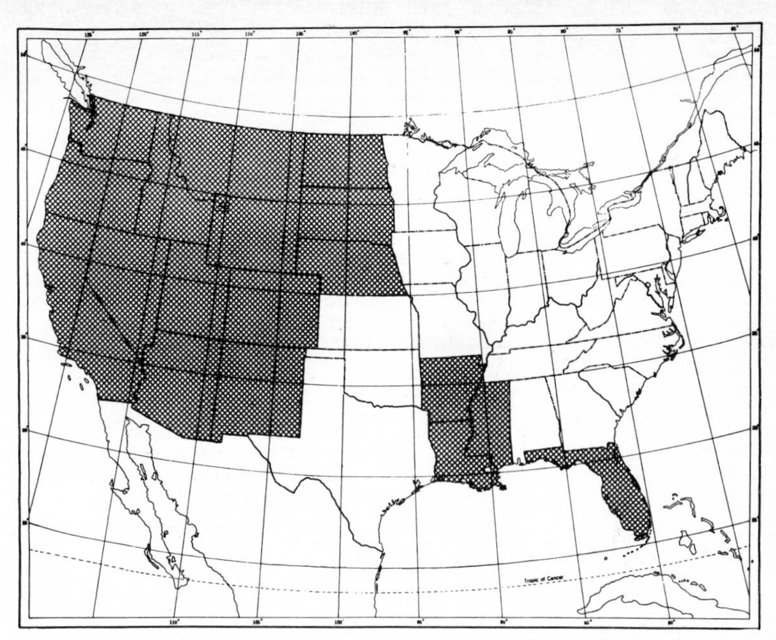

The shaded states above, and Alaska, are the so-called public land states, to which the basic mining law of land claims applies. *(American Map Company)*

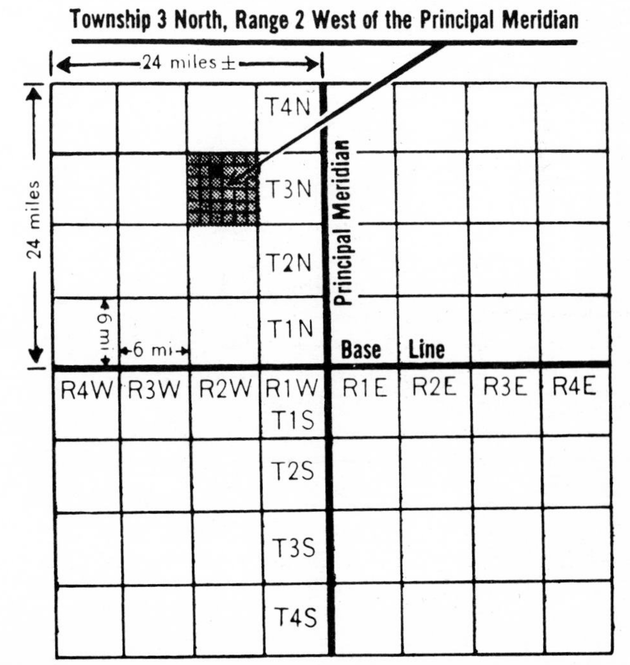

Staking a Mining Claim:
Dividing an area into townships. *(U.S. Department of Agriculture)*

7. Each township is (or can be) subdivided into 36 squares measuring 1 mile on each side. These squares are called sections and are numbered 1 to 36, beginning at the upper right (northeast) and going back and forth, ending at the lower right (southeast). This zigzag method of numbering serves only to put any even-numbered section next to an odd-numbered one, making them easier to remember.

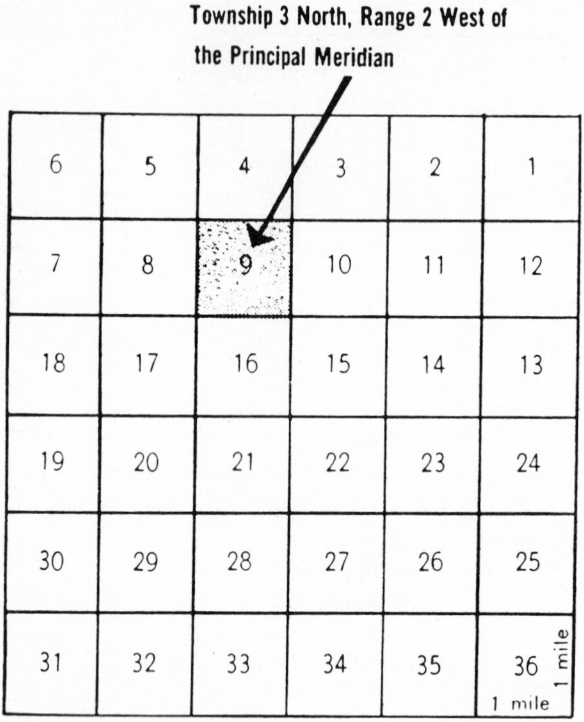

Township 3 North, Range 2 West of the Principal Meridian

6	5	4	3	2	1
7	8	9	10	11	12
18	17	16	15	14	13
19	20	21	22	23	24
30	29	28	27	26	25
31	32	33	34	35	36

1 mile

1 mile

Sections 1 through 6 on the north side and 7, 18, 19, 30, and 31 on the west side are fractional sections.

A township divided into sections. *(U.S. Department of Agriculture)*

8. Each section is (or can be) subdivided into halves, these again into halves or quarters, and so on, into smaller and smaller fractions. (Tracts that are subdivided into blocks and lots seldom concern mining claims.)

The Dominion Lands System of Canada, used in most parts of western

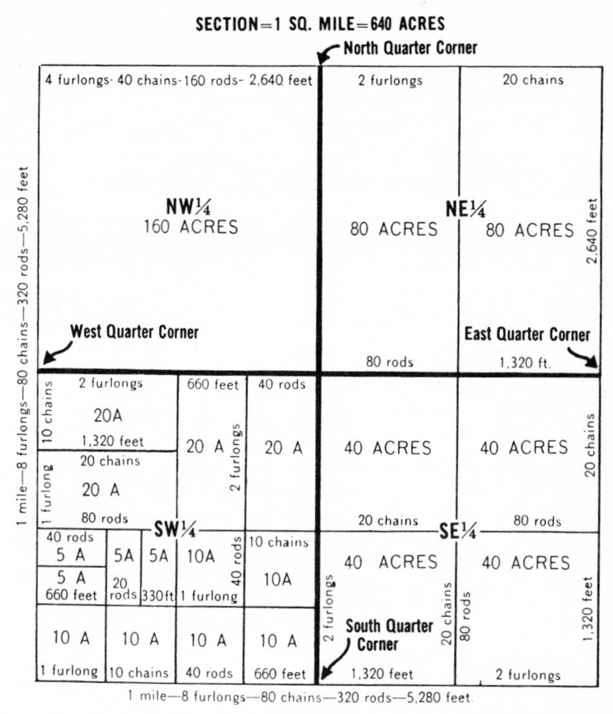

Section of land showing acreage and distances. *(U.S. Department of Agriculture)*

Canada, is much the same as in the United States, but the base line is the 49th parallel of latitude (the international boundary). The sections, however, are numbered exactly in reverse (beginning at the lower right—southeast), and they are subdivided into quarters and then into 16 "legal subdivisions," numbered in the same fashion (beginning at the lower right—southeast). Roman numerals are used for ranges, so that a 40-acre tract might be identified as LS 10, Section 5, Township 4, Range VI, west of the 5th Meridian. In British Columbia, the subdivisions of the townships are different.

Whereas the older, metes-and-bounds system of subdividing land, as used in the eastern part of the United States, has no interest to the prospector on federal land, because mining claims cannot be staked in these states, the situation in Canada is different. In that country, some areas that can be staked are subdivided according to older systems. Counties, or districts, are then subdivided into townships (parishes in parts of Quebec) that are usually 10 miles square. These may be laid out regularly or irregularly and are either named or numbered. Townships are generally divided into 0.1 mile east-west strips, numbered (in Roman numerals) from south to north, and called ranges in Quebec and concessions in Ontario. These strips are divided into lots, which are numbered from east to west.

Having identified the land by its coordinates, the next step is to find out who owns it and whether it is open to prospecting. The U.S. Bureau of Land Management will furnish this information on request, although the best plan is to visit the district office of the U.S. Land Office in which the land is situated and have the land checked for you in your presence. The many complications in land laws—exceptions, withdrawals, patents, etc.—make this step vital. A preliminary idea may perhaps be obtained from a state land-status map of the sort published from time to time by the same bureau.

Claim maps are sold in Canada, each covering a township or a larger area if township subdivision has not been made.

Please understand that even this procedure will not tell you whether open land has already been claimed by someone else, because mining claims are filed at the county seat, and nobody informs the federal government about it. Either you (or an agent) will have to examine the ground in person for the required stakes and notices—but the requirement is not always enforced rigidly—or you will have to inquire at the county courthouse, or preferably both. (Alaska has no counties, but four judicial districts; parishes in Louisiana are like counties.)

Furthermore, even on available ground, certain mineral products cannot be obtained through a mining claim but must be leased. This aspect of mining does not, however, affect the ordinary prospector. After a mining claim is secured, it may be converted to patent, which gives permanent (but taxable) ownership; this formerly common procedure has been made difficult in recent years owing to its having been misused to acquire valuable land for nonmining purposes.

Another essential point to know is that each public-land state has its own laws that modify the federal regulations. Although generally alike, they differ in some respects. Therefore, the state rules should be consulted; several of the states issue simplified instructions. It may be of interest in this connection to recall how the long-uncertain boundary between Utah and Colorado finally came to be noticed during the uranium

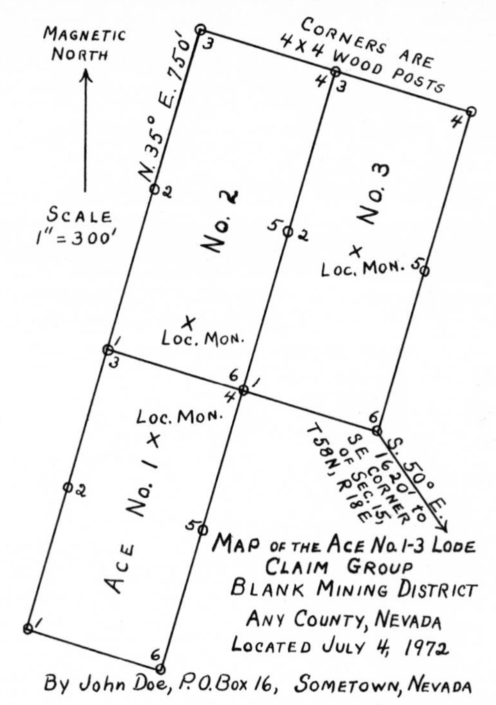

Sample claim location map. *(Nevada Bureau of Mines and Geology)*

boom, when prospectors in this once-remote area realized they had to know whether to follow Utah or Colorado mining law in staking their claims.

It seems best that the above information be followed by a brief discussion of the changes that the prospecting and mining laws are undergoing in response to the social, economic, and political changes of the last decades of the 20th century. A fast-growing population, the depletion of the best vacant land, the rapid consumption of earth's irreplaceable resources, the continued pollution of air and water, the largely unnecessary littering of the ground—these and other unfavorable developments require a new look at the laws governing the use of public land and resources.

It would be well for all citizens to be informed of the problems involved. Neither the notion that the earth is ours to despoil without regard to future generations nor the idea that it can be largely conserved as wilderness and solitude without returning man back to savagery can be accepted. Civilized life is not simple, and compromises must be made. Without any attempt at preachment, here are a few principles of mineral economics that should be taken into account:

1. Mining is a primary source of wealth. In this sense, it is analogous to farming, lumbering, and fishing. Many a miner takes pride in the new, "untainted" wealth that he adds to society: wealth that is a foundation for a wide range of secondary activities.

2. Mineral deposits are nonrenewable. This contrasts with the repeated production of crops (even trees) and animals. A few exceptions are known—such as iron bogs that replenish themselves within a few decades, and the horsetail rush that is supposed to absorb gold from the soil. So-called gold farms in South America are deposits of very finely divided gold that are renewed after times of flood, but these are all more or less curiosities. On this nonrenewability of mineral resources—which are a wasting asset—is based the justification for depletion allowance as a principle of taxation.

3. Mines take a long time to put into production. The time that is required compares unfavorably with the time that is needed to start a factory, for each mine has its own particular set of natural conditions, whereas the basic principle of most factory operations is mass production, and this concept can be carried over into making one factory much like another. The significance of this to national defense should be obvious.

4. Mines must be maintained if they are to remain useful. The International Nickel Company of Canada announced in 1970 that it was pumping 100 million gallons of water from the Victoria mine (abandoned in 1923) in order to learn if further mining is justified. This amount of water would supply the entire needs of a large city for a day. The filling of underground workings with water is a natural phenomenon that is not familiar to outsiders; so too is the tendency to cave when not in use. These likewise have import to national defense. Both of the last two entries (3 and 4) suggest that it is not advisable for a nation to depend too entirely on foreign sources of minerals, for they may be restricted by war or other interference.

5. Mineral resources must be concentrated by nature if they are to be utilized. Seawater (geologically, a rock body) is a vast source of magnesium, but common rock is otherwise seldom used to supply needed minerals. Of the 92 natural elements that are known, only 8 are present in the earth's crust in amounts exceeding 1 percent. Only special mineral occurrences are of value, and these must be exploited where they are found.

Perhaps 1,000 square miles of the entire earth has yielded nearly all its mineral wealth to date. Only 91 mining districts contain about 98 percent of the total value (including unmined reserves) of all metal deposits in the western United States, excluding iron, uranium, and vanadium.

6. The exhaustion of rich, near-surface mineral deposits raises the cost of production. Cost is an economic factor in any type of society. Many of the world's richest mines have already been depleted. Even though a given mineral may never disappear completely from use, increased price becomes an economic burden that may drastically retard industrial progress. For this reason, and in contrast to the fourth entry above, a nation should not use up its richest mineral deposits while letting other countries save theirs.

7. Mining is a specialized vocation, requiring training and experience. Miners turn to other occupations, often in other communities, if they are not needed, but supermarket clerks cannot be converted into miners as quickly as a national emergency may arise.

8. Waste is especially undesirable in regard to mineral resources, which are irreplaceable. Conservation is a necessity—internationally as well as domestically—not merely an ideal.

9. All these considerations are interrelated. There is here no tenth commandment.

Anyone familiar with the history of the United States and Canada will understand that favorable laws to encourage prospecting and mining have promoted the settlement of the West and the building of two great nations "from sea to shining sea." The United States Mining Code of 1872—still in effect—is one of the major pieces of constructive legislation in the American experience. The extent to which it needs renovation, in view of the changing environment, is a matter of deep concern and earnest debate. Such profound students of the subject as Herbert Hoover have discussed it widely. Not until 1969, however, was a bill enacted in the Congress of the United States to establish a national mining and minerals policy. Much remains to be done, however, to coordinate and implement it. The whole subject, complex as it is, requires well-thought-out national mineral policies in each of the various chief producing and consuming countries of the world. This subject, however, is not within the scope of this book. The report on the U.S. Senate hearing on the "National Mining and Minerals Policy" (S. 719, 91st Congress, 1st Session) can be obtained from your congressman.

The bill signed by President Nixon as PL-91-631 in 1971 establishes, as a continuing national policy, the

> need to foster and encourage private enterprise in (1) the development of economically sound and stable domestic mining, minerals, metals and mineral reclamation industries; (2) the orderly and economic development of domestic mineral resources, reserves, and reclamation of metals and minerals to help assure satisfaction of industrial and security needs; (3) mining, mineral, and metallurgical research, including the use and recycling of scrap, and (4) the study and development of methods for the disposal, control, and reclamation of mineral waste products, and the reclamation of mined land, so as to lessen any adverse impact of mineral extraction and processing upon the physical environment that may result from mining or mineral activities.

Virtually all conceivable changes in the mining laws are examined in the 1970 report of the Public Land Law Review Commission. Bills have been introduced into Congress to control mechanized prospecting and other problems.

A comparable examination of the mining laws of Canada, extending into the individual provinces, is presented in *A Study of Claiming and Surveying Procedures in Relation to Mineral (Hardrock) Properties in Canada,* Don W. Thomson, Surveys and Mapping Branch, Department of Energy, Mines and Resources, Ottawa, 1971.

The prospector needs to be aware of changes that may occur, for they will affect his activities in many real ways. The American Mining Congress has recommended certain proposals, and the Colorado Mining Association particularly endorses the following:

1. Retain the basic principles of the Mining Law of 1872, without recodification.
2. Retain the concept of multiple use.
3. Retain the Prudent Man Doctrine as the basis for discovery.
4. Redefine "common variety materials."
5. Establish a form of exploration claim to protect pre-discovery investments.
6. Eliminate distinction between lode and placer claims, establishing a single type of mining claim.
7. Eliminate extralateral rights.
8. Eliminate association placers.
9. Permit location of 20-acre mining claims by reference to the public lands survey.
10. Establish rules for presumption of abandonment of an unpatented mining claim.
11. Establish a uniform system for locating claims and eliminate any required location work.

12. Reform the quasi-judicial procedure now used in the administration of land laws.

13. Limit the length of time which public land may be withdrawn from entry and provide for continuing review of withdrawals and procedures.

14. Establish a procedure for acquiring suitable surface area for disposal of waste or other necessary uses.

15. Insure access across public lands (even if withdrawn) to mining or milling operations.

The nine principal changes in the mining law that seem to have the strongest likelihood of adoption are as follows:

1. To transfer the recording of claims from the many country courthouses to the few federal land offices.

2. To do away with the distinctions between lode and placer claims.

3. To make all claims of even size (perhaps 20 or 40 acres) and have them correspond to the legal subdivisions.

4. To allow present claim owners time (perhaps 2 years) to record their old claims in order to retain them.

5. To abolish extralateral rights, which are explained on page 184.

6. To require assessment work at an increased value (perhaps $10 per acre per year).

7. To require patenting a claim soon after discovery, with the right to acquire additional land for mining and milling purposes.

8. To increase the cost of patenting a claim according to its surface value.

9. To provide an exploration claim, allowing a period of exploration before the required discovery, thereby giving exclusive prospecting rights to fairly large areas for a considerable length of time and permitting the use of advanced exploration techniques.

To substitute leasing for claim staking, to open or close to prospecting large areas of public land, to require a prospector's license, to provide a development contract claim (for a government contract to mine a limited area), to tax mining claims, to restrict the activities of corporate locators, to reclassify the kinds of mineral products—these are just some of the other suggestions that have been advanced, any one of which can be of serious import to the man in the field. It is to be presumed that he will maintain an interest in this aspect of legislation. In order to reduce unnecessary disturbance of the land, Idaho in 1970 and Nevada in 1971 abolished the requirement for location work on newly located mining claims on federal land. Other important changes can be expected in the mining laws of any of the states, and whatever federal regulations develop from the Public

Lands Organic Act of 1971 will likewise affect the situation. For these reasons, no outline of state laws is given in this book; a 1967 summary appeared in *Mineral Industries Bulletin,* Colorado School of Mines, vol. 10, no. 1, 1967.

LAW OF APEX AND EXTRALATERAL RIGHTS

A curious and unique feature of the mining laws of the United States is the law of the apex and the extralateral rights that it confers. Its origin is rather obscure, and it has probably caused more trouble—in and out of litigation—than any other aspect of American mining law. Nevertheless, it exists in the public-land states except where abandoned by agreement, and it will persist at least until the basic laws of 1872 are superseded; it should therefore be understood.

According to this law, a vein that outcrops on your claim, or that comes closest to the surface on your claim—this is its apex—belongs to you and can be mined downward even when it extends beneath adjacent claims. This is the extralateral right, which requires that the end lines of your claim be parallel to each other. In reverse, a vein that is found under-

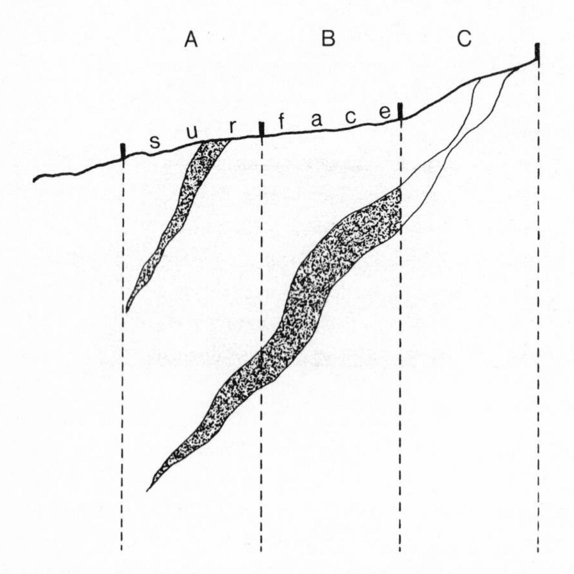

Extralateral rights of a lode claim. *C* owns entire vein if he has these rights; otherwise it is shared by *A* and *B;* *A* owns all of smaller vein. *(After Wolff)*

ground can be traced back to the surface and a claim established at that place. Furthermore, you have no right to mine, even within your own claim, a vein that outcrops on your neighbor's ground.

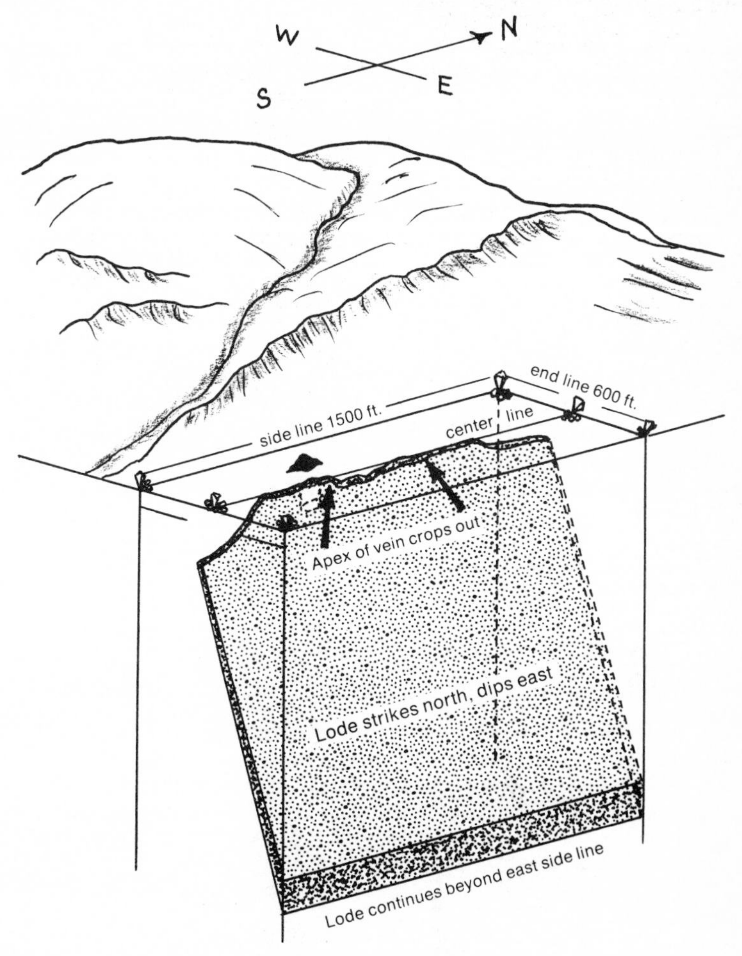

A lode claim under the apex law. *(After Flawn)*

Such a concept implies that veins are simple, not branching, and that they are more or less parallel to one another. However, neither supposition is true, for, as the California courts admit, "veins are irregular in their course and strike, and may be found running through the earth in almost every conceivable direction"—even in the Mother Lode, where veins were relatively "simple and well-behaved." The Bunker Hill mine, in the Coeur d'Alene district of Idaho, contains at least sixteen distinguishable veins and ore bodies.

The General Mining Laws of 1872 are, as modified since, in effect now and will continue to be unless substantially superseded by new laws like those that have been discussed in recent years. As contained in Title 43 of the Code of Federal Regulations, these rules are available free from the Bureau of Land Management, and brief circulars summarizing them, as well as popular leaflets on the subject, can also be had. A number of the states distribute such literature, together with the supplemental rules that apply within the given state. Be sure to get the latest revision.

Title 43, Group 3000 of the law is outlined below. Where an entire section is of no practical concern to the prospector but chiefly of interest to his attorney, the subparts are omitted. Selected statements are followed by questions of the sort most often asked by prospectors and the informal answers to them. These are not to be considered a substitute for qualified legal advice.

3001 *Introduction; General*

Q. What is a mining claim?

A. The words "claim" or "mining claim" have a definite meaning when used in connection with United States mining laws. These words refer to a particular piece of land, valuable for specific minerals, to which an individual has asserted a right of possession for the purpose of developing and extracting discovered minerals. This right is granted the miner if he meets the requirements of the mining laws, and these same laws guarantee him protection for all lawful uses of his claim for mining purposes. However, if the requirements of the law have not been met, no rights are granted.

Q. Who has jurisdiction over mining claims?

A. The Bureau of Land Management, an agency of the U.S. Department of Interior, has the primary responsibility for administering the laws and regulations governing the disposal of most minerals on public lands. The administration of the United States mining laws to "promote the development of the mining resources of the United States" is one of that Bureau's objectives. A Memorandum of Understanding of April 1957, as amended, between the Bureau of Land Management and the Forest Service provides for cooperative procedures in the administration of the mining laws on National Forest System lands, including the examination of mining claims in national forests by Forest Service examiners.

Q. When should I stake a mining claim?

A. Be sure that a mining claim is what you really want. A mining claim is for one purpose only: to permit the development and extraction of certain valuable mineral deposits. Staking a mining claim is neither a simple nor an inexpensive way to obtain a piece of land. The requirements of the mining laws are not easy to meet. Unless you are convinced that you can meet these requirements, a mining claim is not what you are looking for. If you want land to build a house or summer cabin, a swimming pool, a filling station, or any other kind of structure or an entire resort, staking a mining claim is not the way to obtain the land. If you meet the requirements, there are other means, such as purchase under other acts or through a special-use permit, by which it may be possible to occupy land for personal, business, or recreational use.

3001.0–5 *Definitions:* Explained here where necessary.

3001.0–6 *Place of filing:* District land office in each state; in Washington, D.C., if no state office; except, in Montana for North Dakota and South Dakota; in Wyoming for Nebraska and Kansas; in New Mexico for Oklahoma.

3001.0–7 *False statements*

3001.0–8 *Appeals and contests:* As provided by law.

3400 *Mining Claims under the General Mining Laws of 1872; General*

3400.1 *Lands subject to location and purchase.* Vacant public surveyed or unsurveyed lands are open to prospecting and, upon discovery of mineral, to location and purchase, as are also lands in national forests in the public-land states (forest regulations must be observed), lands that are entered or patented under the stockraising homestead law (title to minerals and the use of the surface necessary for mining purposes can be acquired), lands that are entered under other agricultural laws but not perfected, where prospecting can be done peaceably.

Mining locations may be made in the states of Alaska, Arizona, Arkansas, California, Colorado, Florida, Idaho, Louisiana, Mississippi, Montana, Nebraska, Nevada, New Mexico, North Dakota, Oregon, South Dakota, Utah, Washington, and Wyoming.

National parks and monuments. With the exception of Mt. McKinley National Park and Glacial Bay National Monument in Alaska, Organ Pipe Cactus National Monument in Arizona, and Death Valley National Monument in California, mining locations may not be made on lands in national parks and monuments after their establishment.

Minerals in Indian lands. In general, the mineral deposits in Indian reservations are subject to leasing and are under the administration of the Office of Indian Affairs.

Withdrawals. Withdrawals usually bar location under the mining laws, but withdrawals that are made under the Act of June 25, 1910 (36 Stat. 847), as amended by the Act of August 24, 1912 (37 Stat. 497), permit locations of the withdrawn lands that contain metalliferous minerals. Lands withdrawn for waterpower purposes are not subject to location unless first restored under the provisions of section 24 of the Federal Water Power Act.

3400.2 *Minerals under the mining laws.* Whatever is recognized as a mineral by the standard authorities, whether metallic or other substance, when found in public lands in quantity and quality sufficient to render the lands valuable on account thereof, is treated as coming within the purview of the mining laws. Deposits of oil, gas, coal, potassium, sodium, phosphate, oil shale, native asphalt, solid and semisolid bitumen, and bituminous rock (including oil-impregnated rock or sands from which oil is recoverable only by special treatment after the deposit is mined or quarried), and the deposits of sulphur in Louisiana and New Mexico that belong to the United States can be acquired under the mineral leasing laws and are not subject to location and purchase under the United States mining laws. The so-called "common variety" mineral materials and petrified wood on the public lands may be acquired under the Materials Act, as amended.

Q. What types of minerals are there?

A. Minerals are of three types:

1. Locatable. Both metallic (gold, silver, lead, etc.) and nonmetallic (fluorspar, asbestos, mica, etc.) minerals may be located under the mining laws.

2. Salable. By law, certain materials may not be located under the mining laws but may be purchased under the Materials Sale Act of 1947. These include the common varieties of sand, stone, gravel, pumice, pumicite, cinders, and clay. These may be purchased for their fair market value, either at competitive or noncompetitive sales. Petrified wood is not subject to location under the mining laws; small amounts may be removed free of charge by hobbyists for noncommercial use, and larger amounts may be purchased.

3. Leasable. A few other minerals and fuels may be leased from the government and may not be claimed under the mining laws. These are oil and gas, oil shale, potash, sodium, native asphalt, solid and semisolid bitumen, bituminous rock, phosphate, coal, and (in Louisiana and New Mexico) sulfur. All minerals on certain land, such as acquired lands, lands under the jurisdiction of the Department of Agriculture, and areas offshore, are subject to special leasing laws and regulations, which are covered in Part 3220 of Title 43. Senate Bill 2726 was introduced in 1971 to reform the mineral leasing laws.

Q. What is a discovery of valuable mineral?

A. A mining claim may be validly located and held only after the discovery of a valuable mineral deposit. It is a common misunderstanding among prospectors that if they sink a "discovery" shaft or make other mining improvements, and then put up their corner monuments to identify the land, they automatically acquire an interest in the land even though there may be absolutely no indication of valuable minerals within the claim. They also often mistakenly believe that the performance of annual assessment work will perpetuate their "right" to such a claim. Regardless of the prevalence of this belief, such a location is worthless, and no rights to the land have been established according to the law.

The courts have established, and the government follows, the "prudent-man" rule to determine what is a "discovery of a valuable mineral." Under that rule, "where minerals have been found and the evidence is of such a character that a person of ordinary prudence would be justified in further expenditure of his labor and means, with a reasonable prospect of success, in developing a valuable mine," the requirements of the statute have been met. Economic factors, such as market, cost versus returns, etc., are important considerations in applying the prudent-man test. Many people have misunderstood this to mean that merely any showing of a mineral, or a hope or wish for future discovery, is sufficient. This is not correct. There must be an actual physical discovery of the mineral on each and every mining claim, and this discovery must satisfy the prudent-man rule. Traces, isolated bits of mineral, or minor indications are not sufficient to satisfy the prudent-man rule.

3400.3 *Mineral locations in stock driveway withdrawals.* Avoid interference with animals.

3400.4 *Mineral locations in reclamation withdrawals.* Allowed under special application.

3401 *Lands and Minerals Subject to Location*

3401.1 *Manner of initiating rights under location*

Initiation of rights to mineral land. Rights to mineral lands owned by the United States are initiated by prospecting for minerals thereon, and, upon the discovery of mineral, by locating the lands upon which such discovery has been made. A location is made by staking the corners of the claim, posting notice of location thereon, and complying with the state laws regarding the recording of the location in the county recorder's office, discovery work, and other regulations. As supplemental to the United States mining laws, there are state statutes relative to location, manner of recording of mining claims, etc., in the state, which should also be observed in the location of mining claims.

Q. What is the difference between a patented and unpatented claim?

A. A patented mining claim refers to a piece of ground for which the federal government has given a deed or has passed its title to an individual. An unpatented claim is one on which an individual, by the act of valid location under the mining laws, has obtained a right to remove and extract minerals from the land, but where full title has not been acquired from the government. The rights under each claim are somewhat different. You may apply for a patent to a mining claim if you wish, but it is not necessary to have a patent to mine and remove minerals from a valid claim.

Q. What rights do I obtain from a mining claim?

A. If you establish a valid claim, perform and record annual assessment work required by state law, and meet all other requirements of federal and state mining laws and regulations, you establish a possessory right to the area covered by the claim for the purpose of developing and extracting minerals. This possessory right may be sold, inherited, or taxed according to state law. No one else can mine, without your consent, the minerals that you have claimed.

But until you obtain patent to the claim from the government, you do not hold full title to the land. Your possession is based upon discovery of a valuable mineral, and your right to the claim may be questioned or challenged by the government if it appears that your claim lacks discovery, the minerals have been mined out, or the claim does not meet other requirements of the law. If the government's challenge of your claim is successful, the claim is canceled, and you have no rights to the lands.

On unpatented claims, you may use as much of the surface and surface resources of the claim as are reasonably necessary to carry out your mining operations. These uses, however, must be connected with and necessary for mineral development.

On all mining claims located since July 23, 1955, the government, prior to issuance of patent, has the right to manage the surface so far as it does not interfere with mining. The government may manage and dispose of vegetative resources, such as timber and grass, and may manage other resources except minerals subject to location. For claims located prior to 1955, the government may also obtain surface management rights, under the Act of July 23, 1955, except where the claimant proves that his location was valid prior to July 23, 1955.

Q. Is there a way that I can own the surface as well as the minerals?

A. A mining patent or deed that is received from the government gives you the exclusive title to the locatable minerals. In most such cases, you will also obtain full title to the land surface and all other resources. If you do obtain full title to the land and minerals, you of course have the exclusive right to use the surface, subject to local law. You should contact the local Bureau of Land Management or Forest Service office to find if your mining claim is in an area where you may receive full title to both surface and mineral rights.

3401.2 *Who may make locations.* Citizens of the United States, or those who have declared their intention to become such, including minors who are bona fide locators and corporations organized under the laws of any state. Agents may make locations for qualified locators.

Q. How many claims may I stake?

A. There is no limit to the number of claims you may hold, as long as you have made a discovery of a valuable mineral on each one and meet other requirements.

3410 *Nature and Classes of Mining Claims; General*

3410.0–6 *Classes of mining claims*

Q. What types of mining claims are there?

A. Mining claims are of four types:

1. Lode claims. Deposits subject to lode claims include classic veins or lodes having well-defined boundaries. They also include other rock bearing valuable minerals in place and may be broad zones of mineralized rock. Examples include quartz or other veins bearing gold and other metallic minerals and low-grade disseminated copper deposits.

2. Placer claims. Deposits that are subject to placer claims are all those that are not subject to lode claims. These include the "true" placer deposits of sand and gravel that contain free gold and they also include many nonmetallic deposits.

3. Mill site. A mill site is a plot of unappropriated public-domain land of a nonmineral character, which is suitable for the erection of a mill or reduction works. A mill site may be located under either of the following circumstances:

a. When used or occupied distinctly and explicitly for mining and milling purposes in connection with the lode or placer location with which it is associated.

b. For a quartz mill or reduction works that is unconnected with a mineral location.

4. Tunnel sites. A tunnel site is located on a plot of land where a tunnel is run to develop a vein or lode. Tunnel sites cannot be patented.

Q. If I buy public land, do I get the minerals too?

A. Ordinarily, public lands valuable for minerals are not sold. However, in some instances, such lands are sold, and certain minerals are reserved to the government. The notice of sale will clearly state whether minerals that relate to a particular parcel are to be reserved.

3411 *Lode Claims*

3411.1 *Lodes located prior to May 10, 1872.* Present inheritors of those rights can take advantage of the apex law.

3411.2 *Lodes must not have been adversely claimed.*

3411.3 *Length of lode claims.* Not more than 1,500 feet in length along the vein; the side lines are usually parallel and are to be located by length and direction.

3411.4 *Extent of surface ground.* Not more than 300 feet on each side of the middle of the veins. The end lines of the location are to be parallel to each other; the discovery shaft marks the middle of the vein at the surface.

3411.5 *Restriction in width of claims by local laws.* State laws cannot limit the length to less than 1,500 feet or the width to less than 50 feet (except for adverse claims).

3412 *Describing Locations*

3412.1 *Defining of locations by claimants.* Except for placer claims described by legal subdivision, all mining claims must be distinctly marked on the ground so that their boundaries may be readily traced, and all notices must contain the name or names of the locators, the date of location, and such a description of the claim by reference to some natural object or permanent monument as will serve to identify the claim.

3413 *Discovery*

3413.1 *Discovery required before location.* No lode claim shall be located until after the discovery of a vein or lode within the limits of the claim. The object is to prevent the appropriation of presumed mineral ground for speculative purposes, to the exclusion of bona fide prospectors, before sufficient work has been done to determine whether a vein or lode really exists.

3413.2 *Discovery work.* The claimant should, prior to locating his claim (unless the vein can be traced upon the surface), sink a shaft or run a tunnel or drift to a sufficient depth therein to discover and develop a mineral-bearing vein, lode, ledge, or crevice. He should determine, if possible, the general course of such vein in either direction from the point of discovery, by which direction he will be governed in marking the boundaries of his claim on the surface.

3414 *Location*

3414.1 *Location notice: monumenting.* The location notice should give the course and distance as nearly as practicable from the discovery shaft on the claim to some permanent, well-known points or objects, such as stone monuments, blazed trees, the confluence of streams, point of intersection of well-known gulches, ravines, or roads, prominent buttes, hills, etc., that may be in the immediate vicinity. These will serve to perpetuate and fix the site of the claim and render it possible to be identified from the description that is given in the record of locations in the district. They should be duly recorded.

In addition to the foregoing data, the claimant should state the names of adjoining claims, or, if none adjoin, the relative positions of the nearest claims; he should drive a post or erect a monument of stones at each corner of his surface ground; and at the point of discovery or discovery shaft, he should fix a post, stake, or board, upon which should be designated the name of the lode, the name or names of the locators, the number of feet claimed, and in which direction from the point of discovery.

It is essential that the location notice that is filed for record, in addition to the foregoing description, should state whether the entire claim of 1,500 feet is taken in on one side of the point of discovery or whether it is partly upon the other side; and in the latter case, the notice should state how many feet are claimed upon each side of the discovery point.

3414.2 *Location notices to be recorded.* The location notice must be filed for record in all respects as required by the state laws and by local rules and regulations if there be any.

3415 *Tunnel Sites*

3415.1 *Possessory right of tunnel proprietor.* Additional mining rights are granted to claimants of mining-tunnel sites if the tunnel intersects lodes that were not visible at the surface.

3415.2 *Location of tunnel claims.* These are usually square, not more than 3,000 feet on each side, and the lines are to be located by length and direction. Regulations are given as to notices and stakes or monuments required to hold the claim.

3415.3 *Recording of notices.* Regulations as to tunnel claims.

3416 *Placer Claims*

3416.1 *Maximum allowable acreage.* Placer locations, which include all minerals that do not occur in vein or lode formation, may be for areas of not more than 20 acres for each locator, no claim to exceed 160 acres made by not less than eight locators. Placer locations must conform to the public surveys wherever practicable.

3416.2 *Discovery.* Only one is needed, whether by an individual (on 20 acres or less) or by a group (on 160 acres or less).

3416.3 *Locations authorized in 10-acre units.* 40-acre legal subdivisions can be split into square 10-acre tracts, which will have legal standing.

3416.4 *Manner of describing 10-acre units.* Standard geologic description may be used, but tract must be fully identified.

3416.5 Conformity of placer claims to the public land surveys: provides rectangular tracts, avoiding peculiar shapes wherever possible (except on unsurveyed land and in a few other places).

3416.6 *Annual expenditure.* $100 worth of assessment work on either placer or lode claims.

3416.7 *Building-stone placers.* These were acceptable for extracting building stones, but under the Materials Sale Act of 1947, common varieties of stone may be purchased but not claimed.

3416.8 *Saline placers.* Permitted in 1901, but superseded in 1920 by the Mineral Leasing Act.

3416.9 *Petroleum placers.* Permitted in 1897, but superseded in 1920 by the Mineral Leasing Act.

3417 *Millsites*

3417.0–3 *Authority.* Additional rights to land are granted for millsite purpose.

3417.1 *Required use.* In connection with a lode or placer claim. They are limited to 5 acres per claim and the lines are to be located by length and direction.

3420 *Assessment Work; General*

3420.1 *Annual assessment work.* In order to hold the possessory right to a lode or placer location, not less than $100 worth of labor must be performed or improvements made thereon annually. The period within which the work that is required is to be done shall commence at 12 o'clock noon on the first day of September after the date of location of each claim. Where a number of contiguous claims are held in common, the aggregate expenditure that would be necessary to hold all the claims may be made on any one claim. Cornering locations are held not to be contiguous.

Q. May I build a house on a mining claim?

A. You may build a house or cabin or other improvements such as tool sheds or ore storage bins, etc., on a valid mining claim if such structures are reasonably necessary for your use in connection with your mining operations.

Q. May I buy a mining claim?

A. A valid mining claim can be bought and sold, willed or inherited. However, if you buy or sell a claim, remember that you acquire or convey only such rights as you possess under the mining law. If the claim is without a valid discovery or is otherwise defective, it is worthless and is not made valuable by being bought and sold.

A great deal of unwise speculation has resulted from the activities of unethical or misinformed "promoters" who, for a fee, purport to stake mining claims and do annual assessment work for others. Most of these claims are located in areas of rapid expansion and rapidly changing land values. More often than not, these claims have absolutely no value for minerals and are invalid. These promoters are not a part of the mining industry and must not be confused with the legitmate miners or prospectors who are diligently prospecting for minerals and who may, on occasion, wish to sell a valid claim to others for development. Through the inducements of these unethical promoters who misrepresent a mining claim as a possible site for a weekend cabin or a hunting lodge, many people have invested their money in worthless claims and perhaps built a house only to discover that the claim that they purchased was not valid. Remember, an unpatented mining claim is to be used for mining purposes only.

3420.2 *Inclusion of surveys in assessment work.* In addition to the several types of work that may fulfill the annual labor requirement, the requirement can also be satisfied by conducting geological, geochemical and geophysical surveys. Such surveys must be conducted by qualified experts and verified by a detailed report filed in the county or recording district office in which the claim is located. This report must set forth fully the following:

(1) The location of the work that was performed in relation to the point of discovery and boundaries of the claim.
(2) Nature, extent, and cost of the work that was performed.
(3) The basic findings of the surveys.
(4) The name, address, and professional background of the person or persons who have conducted the work.

Such surveys may not be applied as labor for more than 2 consecutive years or for more than a total of 5 years on any one mining claim. Each survey shall be nonrepetitive of any previous survey of the same claim. Such surveys will not apply toward the statutory provision that requires the expenditure of $500 for each claim for mineral patent.

3420.3 *Failure to perform annual assessment work.* The work can be resumed if it is done before someone else relocates the claim.

3420.4 *Determination of right of possession between rival claimants.* The courts will decide.

3420.5 *Annual assessment work not required after patent certificate.* The land then becomes taxable as local property.

3420.6 *Failure of a coowner to contribute to annual assessment work.* Adequate notice ends his interest in the claim.

3421 *Deferments*

3421.0–3 *Authority.* Assessment work may be deferred in certain cases of necessity.

3421.1 *Conditions under which deferment may be granted.* Difficulty of access or other legal impediments.

3421.2 *Filing of petition for deferment, contests.* Procedure.

3421.3 *Nature of action on petition to be recorded.*

3421.4 *Period for which deferment may be granted.* One year, renewable, or less.

3421.5 *When deferred assessment work is to be done.* The usual period after the end of the deferment, in addition to the regular work.

3430 *Disposal of Reserved Minerals; General*

3431 *Under Act of July 17, 1914*

3432 *Under Stockraising Homestead Act* patents to stockraising or grazing homesteads reserve all mineral rights to the government, and any qualified locator may prospect there, although he is liable for damage to crops or permanent improvements. Actual mining requires a written agreement, payment for property, and posting of bond.

3440 *Surveys of Mining Claims; General*

3440.1 *Application for survey.* A correct survey by a cadastral engineer is required for a patent application of a lode claim, of a claim not laid out with the rectangular system, or of a claim that does not conform to the legal subdivisions.

3440.2 *Survey must be made subsequent to recording notice of location*

3440.3 *Plats and field notes of mineral surveys*

3441 *Surveys*

3441.1 *Particulars to be observed in mineral surveys.* Obligations of the surveyor.

3441.2 *Certificate of expenditures and improvements.* $500 in labor or improvements are required for each claim, and other necessary statements must be made in an application for patent.

3441.3 *Mineral surveyor's report of expenditures and improvements*

3441.4 *Supplemental proof of expenditures and improvements.* For late filing.

3441.5 *Amended mineral surveys*

3442 *Mineral Surveyors*

3443 *Contract for Surveys*

3444 *Appointment and Employment of Mineral Surveyors*

3445 *Plats and Notices*

3446 *Posting*

3450 *Lode Claim Patent Application; General*

3450.1 *Application for patent.* Information required.

3450.2 *Service charge.* $25.

3450.3 *Evidence of title.* Manner of supporting the proof of title.

3450.4 *Evidence relating to destroyed or lost records*

3450.5 *Statement required that land is unreserved, unoccupied, unimproved, and unappropriated.* Applicable in Alaska only.

3451 *Citizenship*

3451.1 *Citizenship of corporations and of associations that act through agents.* Manner of establishing proof of citizenship.

3451.2 *Citizenship of individuals.* Statements necessary.

3451.3 *Trustee to disclose nature of trust*

3452 *Possessory Rights*

3452.1 *Right by occupancy.* Relates to old claims when records are lost.

3452.2 *Certificate of court required*

3452.3 *Corroborative proof required*

3453 *Publication of Notice*

3453.1 *Newspaper designation.* Frequency.

3453.2 *Contents of published notice*

3453.3 *Manager to designate newspaper*

3453.4 *Proof by applicant of publication and notice*

3453.5 *Charges for publication*

3453.6 *Payment of purchase and statement of charges and fees*

3455 *Entry and Transfers*

3456 *Diligent Prosecution*

3457 *Application Processing upon Contest or Protest*

3458 *Patents for Mining Claims*

3460 *Mill Sites, Patents; General*

3460.1 *Application for patent*

3460.2 *Mill sites applied for in conjunction with a lode claim*

3460.3 *Mill sites for quartz mills or reduction works*

3460.4 *Proof of nonmineral character*

3470 *Placer Mining Claim Patent Applications; General*

3470.1 *Application for patent*

3470.2 *Proof of improvements for patent*

3470.3 *Data to be filed in support of application*

3470.4 *Applications for placers that contain lodes*

3480 *Adverse Claims, Protests and Conflicts*

3481 *Adverse Claims*

3482 *Protests, Contests, Conflicts, and Segregations*

3483 *Segregation (of mineral from nonmineral land)*

3500 *Multiple Use*

3510 *Public Law 167; Act of July 23, 1955.* Designed to provide for multiple use of the surface of public-land tracts, this law closely affects the holder of a mining claim. It removes common varieties of minerals and rocks from claim staking, and it prohibits the use of mining claims for any other purpose. The significant provisions follow:

3511 *Common Varieties.* A deposit of common varieties of sand, stone, gravel, pumice, pumicite, or cinders is not available for claim staking unless it contains "some other mineral," or "has some property giving it distinct and special value," or contains block pumice in pieces of 2 inches or more, or limestone suitable for use in the production of cement, or metallurgical or chemical grade limestone, gypsum, and other materials. Various factors are considered in determining commercial value.

3512 *Proceedings under the Act*

3512.1 *Restriction on use of unpatented mining claims.* Mining claims do not include rights to vegetation unless necessary, but access to other timber is provided; state water rights are not otherwise interfered with; use of claims is forbidden for "filling stations, curio shops, cafes, tourist or fishing and hunting camps"; access for government purposes cannot be blocked.

3512.2 *Request for publication of notice to mining claimant*

3512.3 *Evidence necessary to support a request for publication*

3512.4 *Publication of notice*

3512.5 *Contests of published notice*

3512.6 *Service of notice*

3512.7 *Service of copies; failure to comply*

3512.8 *Proof of publication*

3512.9 *Failure of claimant to file verified statement*

3513 *Hearings*

3514 *Rights of Mining Claimants*

3530 *Public Law 359; Mining in Power-site Withdrawals; General.* These laws were designed to permit the mining of mineral resources on public lands that were previously withdrawn for power development.

3531 *Power Rights*

3532 *Mining Operations*

3533 *Surface Protection*

3534 *Withdrawals Other Than for Powersite Purposes*

3535 *Operations Risk*

3536 *Location and Assessment*

3537 *Prior Existing Mining Locations*

3538 *Use*

3540 *Public Law 585; Multiple Mineral Development; General.* The purpose of the law was to amend the mining and mining-leasing laws to provide for multiple mineral development of tracts of public land. These provisions refer only to action that is required of the government.

3541 *Claims, Locations, and Patents*

3542 *Procedure to Determine Claims.* Detailed rules are given whereby claims to Leasing Act minerals under unpatented mining locations are determined. These involve the issuing of notices and their publication.

3543 *Hearing.* These regulations pertain to hearings on disputed matters by the Bureau of Land Management.

3544 *Helium.* Previously withdrawn lands containing helium were opened to mining locations and mineral leasing.

3545 *Fissionable Source Materials.* Previous reservations of radioactive minerals were removed, permitting claims to be located and other procedures as for nonfissionable materials.

3630 *Areas Subject to Special Mining Laws.* Certain exceptions to the general mining laws pertain to Oregon and California Railroad and Reconveyed Coos Bay Wagon Road Grant Lands, Olympic National Park, Organ Pipe Cactus National Monument, City of Prescott Watershed, Papago Indian Reservation, Glacier Bay National Monument, and lands patented under the Alaska Public Sale Act.

BUREAU OF LAND MANAGEMENT LAND OFFICES

Alaska

SOUTHERN ALASKA
Anchorage Land Office
555 Cordova Street
Anchorage, Alaska 99501

NORTHERN ALASKA
Fairbanks District & Land Office
516 Second Avenue
Fairbanks, Alaska 99701

Arizona

Arizona Land Office, Federal Building
Phoenix, Arizona 85025

California

SOUTHERN CALIFORNIA
Riverside District & Land Office
1414 Eighth Street
Riverside, California 92502

NORTHERN CALIFORNIA
Sacramento Land Office
Federal Building, Room 4017
Sacramento, California 95814

Colorado

Colorado Land Office
Federal Building, 1961 Stout Street
Denver, Colorado 80202

Idaho

Idaho Land Office
Federal Building
Boise, Idaho 83701

Montana (North Dakota, South Dakota)

Montana Land Office
316 North 26th Street
Billings, Montana 59101

Nevada

Nevada Land Office, Federal Building & U.S. Nevada Courthouse
300 Booth Street
Reno, Nevada 89502

New Mexico (Oklahoma, Texas)

New Mexico Land Office, Federal Building & U.S. Post Office
South Federal Place
Santa Fe, New Mexico 87501

Oregon (Washington)

Oregon Land Office
729 Northeast Oregon Street
Portland, Oregon 97232

Utah

Utah Land Office,
Federal Building
Salt Lake City, Utah 84111

Wyoming (Kansas, Nebraska)

Wyoming Land Office
U.S. Post Office & Courthouse
2120 Capitol Avenue
Cheyenne, Wyoming 82001

Eastern
Arkansas, Iowa, Louisiana, Missouri, Minnesota (minerals only), and all states east of the Mississippi River

Eastern States Land Office
7981 Eastern Avenue
Silver Spring, Maryland 20910

For more information about mining on National Forest System lands, including wilderness, you may contact any of the following Regional Offices of the U.S. Forest Service:

Northern Region

Federal Building
Missoula, Montana 59801

Rocky Mountain Region

Federal Center, Building 85
Denver, Colorado 80225

Southwestern Region

517 Gold Avenue S.W.
Albuquerque, New Mexico 87101

Intermountain Region

324 25th Street
Ogden, Utah 84401

California Region

630 Sansome Street
San Francisco, California 94111

Pacific Northwest Region

Post Office Box 3623
Portland, Oregon 97208

Southern Region

Suite 800
1720 Peachtree Rd. N.W.
Atlanta, Georgia 30309

Alaska Region

Post Office Box 1628
Juneau, Alaska 99801

REFERENCES

United States

Especially clear explanations of the United States mining laws are given by Marion Clawson in U.S. Bureau of Mines Information Circular 7535 (1950, obtainable free from the U.S. Bureau of Mines, 4800 Forbes Avenue, Pittsburgh, Pa. 15213) and by Lorraine Burgin in Colorado School of Mines Mineral Industries Bulletin, vol. 10, no. 1, 1967. The Bureau of Land Management (addresses on page 199) distributes free its simplified pamphlet *Mining Claims, Questions and Answers* (1963).

Further details are given in *Regulations Pertaining to Mining Claims under the General Mining Laws of* 1872, *Multiple Use, and Special Disposal Provisions,* Circular no. 2149, Bureau of Land Management, 1964. The complete laws appear in *Code of Federal Regulations,* Title 43, Public Lands, sold (with annual revisions) by the U.S. Government Printing Office, Washington, D.C. 20402.

The subject is discussed at length in the following books:

American Law of Mining, edited by Rocky Mountain Mineral Law Foundation, Matthew Bender & Company, Inc., New York, various dates; and a successor to *Mining Rights on the Public Domain* (previously *Morrison's Mining Rights*), 6th edition, Emilio D. DeSoto and Arthur R. Morrison, Bender-Moss Company, San Francisco, 1936.

American Mining Law, A. H. Ricketts, California Division of Mines Bulletin 123, 1934 (2 volumes).

A Treatise on the American Law Relating to Mines and Mineral Lands within the Public Land States and Territories and Governing the Acquisitions and Enjoyment of Mining Rights in Land of the Public Domain, 3d edition, C. H. Lindley, Bancroft-Whitney Company, San Francisco, 1914 (3 volumes). This is commonly known as "Lindley on Mines."

Comparative Analysis of American and Canadian Hard Mineral Laws, Harry Macdonell, Matthew Bender and Company, New York, 1965.

In addition, certain of the states (especially Arizona, California, Colorado, and Nevada) publish legal guides for prospectors and miners; some are free, and some are sold. Some states (such as Nevada) issue free leaflets that summarize the regulations. A directory of the state agencies concerned with mining and public lands is given on pages 43–46 of this book.

Canada

Canada does not have uniform prospecting and mining laws on a national scale. Certain aspects, however, are subject to federal restrictions; these are related to national parks and Indian reserves, nuclear-energy materials, explosives, and income taxes. The provincial departments of mines (or their equivalents) administer the regulations in the respective provinces, and the federal Department of Indian Affairs and Northern Development has authority in the Northwest Territories and Yukon Territory. Mining districts, or divisions, have been set up, each with a mining recording office, and some are further subdivided (with suboffices) for convenience. Information and literature can usually be obtained in county towns or other places having a courthouse.

Mineral rights, entirely separate from surface rights, can almost always be obtained in Canada by claim staking except where land has been withdrawn or where prior grants state otherwise; in the older parts of Canada, these must first be verified (at the mining recorder or land titles office). The principal difference from the United States is the absence of extralateral rights, involving the so-called law of the apex (see page 184); "protecting the dip" is therefore a matter to be considered. Lode and placer claims are separate, and other types are also provided as in the United States.

The separate laws of the provinces and territories must be followed. These concern the difference between company and individual claims; the number of claims that may be staked in one year; the size and shape of the claims; the manner of staking; marking, designating, and recording claims; the amount of assessment work ("representation"); the means of patenting or leasing claims; and taxing claims. Large prospecting concessions are given, usually in remote regions, under special conditions. Regulations are usually observed more closely in Canada than in the United States.

Uranium and thorium prospecting require no special permit until the stage of advanced exploration is reached. This includes prolonged surface trenching, detailed geological and geophysical surveying, diamond drilling, underground exploration, and bulk sampling. The Atomic Energy Control Board, in Ottawa (P.O. Box 1046), which issues the permits, is to be kept informed of abandonment or the desire to obtain a free mining permit. The Mineral Deposits Division of the Geological Survey of Canada is to be informed of successful finds of radioactive minerals, the result of exploration with a permit, and any published reports. The Director of the Geological Survey of Canada is to be notified of the precise location of places where strong radioactivity is determined from samples or where assays show 0.05 percent or more of uranium or thorium.

Canadian mining laws are condensed in *Digest of Canadian Mining Laws,* 6th edition, published by the Department of Energy, Mines & Resources, Ottawa ($4 from the Queen's Printer, Ottawa). A clear summary is given on pages 264–268 of *Prospecting in Canada,* 4th edition, by A. H. Lang (Geological Survey of Canada, Ottawa, 1970). Several of the provinces issue free pamphlets.

Following is a directory of the mines departments of the provinces and territories of Canada:

ALBERTA
Department of Mines and Minerals
Edmonton, Alta.

BRITISH COLUMBIA
Department of Mines and Petroleum Resources
Victoria, B.C.

MANITOBA
Mines Branch
Department of Mines and Natural Resources
Winnipeg, Man.

NEW BRUNSWICK
Department of Natural Resources
Fredericton, N.B.

NEWFOUNDLAND
Mineral Resources Division
Department of Mines, Agriculture and Resources
St. John's, Nfld.

NORTHWEST TERRITORIES AND YUKON TERRITORY
Northern Economic Development Branch
Department of Indian Affairs and Northern Development
Ottawa, Ont.

Resident Geologist for Yukon
Federal Building
Whitehorse, Y.T.

Resident Geologist for Northwest Territories
Federal Building
Yellowknife, N.W.T.

NOVA SCOTIA
Department of Mines
Halifax, N.S.

ONTARIO
Department of Mines
Toronto, Ont.

PRINCE EDWARD ISLAND
Deputy Provincial Secretary
Provincial Government Offices
Charlottetown, P.E.I.

QUEBEC
Mineral Resources Branch
Department of Natural Resources
Quebec, Que.

SASKATCHEWAN
Department of Mineral Resources
Regina, Sask.

Australia

The individual states in Australia have their own laws on prospecting and mining. Following is a directory of the mines departments of the states, beginning with the federal agency in the national capital:

Bureau of Mineral Resources
Geology and Geophysics
Constitution Avenue
Parkes, Canberra, A.C.T., 2600

NEW SOUTH WALES
Department of Mines
State Office Block
Phillip Street
Sydney, New South Wales, 2000

QUEENSLAND
Department of Mines
Mineral House
2 Edward Street
Brisbane, Queensland, 4000

SOUTH AUSTRALIA
Department of Mines
169 Rundle Street
Adelaide, South Australia, 5000

TASMANIA
Department of Mines
Public Buildings
Davey Street
Hobart, Tasmania, 7000

VICTORIA
Department of Mines
Treasury Place
Melbourne, Victoria, 3000

WESTERN AUSTRALIA
Department of Mines
St. George's Terrace
Perth, Western Australia, 6000

World and International

World mining laws are presented and explained in historical fashion by Peter T. Flawn in chapter 7 of *Mineral Resources: Geology, Engineering, Economics, Politics, Law* (Rand McNally & Company, Chicago, 1966). A discussion of this subject is also given by Northcutt Ely in chapter 3 of *Economics of the Mineral Industries,* 2d edition, edited by Edward H. Robie (American Institute of Mining, Metallurgical, and Petroleum Engineers, New York, 1964). An older but interestingly written book is

International Mining Law, by Theo. F. Van Wagenen (McGraw-Hill Book Company, New York, 1918).

The laws and their interpretation are given in the following publications:

Summary of Mining and Petroleum Laws of the World, by Northcutt Ely, U.S. Bureau of Mines Circular 8017, 1961. Revised: Part 1, Western Hemisphere, Information Circular 8482, 1970. Part 2, East Asia and the Pacific, Information Circular 8514, 1971. Part 3, Near East and South Asia, Information Circular 8544, 1972. Two more parts are still to be published; later revisions will appear in *Mineral Trade Notes,* issued free by the U.S. Bureau of Mines.

Survey of Mining Legislation with Special Reference to Asia and the Far East, by Office of Legal Affairs, United Nations, published by Columbia University Press, New York, 1950. It deals with the following countries: Australia, Brazil, Burma, Canada, Ceylon, Republic of China, Congo, Egypt, Hong Kong, India, Indonesia, Japan, South Korea, Laos, Malayasia, Mexico, Pakistan, Peru, Philippines, Sarawak, Thailand, Turkey, Viet Nam.

Mining and Petroleum Legislation in Latin America, 2d edition, published by The Pan American Union, Washington, 1969. It deals separately with mining and petroleum legislation in the following countries: Argentina, Barbados, Bolivia, Brazil, Chile, Colombia, Costa Rica, Dominican Republic, Ecuador, El Salvador, Guatemala, Haiti, Honduras, Mexico, Nicaragua, Panama, Paraguay, Peru, Trinidad and Tobago, Uruguay, and Venezuela.

Part 3

14

Studying the Earth

The advantage of being able to recognize minerals, rocks, and geologic structures can scarcely be exaggerated. The history of prospecting is filled with true stories of men who had fortune in their hands but failed to realize it. Later, men passing that way knew what they saw and took advantage of their knowledge. Even when they were wrong about the exact nature of their finds, they had an inkling of the true value and were encouraged to make tests and inquire further.

Thus, the purple quartz at Creede, Colorado, was thought at first to be the same as the purple fluorite at Cripple Creek; nevertheless, both belonged to rich mines. The pink rhodonite of the Camp Bird mine misled Thomas Walsh into believing that he had pink fluorite, as at Cripple Creek, but it also led him into an abandoned tunnel, where he took the sample that made him a millionaire and bought the Hope diamond for his daughter.

Conrad Reed, a boy about 12 years old, found in 1799 a 17-pound nugget, which opened the gold fields of North Carolina. The silversmith to whom it was taken did not recognize its value, because he was thinking only of silver; only later did a jeweler identify it correctly.

In his presidential address to the Mineralogical Association of Canada, M. H. Frohberg spoke of the rewards to the prospector of a knowledge of practical mineralogy. He told of the discovery of uranium in 1929 at Great Bear Lake—one of history's richest mineral finds—by Gilbert La Bine, an unknown prospector, who recognized pitchblende and the bright coatings of secondary uranium minerals by their color. Only 2 years before, an exploration party had spent almost $2 million in a fruitless search for gold in the same area; members of the group had observed coatings due to cobalt, nickel, and copper, but they had passed them by and had not thought to identify the uranium minerals. Mr. La Bine acquired both wealth and prestige from the Eldorado uranium and silver mine.

Practical, working miners had not always been enamored of scientific knowledge. One of them said in 1849: "The mines of California have baffled all science, and rendered the application of philosophy entirely nugatory. Bone and sinew philosophy, with a sprinkling of good luck, can alone render success certain. We have met with many geologists and practical scientific men in the mines and have invariably seen them beaten by unskilled men, soldiers and sailors, and the like." When, however, the nature of the deep but accessible Tertiary gravels in the northern part of the Mother Lode was understood, California mining flowered into new life.

In his address as retiring president of the Colorado Scientific Society, Richard Pearce, one of the most brilliant of pioneer Western metallurgists, stated:

> The ambitious amateur prospector who is generally ignorant of the science of mineralogy, or the laws which appear to regulate the association of minerals, and who is constantly blundering over imaginary riches, which he thinks are in store for him, might well be saved disappointment and loss of time were he better acquainted with what the old Cornish miner would call the "kindly" or "unkindly" appearance of the ore or rock he discovers.
>
> It is a well-established fact that certain metals have each their own particular associates, or kindred surroundings; this is particularly the case with such metals as gold, silver, tin, etc.; as, for example, it would be utter waste of time to look for gold in paying quantities in a vein of magnetite or a lode of pure calcite, and no one in his senses would dream of looking for tin in a vein of calcite or barite.

On the positive side, you can reasonably expect to find platinum and chromium only in areas of peridotite or serpentine. More specifically, in Western Australia, big masses of barren quartz act as large-scale guides to ore. Barite that is found in swamps in the Ural Mountains indicates the presence of gold ore immediately below. Countless other such associations are known and will repay the thoughtful observer.

This is not an old-fashioned notion. As a result of the study of moon rocks brought back by Apollo 11, Edwin Roedder, of the U.S. Geological Survey, stated in 1971:

> The relevance of any understanding of earth processes is that we need mineral resources. And we're going to get mineral resources by finding ore deposits. We find ore deposits essentially not by just looking for them, but by knowing how they got to be, how they occurred, why they are where they are. Once we know these particular features, we can be much better in prospecting for ore deposits. And, so, the more we understand about geological processes, the better we know where to go to look for ore.

In the first book on geology published in the United States, William Maclure, the "Father of American Geology," in 1817 said, "Great sums of money have been lost in the United States, and in other countries, by digging for substances among classes of rocks, which have never been found to contain them elsewhere; and of course the probability was against their being found in that class of rock here."

It was in Colorado—first at Leadville, then at Cripple Creek—that the attitude of the mining industry toward science began to change. The classic monograph by Samuel F. Emmons on the *Geology and Mining Industry of Leadville* caused the publisher of that city's leading newspaper to say that this report had reduced "the pursuit of mining to a fixed science, . . . enabling the miner to sink with intelligence, to drift with knowledge, to cross-cut with certainty, to discover, extract, and hoist the ore with economic appliances, securing the maximum results with the minimum of labor cost."

By applying four geologic guides to prospecting for carnotite deposits in the Colorado Plateau, the U.S. Geological Survey (on behalf of the Atomic Energy Commission) obtained results that were at least twice as favorable as by drilling without geologic guidance. These four guides were: (1) thickness of the ore-bearing sandstone, (2) color of the sandstone, (3) altered mudstone associated with the sandstone, and (4) presence of abundant carbonaceous material within the sandstone. These are discussed further in chapter 22.

The application of knowledge thus is resolved into three subjects, as mentioned in the first sentence of this chapter: One subject is minerals and their identification; a second is rocks and their recognition; the third is geologic structures, the ways in which rocks are disposed in position and relationship. For a logical presentation, these subjects are discussed in this book in the same order. Emphasis is placed on the rock and mineral associations that have proved to be most typical of the world's important mineral deposits, and the drawings have been made for the purpose of showing structures clearly.

Before these are given, however, a brief survey is presented of the best opportunities for prospectors around the world. Large companies that have ample funds can doubtless negotiate favorable prospecting agreements almost anywhere, but this aspect of the subject is not appropriate here. The lone prospector or small group having a modest financial backing must of necessity operate where the government of the country offers a genuine welcome and encouragement. Canada allows equal rights to citizens of every nation. Mexico invites exploration and development by American and Canadian prospectors and miners within the limits of its laws. The opportunities in Australia are tremendous, the country is expanding as a producer of minerals faster than any other, and the conditions of climate and terrain are reasonably favorable. Such is not the case everywhere in Latin America south of Mexico, although huge areas there have not been explored thoroughly. Von Bernewitz believed that the best opportunities for the prospector lie in Canada, where an abundance of water, timber, fish, game, and rock outcrops offsets the disadvantage of dense growth and make prospecting easier than in the mountains and deserts of the western United States. He also strongly favored Australia and considered that Alaska was just beginning to open up its "great mineral possibilities."

In accordance with the Von Bernewitz subdivision of the conterminous United States into western, central, and eastern sections, here are lists of the main mineral products of these three divisions, suggesting what to look for:

First, the Western States: Alaska, Arizona, California, Colorado, Idaho, Montana, New Mexico, North Dakota, Oregon, South Dakota, Utah, Washington, Wyoming. The mineral resources are dominantly metallic.

Next, the Central States: Arkansas, Illinois, Kansas, Kentucky, Louisiana, Michigan, Minnesota, Missouri, Oklahoma, Tennessee, Texas, Wisconsin. The mineral resources are both metallic and nonmetallic.

Finally, the Eastern States: Connecticut, Florida, Georgia, Maine, Maryland, Massachusetts, New Hampshire, New Jersey, New York, North Carolina, Pennsylvania, Rhode Island, South Carolina, Vermont, Virginia, West Virginia. The mineral resources are dominantly nonmetallic.

In addition to the experience gained by seeing and handling specimens of rocks and minerals, you may gain much by subscribing to a mining newspaper or magazine (see pages 405–406) for inspiration as well as information, both current and standard.

The following is a list of the colleges and universities in the United States, arranged alphabetically by states, that offer degrees (and hence maintain facilities) in mineral engineering, mineral management, mining engineering, or mining technology:

Alabama	University of Alabama (Huntsville, Birmingham)
Alaska	University of Alaska (College)
Arizona	Eastern Arizona College (Thatcher)
	University of Arizona (Tucson)
California	Santa Ana College (Santa Ana)
	Stanford University (Stanford)
	Taft College (Taft)
	University of California (Berkeley)
Colorado	Colorado School of Mines (Golden)
Idaho	University of Idaho (Moscow)
Illinois	University of Illinois (Urbana-Champaign)
Kentucky	Madisonville Community College (Madisonville)
	University of Kentucky (Lexington)
Massachusetts	Massachusetts Institute of Technology (Cambridge)
Michigan	Michigan Technological University (Houghton)
Minnesota	Brainerd State Junior College (Brainerd)
	Hibbing State Junior College (Hibbing)
Missouri	University of Missouri (Rolla)
Montana	Montana College of Mineral Science and Technology (Butte)
Nevada	University of Nevada (Reno)
New Mexico	New Mexico Institute of Mining and Technology (Socorro)
New York	Columbia University (New York)
Ohio	Ohio State University (Columbus)
Pennsylvania	Pennsylvania State University (University Park)
	University of Pittsburgh (Pittsburgh)
South Dakota	South Dakota School of Mines and Technology (Rapid City)
Utah	College of Eastern Utah (Price)
	University of Utah (Salt Lake City)
Virginia	Virginia Polytechnic Institute (Blacksburg)
Washington	Everett Commercial College (Everett)
	University of Washington (Seattle)
	Washington State University (Pullman)
West Virginia	Bluefield State College (Bluefield)
	West Virginia University (Morgantown)

Wisconsin	University of Wisconsin (Madison)
	Wisconsin State University (Platteville)
Wyoming	Central Wyoming College (Riverton)

A proposed 2-year course in mineral technology for the Community College of Denver would educate qualified persons to become technicians for the mineral industry. The hundreds of institutions that give degrees (and hence maintain departments) in geology are listed in the *Directory of Geoscience Departments,* published at intervals by the American Geological Institute, 2201 M Street NW, Washington, D.C. 20037.

Following is a list of the colleges and universities in Canada, arranged alphabetically by provinces, that offer degrees in mining and mineral engineering:

Alberta	University of Alberta (Edmonton)
	University of Calgary (Calgary)
	University of Lethbridge (Lethbridge)
British Columbia	University of British Columbia (Vancouver)
	Notre Dame University of Nelson (Nelson)
	University of Victoria (Victoria)
	Simon Fraser University (Burnaby)
	Seminary of Christ the King (Mission City)
Manitoba	University of Manitoba (Winnipeg)
	University of Winnipeg (Winnipeg)
	Brandon University (Brandon)
	St. John's College (Winnipeg)
	St. Paul's College (Winnipeg)
New Brunswick	University of New Brunswick (Fredericton)
	St. Thomas University (Fredericton)
	Université de Moncton (Moncton)
	Mount Allison University (Sackville)
Newfoundland	Memorial University of Newfoundland (St. John's)
Nova Scotia	University of King's College (Halifax)
	Nova Scotia Agricultural College (Truro)
	Nova Scotia Technical College (Halifax)
	Dalhousie University (Halifax)
	Acadia University (Wolfville)
	Saint Mary's University (Halifax)
	St. Francis Xavier University (Antigonish)
	Mount Saint Vincent University (Halifax)
	Maritime School of Social Work (Halifax)
	Collège Ste-Anne (Church Pointe)
Ontario	Lakehead University (Port Arthur)
	University of Windsor (Windsor)
	Laurentian University of Sudbury (Sudbury)

	Brock University (St. Catharines)
	University of Waterloo (Waterloo)
	University of Western Ontario (London)
	Waterloo Lutheran University (Waterloo)
	University of Guelph (Guelph)
	Victoria University (Toronto)
	McMaster University (Hamilton)
	University of Toronto (Toronto)
	University of St. Michael's College (Toronto)
	University of Trinity College (Toronto)
	York University (Toronto)
	Knox College (Toronto)
	Wycliffe College (Toronto)
	Osgoode Hall Law School (Toronto)
	Trent University (Peterborough)
	Queen's University at Kingston (Kingston)
	University of Ottawa (Ottawa)
	Carleton University (Ottawa)
	Saint Paul University (Ottawa)
	Royal Military College of Canada (Kingston)
	University of Saint Jerome's College (Waterloo)
	Huron College (London)
	King's College (London)
Prince Edward Island	St. Dunstan's University (Charlottetown)
	Prince of Wales College (Charlottetown)
Quebec	McGill University (Montreal)
	Université de Montreal (Montreal)
	Sir George Williams University (Montreal)
	Université Laval (Québec)
	Université de Sherbrooke (Sherbrooke)
	Bishop's University (Lennoxville)
	Loyola College, 7141 Sherbrooke St. W. (Montreal)
	Marianopolis College, 3647 Peel St. (Montreal)
	Collège Sainte-Marie (Montreal)
	Hautes Etudes Commerciales (Montreal)
	Ecole de Médecine Veterinaire (St-Hyacinthe)
	Collège Jean-de-Brébeuf (Montreal)
Saskatchewan	University of Saskatchewan (Saskatoon)
	University of Saskatchewan (Regina)

Technical courses in mineral education that are intermediate between high school and university education are offered by the following institutions:

Alberta	Northern Alberta Institute of Technology (Edmonton)
British Columbia	British Columbia Institute of Technology (Burnaby)
Newfoundland	College of Trades and Technology (St. John's)

Ontario Provincial Institute of Mining (Haileybury)
Cambrian College (Sault Ste. Marie)
Lakehead University (Thunder Bay)

In Great Britain, details of courses in mining and mineral science are given in *Higher Education in the UK*, sold (10 shillings) by the British Council, 65 Davies Street, London, W.1, and in *Degree Course–Guide Mining Engineering* (7 shillings, 6 pence), by Careers Research and Advisory Centre, Batemans Street, Cambridge. Kingston College of Technology offers a B.Sc. (Special) degree in geology, and 16 other technical colleges give courses that lead to B.Sc. General degrees with an option in geology. Information is given in *A Compendium of Advanced Courses in Technical Geology*, which is available at every school in the United Kingdom.

15

Elementary Mineralogy

The advantage of having at least an elementary knowledge of minerals has been brought out already. As a prospector, you can identify the most common minerals yourself, either by sight or after making a few simple tests, and then obtain aid in having more difficult identifications performed for you by someone who is more experienced. It is of the utmost importance that the correct nature of a mineral be decided upon before money is spent on developing a mine. This seems so obvious that it would not be worth mentioning except for the numerous occasions known to every professional when it was blithesomely overlooked. The author has seen a "topaz" mine opened on a vein of quartz, a "tin" mine opened in a hornblende-rich granite, and a "gem" placer worked in gravel full of colored glass.

Most courses in mineralogy begin with a study of crystallography. Minerals tend to occur in geometric forms, called crystals, when conditions are favorable for free growth. Certain minerals—garnet, for example—form as crystals more often than others. Some minerals occur readily as crystals and as imitative forms (such as icicles), or they may be massive

Probably no other mineral is more familiar in crystal form than quartz. *(Westinghouse Electric Corporation)*

(entirely irregular). The shape of a mineral—whether a crystal or not—is called its habit.

A few mineral habits when recognized by a prospector have yielded vast fortunes. The peculiar "visor twins" of cassiterite, protruding from quartz, have led to rich tin mines. The greasy, curved faces of a diamond crystal, the cogwheel-shaped crystals of bournonite (cogwheel ore), the barrel-shaped crystals of corundum (ruby, sapphire)—these are other familiar examples.

Pictures are seldom of much help in identifying minerals, and colored pictures, though attractive, are largely of illusory value. Experience is the best guide to recognizing minerals. Fortunately, in spite of considerable diversity, most minerals are usually found in relatively few habits; some of these are easily learned. You are strongly advised to buy a standard set of specimens from a dealer who specializes in educational (not tourist) minerals, and to spend an occasional hour looking over a museum collection, of which there are hundreds that are accessible to visitors. By observing separate specimens of minerals, you can learn to recognize them in rocks.

The scientific definition of a mineral is that it is a natural, inorganic

Cross shaped twin crystals of staurolite are among the easiest of all minerals to recognize. *(Field Museum of Natural History)*

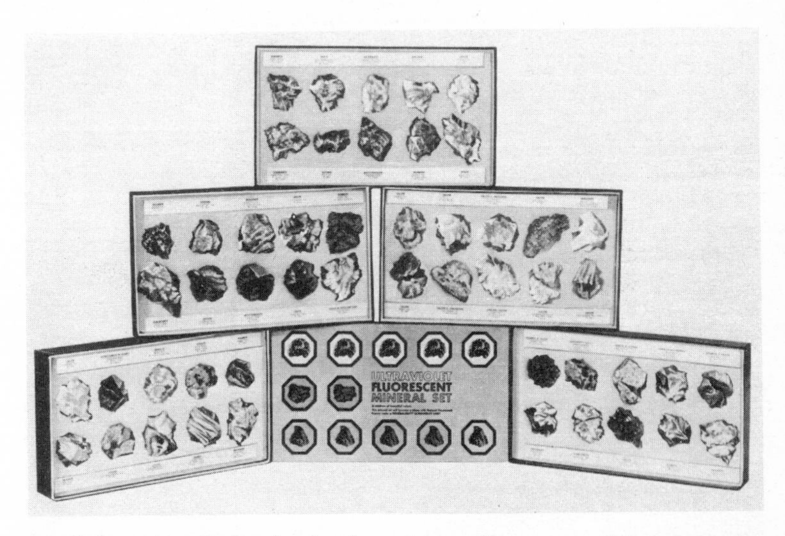

A well-chosen box of mineral and rock specimens will prove a good investment of a few dollars. *(Ultra-Violet Products, Inc.)*

chemical element (such as gold or carbon—diamond or graphite) or a chemical compound (two or more elements in definite proportions). To the prospector, however, a mineral is anything that is commonly considered as such: this is stated in the United States mining law, discussed on page

188 of this book. Thus, coal, limestone, or other rocks are not truly minerals, but they certainly are mineral resources, as is petroleum, which is neither mineral nor rock.

In the rest of this chapter, however, only minerals as they are scientifically considered will be taken into account. Depending on how they are classified, there are today about 2,000 named minerals. Some (such as native gold) are known to everyone, some (such as tanzanite) are extremely rare but well publicized, some (such as diamond) are thought to be rare but actually are not, and some are known by only one or a few specimens. One or two have been carefully described and then have disappeared as specimens.

Minerals have a number of properties, or characteristics, by which they can be identified. The most useful ones to the prospector are the following:

1. Habit: crystal form (cube, for example), imitative shape (as dogtooth spar, which is calcite), or massive.

Among the habits that the collector may especially want to know when he sees them or comes across them in mineral descriptions are the following:

Acicular: needlelike
Amygdaloidal: almond-shaped (in cavities)
Banded: zoned
Bladed: flattened
Botryoidal: grapelike
Capillary: hairlike
Colloform: rounded
Columnar: pillarlike
Concentric: circularly banded
Concretion: a lump
Dendritic: branching
Divergent: radiating

Aragonite may look like coral. *(Field Museum of Natural History)*

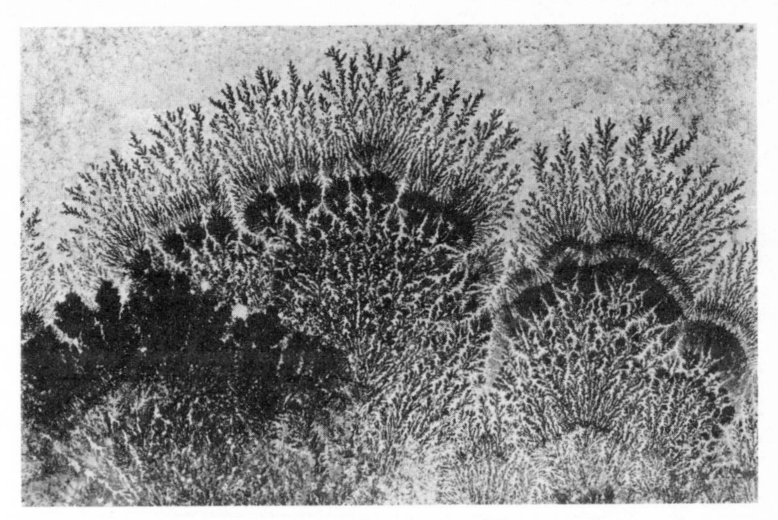

Dendrites occur on rock surfaces and in moss agate.

A geode is typically lined or filled with minerals and crystals. *(Ward's Natural Science Establishment)*

Drusy: in tiny crystals
Felted: matted
Fibrous: threadlike
Filiform: in snarled threads
Foliated: leaflike
Geode: lined cavity in nodule
Globular: rounded
Granular: grainlike
Lamellar: platelike
Laminated: sheetlike
Mammillary: breastlike
Micaceous: sheetlike
Oolitic: in small spheres (fish egg)
Pisolitic: pealike
Plumose: featherlike
Prismatic: elongated
Pyramidal: pointed
Radiated: divergent
Reniform: kidneylike
Reticulated: latticelike
Stalactitic: iciclelike
Stellated: starlike
Tabular: platelike

2. Luster: either metallic or nonmetallic; the latter may be recognized as vitreous (glassy), adamantine (diamondlike), resinous, pearly, silky, etc.

3. Color: very helpful if reliable, but many minerals show a range of color, and impurities effect the color greatly; observe a fresh surface, but even a tarnish can be informative (as on bornite).

4. Streak: the color of the powdered mineral; unless white, it may be helpful in identification (especially with hematite, limonite, covellite, and certain others).

5. Tenacity: most minerals are brittle, but some hold together in special ways—flexible (can be bent, as chlorite), elastic (springs back, as mica), malleable (can be hammered, as native metal), sectile (can be cut, as argentite).

6. Cleavage: regular breakage; note both the ease and the direction of cleavage.

7. Parting: less uniform breakage; pseudocleavage.

8. Fracture: irregular breakage, described as conchoidal (shell-like), even, uneven, etc.

9. Hardness: resistance to scratching. The Mohs scale is standard, as follows:

Streak is tested by rubbing the mineral on a piece of unglazed porcelain. *(Ward's Natural Science Establishment)*

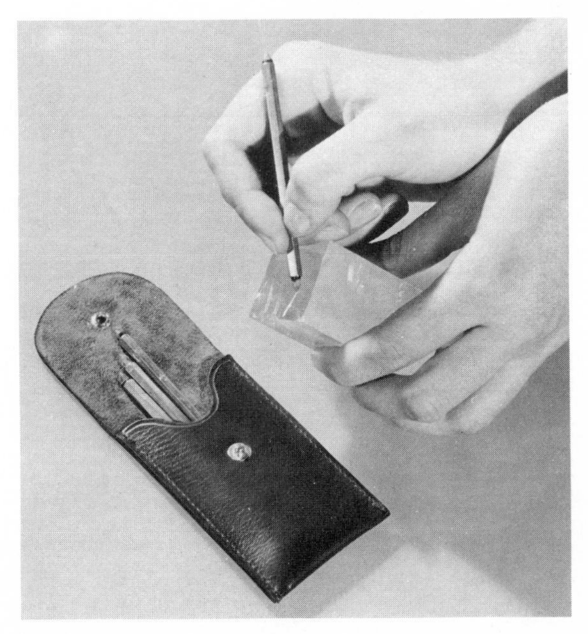

Hardness in mineralogy refers to resistance to scratching. *(Ward's Natural Science Establishment)*

Number 10 Diamond
9 Corundum
8 Topaz
7 Quartz
6 Feldspar (orthoclase)
5 Apatite

4 Fluorite
3 Calcite
2 Gypsum
1 Talc (if pure)

Fresh, pure surfaces should be used, and care must be taken to distinguish a true scratch (on a softer mineral) from a mark (on a harder mineral). For convenience, you can use a good file (6½), a good knife (5½), window glass (5½), a copper coin (3½), or a fingernail (2½).

10. Specific gravity: relative density—the weight compared with an equal volume of water; hefting a specimen in your hand may indicate its specific gravity, after some practice with known minerals; or actual measurements can be made by immersion—the formula is:

$$\frac{\text{(weight in air)}}{\text{(weight in air) minus (weight in water)}}$$

11. Magnetism: magnetite and pyrrhotite are the only common minerals

The Jolly balance has a stretched spring that records specific gravity. *(Standard Oil Company, New Jersey)*

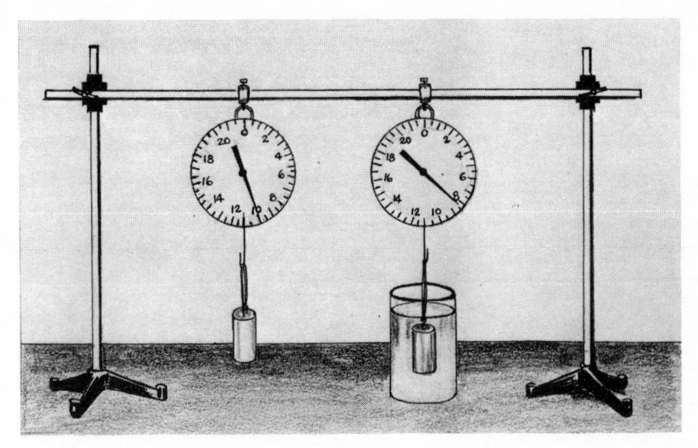

Archimedes found that a substance weighs less in water than in air. Specific gravity is a measure of the difference. *(W. M. Welch Scientific Company)*

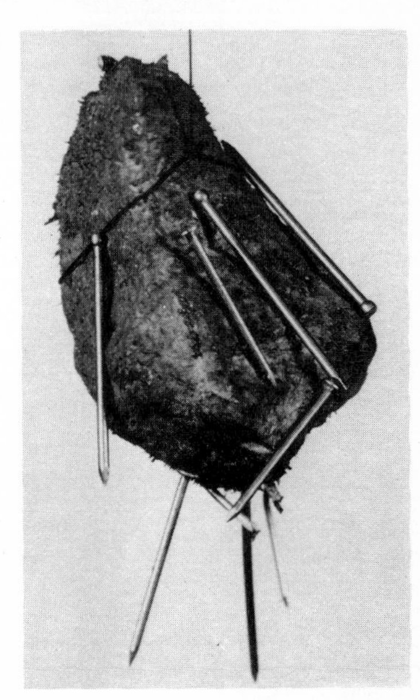

Magnetite is the only common black mineral that is naturally magnetic. This variety is called lodestone. *(Filer)*

that are attracted readily by an ordinary magnet; lodestone, a variety of magnetite, is itself a natural magnet.

12. Minor properties: certain minerals can be identified by taste (halite, being table salt, tastes salty), odor (clay minerals smell earthy when moist), or feel (very soft minerals feel slippery or soapy).

The above physical properties can be supplemented by blowpipe and chemical tests, discussed below.

Blowpipe methods are very useful for a rapid identification of many minerals, even in the field. Combined with a relatively few of the standard chemical tests involving the use of simple reagents, the blowpipe serves the prospector and mineral collector well, providing him with about all the means of identification that he needs, unless he has a fully equipped laboratory and the ability to take advantage of it.

Below is a list of the apparatus that will be found most useful for blowpipe work. Kits of several sizes are sold by dealers in prospecting equipment, as listed on pages 59–60.

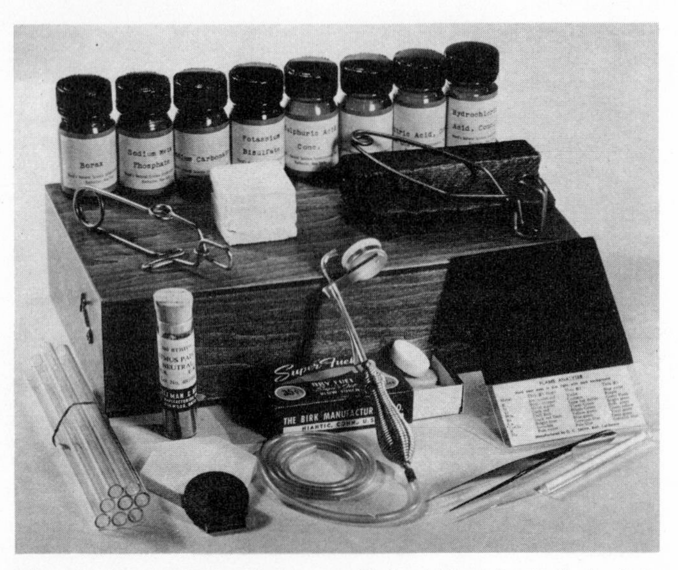

Mineral-testing kits can be bought or assembled. A blowpipe is the basic item of equipment. *(Ward's Natural Science Establishment)*

Mortar and pestle To crush and powder most specimens; only a few tests are made with solid fragments. Agate is best but expensive; steel blocks can be used with a hammer; a steel "diamond mortar" fits closely to prevent loss of material.

Lamp To provide a flame. A bunsen burner that is fitted with a pinched end (either as a cap or inserted into the burner) is most convenient for use with a regular source of gas (natural or artificial), to which it is connected with a rubber or plastic tube. An alcohol lamp, burning alcohol or other fuel (liquid or solid), can be used instead. So can an ordinary candle. Portable torches that burn propane or other fuel also make useful lamps and are preferred by some, especially in the field.

Blowpipe To concentrate and direct the flame and provide reducing and oxidizing environments at high temperature. Various kinds can be used, but a simple tube that is made of brass and blown by the operator will suffice.

Charcoal To support the specimen and also provide a reducing surface of considerable effect. Use blocks or sticks that are prepared especially for blowpipe work. Dig a small hole to hold the material; scrape clean after each use.

Forceps To hold solid fragments; steel or alloy. Cheap forceps made of brass will interfere with flame tests.

Glass tubes To make open-tube or closed-tube tests. Cut into lengths of 5 to 6 inches and bend (when heated) to make an open tube for oxidizing effects; one length is heated and pulled apart to make two closed tubes (open at one end) for nonoxidizing effects. Discard after each use.

Wire To make bead tests (must be platinum, typically 27 gauge) and flame tests (most conveniently Nichrome, used only once and discarded). Insert one end in a closed tube and remelt to close it; the other end can be bent into a loop to hold a bead.

Hammer As mentioned above.

Triangular file or glass cutter

Color screen or blue glass

In addition to these items, there are the small bits of equipment that are useful in testing the physical properties of minerals as described on page 218: a streak plate of unglazed or chipped porcelain, a magnet or magnetized knife blade, hardness points or minerals of standard hardness; magnifying lens.

The chemicals that are customarily used in small amounts in blowpipe work are as follows:

Borax, powdered
Sodium carbonate or bicarbonate, powdered
Potassium bisulfate
Bismuth flux: potassium iodide and sulfur, mixed
Boron flux: potassium bisulfate and powdered fluorite, mixed
Potassium nitrate

Powdered gypsum
Tin chloride
Sodium fluoride or lithium fluoride: to test for uranium, but not useful unless an ultraviolet lamp is available
Litmus paper
Cobalt nitrate solution
Hydrochloric acid, nitric acid, sulfuric acid: to assist flame tests; conveniently kept in small bottles that are fitted with eyedroppers
Fuel for lamp, if needed

Chemical tests require some or most of the following pieces of equipment:

Test tubes, heat-resistant glass
Test-tube holder
Beakers, glass or plastic
Funnel, glass
Filter paper
Watch glasses
Dishes, porcelain

Various stands are convenient for supporting the equipment being used, and additional items can be added from time to time.

It is no easier to teach blowpiping than swimming, but it is just as easy to learn by trying. Your puffed-up cheeks will supply a steady, continuous flow of air, which is inhaled through the nose and exhaled through the mouth into the blowpipe. Only a low flame is needed; the blowpipe is placed close to it (for reducing effects) or barely within it (for oxidizing effects); the flame is directed somewhat downward. This flame consists of three parts, which are important to understand, as follows:

1. The inner flame (yellow in blowpipe, black in candle) consists of unburned gas and moisture; it is not used.

2. The middle flame (blue) consists of carbon monoxide and moisture; called the reducing flame, it takes away oxygen from the mineral and may yield globules (on charcoal) of gold, silver (or both), copper, lead, or tin. Magnetic masses indicate iron, nickel, or cobalt.

3. The outer flame (nearly invisible in blowpipe, yellow in candle) consists of carbon dioxide and moisture; called the oxidizing flame, it is the hottest part and adds oxygen to the mineral.

The three parts of the flame grade into one another; practice will make it possible for you to tell what is happening. Rather than practice routinely to develop skill in blowpiping, it is recommended that the scale of fusibility be tried with small fragments of minerals that are held in forceps, and the desired skill will come naturally. This scale is:

1. Stibnite (or use galena): fuses very easily (even without a blowpipe)
2. Chalcopyrite: fuses easily
3. Almandite garnet (or use gypsum): thin splinters fuse easily to a globule
4. Actinolite (or use albite feldspar): thin splinters fuse to a globule
5. Orthoclase feldspar: thin edges fuse
6. Bronzite: barely fuses on thinnest edges
7. Quartz: infusible

Specimens that break apart (decrepitate) may have to be heated very slowly or else powdered and made into a paste with water if the degree of fusibility is to be determined.

Perhaps the next blowpipe experiments to practice would be the flame tests. Certain elements give characteristic colors to a flame; some of them are especially helpful in identification. The mineral must be vaporized; this is sometimes done more easily with a fragment (held in forceps, but not brass ones) or the powder (picked up on the end or loop of a wire) by moistening it with hydrochloric or sulfuric acid. A few elements require special treatment. The following flame colors are the most useful:

Red: indicates lithium, strontium, calcium; if orange, it means calcium, but all three elements may be much the same color of red.

Yellow: indicates sodium, but this color appears too often to have any particular meaning.

Green: indicates barium, boron, copper (as oxide), phosphorus; boron and phosphorus minerals often require moistening with sulfuric acid. The famous discovery of "20-Mule-Team" borax deposits was made in Death Valley, California, by Aaron Winters, an unsuccessful gold prospector, who, on the advice of a company scout, tested the rock by pouring sulfuric acid and alcohol on it and lighting it, obtaining a green flame.

Blue: indicates copper (as chloride), antimony, arsenic; adding copper to the flame will test it for chlorine.

Violet: indicates potassium, but blue glass or filter is usually needed to cut out the yellow of sodium; potassium minerals may need to be mixed with powdered gypsum.

Some minerals swell up, bubble (intumesce), leaf off (exfoliate), change color, glow, become magnetic, or act in other ways that may give a clue to their identity. Aluminum minerals turn blue when heated, moistened with a drop of cobalt nitrate solution, and heated again; so also does hemimorphite and any fusible mineral—this is therefore a test only for infusible minerals that contain aluminum.

Bead tests are made by heating a flux (almost always borax) in a loop of platinum wire (in a holder). The hot bead is then touched to a tiny amount of powdered mineral, which is then fused in a blowpipe flame. First, expel as much arsenic, antimony, or sulfur as possible by heating the powder on charcoal in the oxidizing and reducing flames (alternately); this process is called roasting and is necessary to prevent destruction of the platinum wire. Certain elements give colored beads, but only a few are distinctive enough to aid in identification. Below are listed the more important ones; the color may be different in the oxidizing and reducing flames, or whether hot or cold. Yellow beads are common but not diagnostic.

Brown in reducing flame: indicates molybdenum

Green in reducing flame: indicates chromium, vanadium (grayish green when hot), copper (cooling to opaque red), iron (bottle green), uranium

Blue at all times: indicates cobalt

Violet in oxidizing flame, cooling to reddish brown: indicates nickel

Brownish violet in reducing flame after cooling: indicates titanium

The following two tests are better made with fluxes other than borax:

Sodium carbonate gives an opaque, bluish-green (turquoise) bead, indicating manganese.

Sodium fluoride or lithium floride gives a yellowish-green bead, which is fluorescent (glows) in ultraviolet light, indicating uranium.

Heated with a blowpipe on charcoal, certain elements in many minerals vaporize, oxidize, and deposit as sublimates (coatings) on the charcoal. In so doing, they may show flame colors, give off recognizable odors, or leave behind some metallic (perhaps magnetic) metal: gold, silver (or both), copper, lead, or tin, perhaps associated with iron, nickel, or cobalt. The most useful charcoal tests are the following:

White coating: indicates arsenic, antimony (deposits farther away), selenium (red tinge, also has peculiar odor), tellurium. Selenium and tellurium coatings color a reducing flame blue and bluish green, respectively.

Yellow coating: indicates lead (also white), bismuth (also white), zinc (cooling to white), molybdenum (cooling to white), tin (cooling to white). Lead and bismuth coatings give different colors (yellow, red, respectively) when mixed with bismuth flux; zinc coating turns green, and tin coating turns bluish green when moistened with cobalt nitrate solution; molybdenum coating colors a reducing flame blue.

When heated in a glass tube, certain elements in many minerals vaporize, oxidize, and deposit as sublimates (coatings) inside the glass. In so doing, they may give off recognizable odors, and their escaping fumes may react acid or alkaline to litmus paper (turning it red or blue, respectively).

The most useful open-tube tests are the following; but certain minerals give various combinations:

Water drops: indicate hydrogen and oxygen
Gray drops: indicate mercury
Dark orange-red coating, cooling to pale yellow: indicates sulfur (also gives sulfur odor)
White coating: indicates arsenic (also gives garlic odor), antimony, molybdenum
Yellow coating: indicates molybdenum
Yellow coating, cooling to white: indicates antimony

The most useful closed-tube tests are the following; but certain minerals give various combinations:

Water drops: indicate hydrogen and oxygen
Black coating: indicates mercury
Black and gray coatings: indicate arsenic
Gray drops: indicate mercury
Reddish-brown coating: indicates antimony
Deep-red coating, cooling to reddish yellow: indicates arsenic
Dark orange-red coating, cooling to pale yellow: indicates sulfur

Simple chemical (wet) tests may use the following reagents (especially) in addition to the acids specified on page 227 for flame tests:

Ammonium carbonate
Ammonium hydroxide ("ammonia")
Ammonium molybdate solution
Ammonium oxalate
Barium chloride
Dimethylglyoxime
Hydrogen peroxide
Potassium ferricyanide
Potassium ferrocyanide
Silver nitrate
Sodium sulfide
Tin metal
Zinc metal

Certain of these reagents can be bought from the dealers in blowpipe equipment (see page 59). Many hobby chemicals are sold at $1 per item by Spectro-Chem Inc., 1354 Ellison, Louisville, Kentucky 40204. Some liquids are best obtained locally from drugstores and chemical dealers. Liquids are most conveniently carried in the field in plastic containers.

Systematic chemical tests are mostly beyond the scope of this book, but certain ones fit readily into the blowpipe procedures and give helpful indications as to the composition of minerals. Many minerals—especially silicates, oxides, tungstates—are insoluble or nearly so, and so these minerals must be fused (as with sodium carbonate on charcoal) before they dissolve in acid to give the desired tests. The following chemical reactions are generally simple enough to be made by the interested prospector or mineral collector. A few test for a specific mineral (cassiterite), some test for specific elements (sulfur, etc.), and some test for the compounds that represent the so-called chemical classes (carbonates, etc.).

Test for cassiterite Various chemical tests can be made for the metallic element tin, but this is a specific test for what is by far the most important mineral containing tin: cassiterite, which is tin oxide. Cassiterite is often difficult to recognize: hence its names wood tin and tin stone. One or many specimens can be placed in a receptacle, together with a piece of metallic zinc and some hydrochloric acid. Boiling the acid and letting it stand afterward may help the reaction, which produces a coating of gray tin. This becomes shiny when the mineral is rinsed in water and rubbed on cloth.

Tests for Elements

1. Powder the mineral.
2. *a.* Put part of the powdered mineral in a receptacle.
   b. Add dilute nitric acid.
   c. Heat the acid to dissolve the mineral.
   d. Filter off any precipitate.
   e. Add dilute hydrochloric acid to a quarter of the liquid.
      (1) A white precipitate indicates silver.
   f. Add silver nitrate to a second quarter of the liquid.
      (1) A white precipitate indicates chlorine, bromine, or iodine.
   g. Add hydrogen peroxide to a third quarter of the liquid.
      (1) A reddish-brown liquid indicates vanadium.
   h. Add dilute ammonium hydroxide to the last quarter of the liquid.
   i. Add dimethylglyoxime solution to the liquid.

(1) A bright-red precipitate indicates nickel.

3. *a.* Put the other part of the powdered mineral in another receptacle.
   b. Add dilute hydrochloric acid.
   c. Add a little dilute nitric acid.
   d. Heat the acid to dissolve the mineral.
   e. Add dilute ammonium hydroxide.
   (1) A deep-blue color indicates copper.
   (2) A white precipitate indicates aluminum; many other precipitates may form, but these are difficult to identify.
   (3) A red precipitate indicates iron; aluminum may also be present.
   f. Filter off any precipitate.
   g. Add sodium sulfide to part of the liquid.
   (1) A white precipitate indicates zinc.
   h. Add ammonium oxalate to another part of the liquid.
   (1) A white precipitate indicates calcium, barium, or strontium.
   i. Filter off any precipitate.
   j. Add sodium phosphate to the liquid.
   (1) A white precipitate indicates magnesium.
   k. Add hydrogen peroxide.
   (1) A reddish-brown liquid indicates vanadium.

Test for carbonates All carbonate minerals and rocks will dissolve in acid with effervescence (fizzing, bubbling). These include calcite, aragonite, dolomite, siderite, smithsonite, azurite, malachite, witherite, cerussite, limestone, and marble as well as other, less valuable ones. The activity varies: some minerals respond energetically only when powdered or in hot acid or nitric acid. Avoid confusing this easy test with the mere bubbling of boiling acid or the escape of hydrogen sulfide, which smells like rotten eggs.

Test for nitrates Heated with potassium bisulfate in a closed tube, these minerals give off red vapors of nitrous oxide.

Test for borates Powdered, mixed with the boron flux, and fused, these minerals give a bright-green flame.

Test for phosphates A simple test for phosphate rock is worth knowing, for this valuable rock looks like various worthless rocks. Put some nitric acid on the rock, add a solution of ammonium molybdate, and the rock turns yellow. When testing phosphate minerals—such as apatite, monazite, turquoise—they should be powdered, and the liquids should be heated in a receptacle and then allowed to cool for several hours, if necessary, until a yellow powder appears.

Test for arsenates Powder the mineral, put it with a splinter of charcoal in a closed tube, and heat the tube. Black and gray sublimates form inside the tube.

Test for vanadates Powder the mineral, put it in a receptacle, add acid, heat the liquid to dissolve the mineral, add hydrogen peroxide. The liquid turns reddish brown from pervanadic acid.

Test for sulfates If the mineral is soluble, powder it, put it in a receptacle, add dilute hydrochloric acid, heat the liquid to dissolve the mineral, and add barium chloride. A white precipitate of barium sulfate forms. This can be done on a minute scale and observed under a lens.

Test for chromates Green bead test.

Test for tungstates These must be powdered and fused first, using sodium carbonate as flux. They are then dissolved in hydrochloric acid, boiled with tin or zinc metal, and the solution goes through a sequence of yellow, blue, and brown colors.

Test for molybdates Bead tests.

Test for uranates Fluorescent bead test.

Test for silicates These minerals are often difficult to determine.

The following books cover blowpiping at length and may be consulted for detailed instructions:

Manual of Determinative Mineralogy, With an Introduction on Blowpipe Analysis, 16th edition, George J. Brush and Samuel L. Penfield, John Wiley & Sons, Inc., New York, 1898.

Identification and Qualitative Chemical Analysis of Minerals, 2d edition, Orsino C. Smith, Van Nostrand-Reinhold Company, New York, 1953.

Descriptive Mineralogy, William Shirley Boyley, Appleton-Century-Crofts, New York, 1917.

A *Textbook of Mineralogy. With an Extended Treatise on Crystallography and Physical Mineralogy,* 4th edition, Edward Salisbury Dana and William E. Ford, John Wiley & Sons, Inc., New York, 1932.

Elements of Mineralogy, Crystallography, and Blowpipe Analysis, Alfred J. Moses and Charles Lathrop Parsons, Van Nostrand-Reinhold Company, New York, 1900.

A *Pocket Handbook of Minerals . . . ,* 2d edition, G. Montague Butler, John Wiley & Sons, Inc., New York, 1912.

A *Manual of Determinative Mineralogy with Tables . . . ,* 4th edition, J. Volney Lewis and A. C. Hawkins, John Wiley & Sons, Inc., New York, 1931.

International Library of Technology: *Blowpiping, Mineralogy, Assaying, Geology, Prospecting, Placer and Hydraulic Mining,* International Textbook Company, Scranton, Pa., 1903.

Manual of Qualitative Blowpipe Analysis and Determinative Mineralogy, F. M. Endlich, Scientific Publishing Company, New York, 1892.

Handbook of Mineralogy, Blowpipe Analysis and Geometrical Crystallography, 2d edition, G. Montague Butler, John Wiley & Sons, Inc., New York, 1918.

In addition, see most of the books on mineralogy and mineral identification that are listed below.

The following books of United States origin are selected from the many that are available on mineralogy and mineral identification. They are listed in approximate order of difficulty, from popular books for the layman-collector to standard American college textbooks. Similar books are published in other countries, especially Great Britain, and in other languages, especially French, German, and Russian.

How to Know the Minerals and Rocks, Richard M. Pearl, McGraw-Hill Book Company, New York, 1955.

Rocks and Minerals, revised edition, Richard M. Pearl, Barnes & Noble, Inc., New York, 1965.

Minerals and How to Study Them, 3d edition, Edward Salisbury Dana, revised by Cornelius S. Hurlbut, Jr., John Wiley & Sons, Inc., New York, 1949.

Minerals and Rocks, Russell D. George, Appleton-Century-Crofts, New York, 1943.

Introduction to the Study of Minerals, 3d edition, Austin Flint Rogers, McGraw-Hill Book Company, New York, 1937.

Elements of Crystallography and Mineralogy, F. Alton Wade and Richard B. Mattox, Harper & Row, Publishers, Inc., New York, 1960.

Principles of Mineralogy, William Dennen, The Ronald Press Company, New York, 1959.

Elements of Mineralogy, Alexander N. Winchell, Prentice-Hall, Inc., New York, 1942.

Dana's Manual of Mineralogy, 18th edition, Cornelius S. Hurlbut, Jr., John Wiley & Sons, Inc., New York, 1971.

An Introduction to the Study of Minerals and Crystals, 5th edition, Edward Henry Kraus, Walter Fred Hunt, and Lewis Stephen Ramsdell, McGraw-Hill Book Company, New York, 1959.

The following magazines deal with the hobby aspects of minerals, gems, and related subjects. Some of the mining periodicals that are listed on page 405–406 discuss minerals, but mostly from the standpoint of extracting, recovering, and treating them.

Earth Science
P.O. Box 1815
Colorado Springs, Colorado 80901

Gems and Minerals
P.O. Box 687
Mentone, California 92359

Lapidary Journal
P.O. Box 2369
San Diego, California 92112

Mineral Digest
P.O. Box 341, Murray Hill Station
New York, New York 10016

Rocks and Minerals
P.O. Box 29
Peekskill, New York 10566

The Mineralogical Record
P.O. Box 783
Bowie, Maryland 20715

Canadian Rockhound
941 Wavertree Road North
Vancouver, British Columbia

New Zealand Lapidary
P.O. Box 19-150
Avondale, Auckland 7, New Zealand

16

Rocks: Elementary Petrology

To the geologist, rock is any substantial portion of the earth, whether solid or loose. To the engineer, rock includes only solid material that cannot be excavated by hand. In the geologic sense, rock includes glaciers of ice and sometimes even bodies of water which, after all, are only liquid ice. At the Kennecott copper mine, in Alaska, the "country rock" surrounding "glacier ore" from the Bonanza vein is actually glacial ice!

Rocks are classified in various ways. The main categories include stratified (sedimentary rock and lava rock)—which are layered and occur in beds (layers, strata)—and unstratified, which are massive, or nonlayered. Rocks are also commonly classified according to their mode of formation as igneous (having cooled from a molten state), sedimentary (deposited under surface or near-surface conditions, by accumulation or precipitation), and metamorphic (definitely altered from an earlier igneous or sedimentary rock). Igneous rock (except lava) and metamorphic rock are typically not layered and generally involve high temperature and pressure: they are often grouped together as crystalline rock because of the way their mineral grains interlock. Intrusive rock is igneous rock that has cooled, solidified, and crystallized deep within the earth's crust; extrusive rock (including lava) has cooled upon the surface of the earth; at shallow depths, the two kinds grade into each other.

Using the features that are mentioned above as the basis for classifying them, rocks are named according to their composition (chemical and mineral) and their texture (the size, shape, and arrangement of the grains—mineral or glass). These relationships will become clear as the main rocks of interest to prospectors are described below. The study of rocks is called petrology and is here confined to observations made only with a hand lens or the unaided eye. A detailed, laboratory classification, made with a petrographic microscope, may apply names that are somewhat different but not seriously so. Many unnecessary names have been given to rocks, especially the igneous ones, and simplification is required.

A few words are in order here about the economic uses of rock. Since the adoption of the law regarding "common varieties" of rock in the United States, ordinary stone cannot be the basis of a mining claim. Hence, the extraction of rock commercially is restricted to leasing public land or making contractual arrangements on private land. Dimension stone is rock that is quarried or used (or both) in specified shapes or sizes. It is used mostly in building (exterior and interior), monuments and memorials, curbstones and flagstones, paving blocks, bridge abutments, and retaining walls. Crushed stone is broken rock that is used mostly for concrete aggregate (concrete is cement plus crushed rock, sand, water, and a binder), road metal (for foundations and surfacing dirt or macadamized roads), ballast (railroad and ship), and many special-purpose sands; smaller amounts of crushed stone are used in filter beds (for purifying water), riprap (for river and harbor construction), and poultry grit. Certain rocks have even more important uses; for example, limestone is the basis of the cement industry, and gypsum is the basis of the plaster industry. The significant occurrences of certain rocks with certain ore and nonmetallic minerals—thus, the best asbestos is associated with serpentine—are emphasized in chapters 18 to 32, where these mineral products are discussed in detail.

IGNEOUS ROCKS

Let us start with igneous rocks, for these were presumably the first rocks to form as the earth cooled. No such original rock is known in the earth's crust, but the moon may have yielded some of its original rock not re-

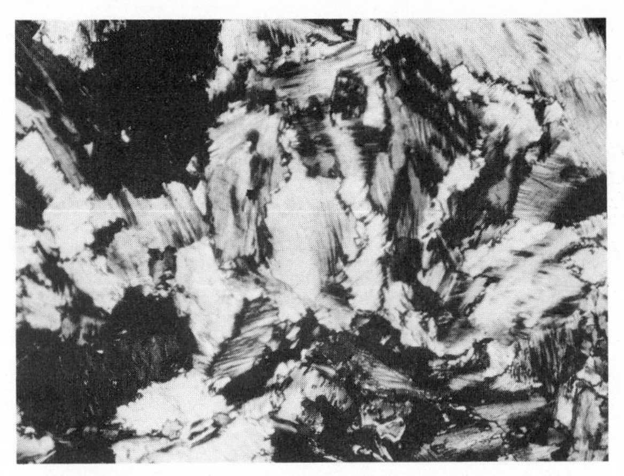

An interlocking texture is characteristic of igneous rocks. *(Buehler Ltd.)*

worked by later geologic (lunar) processes. Igneous rocks are usually recognized by their interlocking textures, their hardness (unless porous or weathered), and their typical minerals. Igneous rocks that have cooled from lava may be porous (and consequently light and somewhat soft), may be glassy, and are fine-grained except for the larger crystals (phenocrysts) enclosed within them. Igneous rocks that have cooled beneath the surface are dense and rather heavy, and the grain size increases with the depth at which they formed. Glass never forms except near or at the surface. Any igneous rock is a porphyry when it contains phenocrysts that are set in a background (groundmass) of smaller grain size.

The chief igneous rocks of interest to the prospector are the following:

1. Granite. This rock makes up the bulk of the world's continents. It is a more or less coarsely grained rock of both igneous and metamorphic origin. Most granite is either white to gray or pink to red, the color depending on the color of the potassium feldspar, which is the main mineral. Quartz is the other essential mineral, and other common minerals are plagioclase feldspar, biotite mica, hornblende, and muscovite mica. Numerous other minerals may occur in small amounts, but ordinary granite has little commercial use except as one of the most important materials for dimension and crushed stone. Tin, tungsten, and molybdenum deposits are found in quartz veins that are associated with granite. Numerous other mineral products come from granitelike rocks.

2. Pegmatite. A very coarse but variably grained rock, pegmatite is

Landscapes like this one in the Wenatchee Mountains of Washington are typical of igneous rock exposures. *(U.S. Geological Survey)*

No feature is more characteristic of pegmatite than the ancient-writing pattern of graphic granite. *(Ward's Natural Science Establishment)*

typically considered a kind of granite, although pegmatite phases of other igneous rocks also occur. The minerals in pegmatite often grow into enormous sizes, many weighing tons. Simple pegmatite consists of the same minerals as granite—feldspar, quartz, mica—but complex pegmatite is noted for its unusual minerals. Some of these are found scarcely anywhere else. Gems, radioactive minerals, tin and tungsten ores, and a host of other valuable minerals are mined from pegmatite; these include feldspar, mica, rose quartz, beryl, tourmaline, spodumene, amblygonite, topaz, apatite, fluorite, monazite, columbite, tantalite, and others. Steam is a major factor in the formation of pegmatite. To the specimen collector, pegmatite is treasure house and happy hunting ground, mecca and paradise.

3. Aplite. Related to granite, aplite is as fine-grained as pegmatite is coarse. Aplite and pegmatite are typically associated with granite in the field.

4. Syenite. Like granite in most respects but containing little or no quartz, syenite is therefore mainly a potassium-feldspar rock of rather coarse texture. Plagioclase feldspar, biotite mica, hornblende, and pyroxene are the other minerals that are mostly present. Syenite occurs with gold and iron.

5. Nepheline syenite (having nepheline) is a source of unusual minerals when it has a pegmatite phase, from which corundum may be mined.

6. Quartz monzonite. When both potassium and plagioclase feldspar are fairly equal in amount, granite grades into quartz monzonite. Hornblende, biotite mica, and pyroxene are the other main minerals. Valuable copper and molybdenum deposits (of the porphyry type) occur with this rock, as do gold and silver.

7. Monzonite. The near absense of quartz changes quartz monzonite to monzonite, which is very similar to syenite. Gold, copper, and iron are found with it.

8. Granodiorite. This is intermediate between quartz monzonite and quartz diorite and yields the same mineral products.

9. Quartz diorite, tonalite. Plagioclase feldspar is much more abundant than potassium feldspar in this rock, and quartz is also present. So are biotite mica, hornblende, and pyroxene. Copper, gold, and silver occur with this rock.

10. Diorite. A reduction in the amount of quartz causes quartz diorite to grade into diorite. Hornblende and biotite mica are usually present. Diorite is used to some extent as dimension stone, and gold, copper, and iron are found with it.

11. Gabbro. The low-silica equivalent of granite, gabbro is an important rock. It is often sold as black granite for building and monument

purposes. Pyroxene is the main dark ("basic" or "mafic") mineral, and olivine is also important, but plagioclase feldspar is the dominant mineral.

12. Norite. A variety of gabbro that contains hypersthene, norite is a source of platinum, gold, silver, copper, and nickel.

13. Anorthosite. This is a light-colored kind of gabbro, because it consists almost entirely of plagioclase feldspar. It is associated with iron and titanium.

14. Peridotite. When pyroxene and olivine make up a coarse-grained rock, it is called peridotite. It is a source of platinum, chromium, and asbestos.

15. Kimberlite. A variety of peridotite, here is the original source of diamond.

16. Pyroxenite. This consists almost wholly of pyroxene.

17. Dunite. It consists almost solely of olivine.

18. Amphibolite. This is composed mainly of amphibole, which is usually hornblende, and so hornblendite is also an appropriate name.

19. Diabase, dolerite. As the grain size of gabbro decreases, it becomes diabase, or dolerite (British). This rock is used as dimension stone and occurs with silver, cobalt, and nickel.

20. Basalt. The extrusive, or volcanic, equivalent of gabbro, basalt makes up the rock of the Pacific Ocean basin and occurs on the continents as well. Basalt is found as lava flows, dikes, and other bodies. It is fine-grained, usually with gas cavities. So-called traprock, used as dimension stone, is almost always basalt.

21. Olivine basalt. Olivine is seen prominently in this kind of basalt.

22. Andesite. This, as a fine-grained extrusive rock, is otherwise equivalent to diorite.

23. Dacite. It is fine-grained, otherwise equivalent to tonalite, or quartz diorite.

24. Quartz latite. It is equivalent to granodiorite, but is fine-grained because of its extrusive origin.

25. Latite. This is a fine-grained rock, otherwise equivalent to monzonite.

26. Quartz latite. The fine-grained equivalent of quartz monzonite.

27. Phonolite. It is the fine-grained equivalent of nepheline syenite and is an important gold-bearing rock at Cripple Creek, Colorado.

28. Trachyte. The fine-grained equivalent of syenite.

29. Rhyolite. This is the fine-grained equivalent of granite. The above

extrusive rocks (basalt to rhyolite) are known as source rocks of gold, silver, copper, nickel, iron, and fluorspar.

30. Obsidian. Volcanic glass is mostly obsidian.

31. Pumice. Porous glass of volcanic origin is known as pumice. It is widely used in abrasives, cement, concrete, and other industrial products.

32. Porphyry. Although this term is well known to prospectors, it actually belongs with the name of any igneous rock that shows visible grains (of any size) set in a background (groundmass) of finer grain size. Thus, we have granite porphyry, basalt porphyry, or any other combination. The extrusive rocks are generally porphyries. As popularly used, porphyry is found as a source rock of many metals, including gold; the porphyry-copper deposits are a special type.

Pumice is volcanic glass made porous by the escape of gas. *(U.S. Geological Survey)*

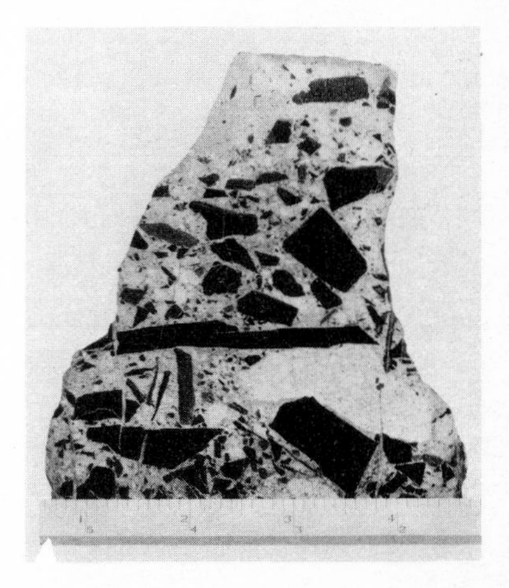

The two-stage pattern shown here is typical of a porphyry. Almost any igneous rock may occur with a porphyritic texture. *(U.S. Geological Survey)*

SEDIMENTARY ROCKS

These are generally classified according to whether they are:

1. Composed of solid fragments that were deposited by running or standing water, or glacial ice, or wind, or
2. Composed of minerals that were deposited as chemical precipitates from a dissolved state, or
3. Composed of organic matter, either plant or animal. Living organisms may also cause some or much of the chemical precipitation of group 2, but this is often difficult to prove.

Sedimentary rocks are generally recognized by their distinct layering, their more or less uniform texture, their typical minerals, and the possible presence of fossils.

The chief sedimentary rocks of interest to the prospector are the following:

1. Conglomerate. When loose rock and mineral matter of gravel size are naturally cemented together to become firm rock, it is called conglomerate. Smaller particles partly fill the spaces between the larger ones. The most common cementing agents in sedimentary rocks are calcium carbonate (calcite), silica (quartz), iron oxides (hematite, limonite), and clay. Copper, gold, silver, and diamond are found in this rock in certain localities of note.

Studying structures in sedimentary rock in Montana, a possible clue to the occurrence of minerals. *(Standard Oil Company, New Jersey)*

Ripple marks left by ancient currents of water are among the characteristic features of sedimentary rock. This outcrop is near Lander, Wyoming. *(Standard Oil Company, New Jersey)*

2. Breccia. Angular rather than rounded fragments of coarse size become breccia instead of conglomerate. Other kinds of breccia (igneous, volcanic, fault, etc.) are also known wherever broken rock is found.

3. Sandstone. Sand-sized particles are cemented together to form sandstone. Typically, it consists of quartz grains, simply because quartz is the most durable of the common minerals; but sand may be composed of other materials, such as coral sand (as on tropical islands), black sand (as in Hawaii), gypsum sand (as the White Sands of New Mexico), and still others. Apart from these special kinds, most sandstone owes its color to the cementing minerals. Sandstone is one of the principal kinds of dimension stone and is associated with both metallic and nonmetallic deposits.

4. Arkose. The presence of a substantial amount of feldspar in sedi-

A landscape of layered rocks like this one near Lander, Wyoming, is typical of sedimentary rock. *(Standard Oil Company, New Jersey)*

mentary rock makes it arkose. Thus, there is arkosic sandstone and arkosic conglomerate.

5. Siltstone. Particles in size between sand and mud constitute siltstone.

6. Shale. Mud or clay accumulates to form shale by compaction and cementation. The layers are usually fine. The color may be variable, and so also may be the other minerals that are present, so that mica flakes, "lime" (calcium carbonate), and silt are common in shale. The lightweight aggregate, ceramic, and cement industries use considerable shale, but it is much too weak to be considered a building stone. In Scandinavia, for example, important sulfide mineralization in certain regions is associated with graphitic shale.

7. Limestone. Formed either by chemical precipitation or by the activity of certain plants and animals, limestone includes many different varieties of rock. It is a major kind of dimension stone and is the foundation of the cement industry; as crushed stone and a chemical raw material, limestone serves a number of uses. Pure limestone consists entirely of calcite. Although it would be easy to list a number of varieties of nearly every rock that is described in this chapter, this has been avoided; how-

ever, the following kinds of limestone deserve to be considered because of their importance and familiar occurrence. Nevertheless, it is ordinary limestone that is most likely to be found in association with a diversity of ore deposits, especially when it is adjacent to igneous intrusions.

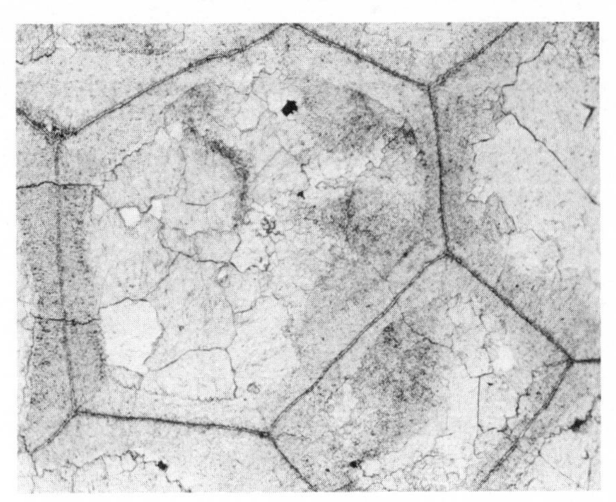

The mosaic texture of coral limestone, a sedimentary rock, is shown here. *(Buehler Ltd.)*

The presence of fossils, often abundantly, is characteristic of much limestone. This specimen is from Austin, Texas. *(Ward's Natural Science Establishment)*

a. Chalk. Fine-grained, porous limestone, chalk is composed mostly of the shells of one-celled animals called foraminifers.

b. Travertine. Limestone that has been deposited by hot or cold springs and in caverns is travertine.

c. Calcareous tufa. This is porous travertine.

d. Coquina. It consists of shells, often broken or rounded, that have been cemented together. Coquina is best known on the east coast of Florida.

e. Coral. A reef limestone of organic origin.

8. Dolostone. The double carbonate of calcium and magnesium is called dolomite as a mineral, and either dolomite or dolostone as a rock. Dolomite is used like limestone except for cement, and in refractories (heat-resistant materials) and insulation. It is also a source of magnesium metal.

9. Gypsum. Calcium sulfate combines with water to form gypsum, which is deposited only by the evaporation of sea water. The mineral and the rock have the same name. Gypsum is the basis of the plaster industry and is also used in minor ways.

10. Anhydrite. Calcium sulfate without water is deposited as anhydrite. The mineral and the rock have the same name. Anhydrite may alter to gypsum or be changed from gypsum. Future uses may include plaster.

11. Rock salt. Halite, the mineral that is common table salt, is deposited by evaporation to form huge beds of rock salt. Although of unequalled importance for man's use—salt is a major raw material for the chemical industry as well as an essential food—deposits of rock salt are not likely to be of much value to the prospector.

12. Chert. Related to quartz (especially to the flint variety), chert is white and dense. Iron and manganese deposits are sometimes associated with it.

METAMORPHIC ROCKS

Apart from weathering at or near the earth's surface, substantial changes of an igneous or sedimentary rock transform it into metamorphic rock. The changes are in chemical or mineral composition, texture, or structure. The factors that cause these changes are heat, pressure, and the presence of hot fluids. Metamorphic rocks are recognized by their often wavy and layered patterns, their hardness and heaviness, and their characteristic minerals. Most metamorphic minerals, however, are the same as those that were in the previous rock from which the new rock was derived.

The chief metamorphic rocks of interest to the prospector are the following:

1. Gneiss. The most coarsely banded (foliated) of all rocks, gneiss typically has alternate layers of light and dark minerals. The light-colored bands are generally feldspar (by some definitions, gneiss must contain feldspar) and quartz. The dark bands consist of such minerals as hornblende and biotite mica. Besides obvious names such as hornblende gneiss, the several varieties may be named according to their composition, as granite gneiss.

2. Schist. As the layers become narrower and more readily separated, gneiss grades into schist. The more important varieties include mica (muscovite or biotite) schist, garnet schist, staurolite schist, and hornblende schist. Gneiss and schist both yield many mineral products, both metallic and nonmetallic.

A thin-section of staurolite-mica schist reveals the oriented pattern typical of many metamorphic rocks. *(Buehler Ltd.)*

3. Quartzite. When metamorphosed, quartz sandstone is changed to quartzite. Tightly cemented sandstone is often (but wrongly) called quartzite. True quartzite is probably the most durable rock known. Occurring with it are gold, copper, lead and zinc. iron, and manganese.

4. Slate. The metamorphism of shale yields slate, which is fine-grained and splits readily into slabs, helping to make it an important dimension stone. It is also found with gold, lead and zinc, and lesser mineral products (antimony, pyrite).

This sort of landscape is typical of several types of metamorphic rock. *(Standard Oil Company, New Jersey)*

5. Marble. Either limestone or dolomite metamorphoses to marble. White when pure, marble usually shows patterns of color in patches or streaks. Marble is one of the most important kinds of dimension stone.

6. Serpentine. As a metamorphic rock, serpentine is derived from peridotite. The mineral content, chiefly serpentine, gives this rock a green color. "Basic" ores—mainly of nickel, chromium, and platinum—are characteristically associated with serpentine, and so are asbestos, diamond, magnesite, mercury, and talc.

BOOKS ON PETROLOGY

The following books are standard American books on the study of rocks, arranged in approximate order of difficulty, from popular books for the layman-collector to those of college level. Similar books are published in other countries and languages, especially French, German, and Russian. Books dealing with the economic aspects of rocks are listed at the end of chapters 24 (for metallic minerals) and 32 (for nonmetallic minerals and rocks).

How to Know the Rocks and Minerals, Richard M. Pearl, McGraw-Hill Book Company, New York, 1955.

Rocks and Minerals, revised edition, Richard M. Pearl, Barnes & Noble, Inc., New York, 1965.

Minerals and Rocks, Russell D. George, Appleton-Century-Crofts, New York, 1943.

A *Handbook of Rocks. . . ,* 6th edition, James Furman Kemp and Frank F. Grout, Van Nostrand-Reinhold Company, New York, 1940.

Rocks and Rock Minerals, 3d edition, Louis V. Pirsson and Adolph Knopf, John Wiley & Sons, Inc., New York, 1947.

Guide to the Study of Rocks, 2d edition, L. E. Spock, Harper & Row, Publishers, Inc., New York, 1962.

Introduction to Petrology, Brian Bayly, Prentice-Hall, Inc., Englewood Cliffs, N.J., 1968.

Petrology, Walter T. Huang, McGraw-Hill Book Company, New York, 1962.

17

Elementary Geology

This chapter describes the geologic structures that the prospector should know about, and it explains how they are related to mineral deposits. The structures themselves may, it is true, be of interest largely to the academic geologist, for they tell him how the earth is made. The mineral deposits, on the other hand, are the actual goal of the prospector. When the relationships between structure and deposit are recognized, finding a mineral deposit becomes easier and more certain. The position of the ore district in the regional structure should also be determined if possible.

So many different kinds of structures and deposits are known to exist that it is not possible to present them all here. (The large mining companies are beginning to use computers to sort out these combinations and determine which are likely to be profitable.) Therefore, it seems best to describe the most significant geologic features, which everyone—scientist, prospector, mineral collector, interested layman—ought to know. Presumably, you have already at least looked through the preceding chapters on elementary mineralogy (chapter 15) and elementary petrology (chapter 16), for minerals and rocks are the building materials of which our planet Earth is composed. In form and position, they constitute the geologic

The veined structure of rock is of vital significance in petrology and economic geology. *(Ward's Natural Science Establishment)*

structures; when present in valuable amounts, they become the mineral deposits.

The most useful generalization about the application of geology to mineral hunting is that each mineral, whether metallic or nonmetallic, has its characteristic associates and is more or less restricted to certain geologic environments. This was brought out in chapter 14, where Richard Pearce was quoted as saying that "it would be utter waste of time to look for gold in paying quantities in a vein of magnetite or a lode of pure calcite."

Perhaps the next most useful statement about geology is to emphasize the importance of contacts: the places where two kinds of rock come together. If the contact is the result of slippage (faulting), the fault surface may or may not have anything to do with mineralization; the importance of faults is discussed later in this chapter (see page 262). Contacts are the boundaries not only between different kinds of rocks but also between two different geologic processes. The shift from one environment to another provides a background in both time and place that concentrates the prospector's attention in this search. Where the contact is gradational, a third (transitional) zone appears, and this may prove of greater interest than the other two. The rule, therefore, is: look for contacts!

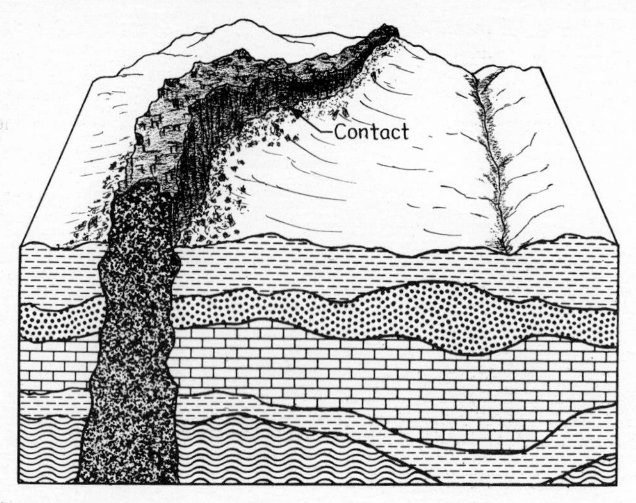

An igneous contact.

The largest bodies of rock in the crust of the earth are those of intrusive (igneous) origin and metamorphic origin. These are formed under conditions of maximum temperature and pressure, deep within the crust and in the upper mantle, which together constitute the lithosphere. Separately or together, they may be found as entire mountain ranges, and they would cover the whole continental part of our globe if the relatively thin covering of other rocks could be removed to expose them. Granite, gneiss, "granite gneiss," and related rocks are the ones occurring here.

The typical sequence in the formation of intrusive rocks, as currently understood, is outlined in the rest of this section. The relationships in the field are as practical to the prospector as they are instructional to the geologist. This brief summary will have to be expanded in later sections.

It is known that no zone of molten rock exists in the upper part of the earth's interior, inasmuch as "secondary" earthquake waves, which do not travel through liquid, move freely across this region. Molten rock (called magma) must therefore originate only in localized pockets where radioactive heat accumulates enough to melt it, or else the pressure is reduced by rock shifting, or both.

This molten rock slowly rises into a position of less pressure and lower temperature, either forcing its way into older rock (hence the term "intrusive") or being allowed to enter it permissively. It then comes to rest, cools, shrinks (leaving joints and cavities), and crystallizes to form recognizable minerals. Some of these minerals may be concentrated enough to be of commercial value; others may be disseminated (scattered) or injected by force into surrounding rock and yet be worth mining.

At the contact with the surrounding rock ("country rock"), a zone of contact metamorphism may develop as the intrusion loses heat and chemi-

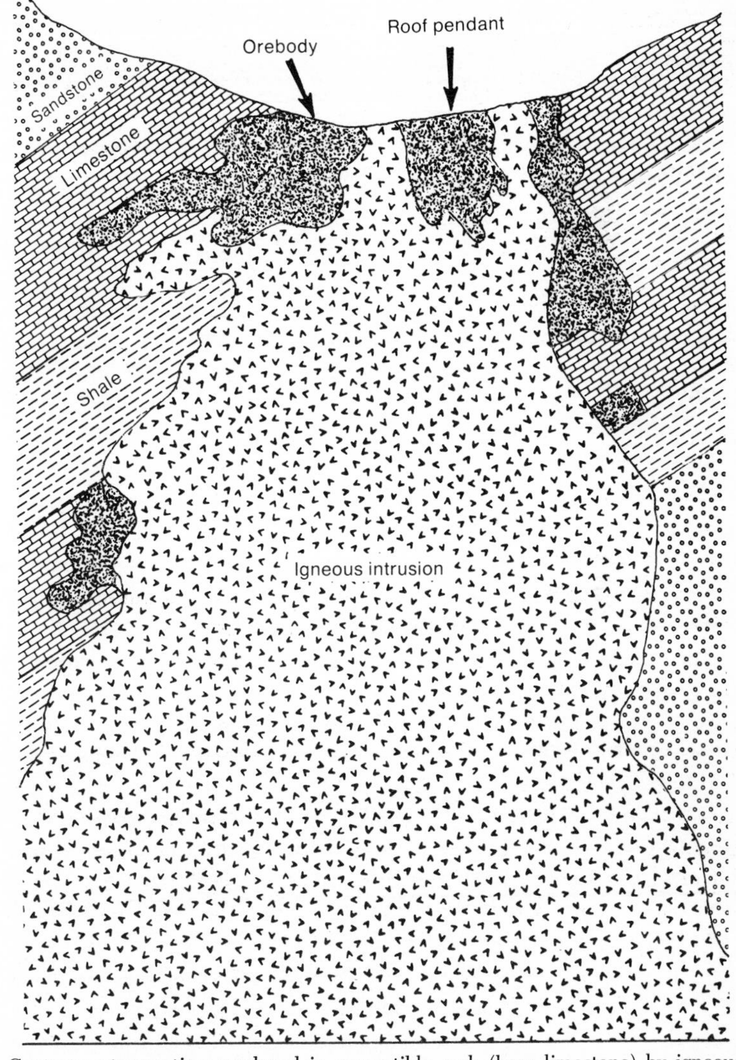

Contact metasomatism produced in susceptible rock (here limestone) by igneous intrusion (quartz monzonite). Roof pendants (altered limestone) are suspended islands, often favorable ore sites. *(After Bateman)*

cals to the adjacent rock. Tungsten is a metal that is typically of contact origin. Metasomatism implies an exchange of material between magma and country rock.

Offshoots of the intrusive rock penetrate structural openings or directions of weakness in the surrounding and overlying rock, becoming dikes and related bodies. Hot liquid and gas expelled from the magma make their way outward and upward as hydrothermal fluids, forming pegmatite first, then quartz veins, then ore deposits of gold and other metals. Where they are not concentrated in veins, the ore minerals may be disseminated, as in the porphyry copper deposits. They may replace the country rock extensively. Finally, nearly depleted of their useful content of minerals, the fluids mix with groundwater and may reach the surface as geysers, hot springs, or cold springs.

The process of igneous intrusion as outlined above may cause the metamorphism of sedimentary rock on a large scale. Or, as now seems more likely, the metamorphism that is associated with mountain building may instead transform sedimentary material into granite, equally on a large scale. The close association of igneous and metamorphic rock in the vast, ancient (Precambrian) regions of the world is remarkable.

Molten rock that rises to the surface of the earth is called lava. It may come from the same source as magma (in the conventional sense), probably losing gas, liquid, and mineral matter as it ascends.

It may, however, have a different place of origin. Nevertheless, lava in Hawaii is known to come from depths as much as 36 miles beneath the surface of the earth, within the mantle.

INTRUSIVE BODIES

Bodies of intrusive rock are named according to shape, size, and relationship to the enclosing rock, as follows:

1. Batholith. The largest igneous bodies are termed batholiths. They are longer than they are wide, have flat bottoms but an unknown depth, and are exposed over an area of 40 square miles or more. Owing to the escape of fluids from a batholith, as mentioned above, ore deposits tend to be concentrated in the higher levels of the intrusion (the summit cupolas) and in the surrounding and overlying rock. Contact deposits also occur nearest the summit cupolas. Where erosion has removed the upper parts of a batholith, the ore deposits are usually gone too.

2. Stock. Circular or elliptical exposures of less than 40 square miles

constitute stocks. A stock may be merely the cupola of a batholith, which would widen below if its shape were known.

3. Dike (dyke, British). Tabular bodies that cut across older rock of any kind, at any angle, are dikes. Being smaller than a batholith or stock, a dike is accompanied by less heat and pressure and a lesser content of min-

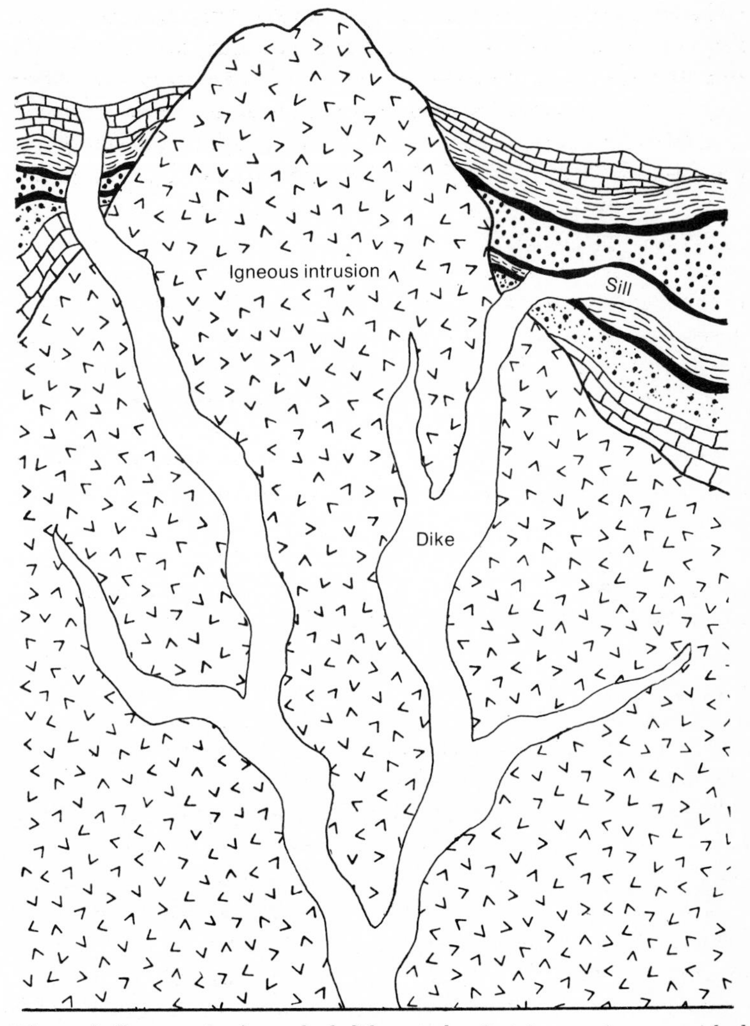

Dikes and sills separating from a batholith or stock. Certain ore veins are enriched where they cross dikes. *(After Landes, Bateman)*

eralizing fluids, and so it produces smaller or fewer ore deposits. The Great Dyke of Rhodesia, the world's largest (330 by 4 miles), is a notable exception, famous for its platinum, chromium, and asbestos. Basic dikes—the dark ones—are most likely to occur with ores.

4. Sill. Like dikes except in their relationship to the surrounding structure, sills are parallel to the enclosing strata or lava flows.

5. Laccolith. When a sill arches up the rock above it, a laccolith is produced.

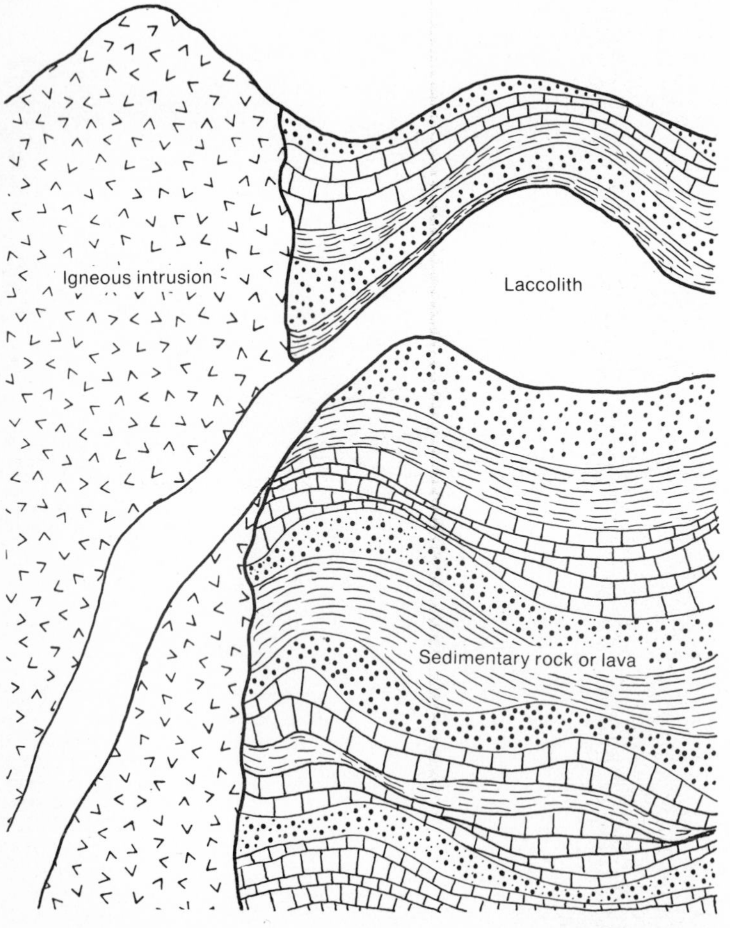

Laccolith derived from adjacent source body. *(After Landes, Bateman)*

6. Lopolith. A sunken surface in the center forms a lopolith rather than a laccolith.

7. Volcanic neck. This is the connecting link between intrusive and extrusive bodies of igneous rock.

Extrusive rocks are those of lava origin, but these cannot be sharply different from those that cool at shallow depths. As noted, a volcanic neck is transitional between intrusive and extrusive activity.

A volcanic neck, such as Shiprock, New Mexico, is transitional between intrusive and extrusive rock. *(Stanolind Oil and Gas Company)*

Sedimentary processes include those that are residual, as caused by weathering, and those that are mechanical, which yield placers.

Metamorphic processes are, as already suggested, closely associated with igneous processes. They, however, may result in drastic changes to sedimentary rock as well as to older igneous rock.

Various classifications of mineral deposits have evolved over the years since they were first studied in a scientific manner in central Europe during the 16th century. Based on the scheme of Waldemar T. Lindgren, as modified by Alan M. Bateman, the modern knowledge of the origin of rocks and their accompanying minerals, as presented in the above paragraphs, can be summarized in the following simplified outline:

1. Magmatic concentration:
   a. Early magmatic
      (1) Disseminated (scattered)
      (2) Segregated (concentrated)
      (3) Injected (forced into other rock)
   b. Late magmatic
      (1) Residual liquid segregation
      (2) Residual liquid injection
      (3) Immiscible (unmixible) liquid segregation
      (4) Immiscible liquid injection
2. Sublimation (deposition from a vapor)
3. Contact metasomatism
4. Hydrothermal processes
   a. Cavity filling
   b. Replacement
      (1) Massive
      (2) Lode
      (3) Disseminated
5. Sedimentation
6. Evaporation
7. Secondary concentration
   a. Residual concentration (by weathering)
   b. Mechanical concentration (placers)
8. Oxidation and enrichment (of igneous and hydrothermal deposits)
9. Metamorphism
   a. Primary
   b. Secondary

STRUCTURES FAVORABLE TO MINERAL DEPOSITION

Recent studies—becoming known to the large mining companies but not as yet to a much wider public—have shown the close relationship between ore bodies and areas of volcanic rock. As a result, volcanic regions are being given new attention, and a number of important ore deposits have been discovered—especially in the United States, Canada, and Australia—around volcanic necks, where intrusive and extrusive igneous activities merge.

Batholiths are now believed to be shallower than was formerly thought and to have produced the volcanic rock that is found deposited above them.

Rising slowly into this volcanic cover, the batholith thus intrudes material of its own making. The fluids (gas and liquid) that are given off yield the ores, whose presence is suggested by breccia (broken rock) and porphyry (visible grains in a finer background).

Where the volcanic cover is thick, the existence of mineral bodies below

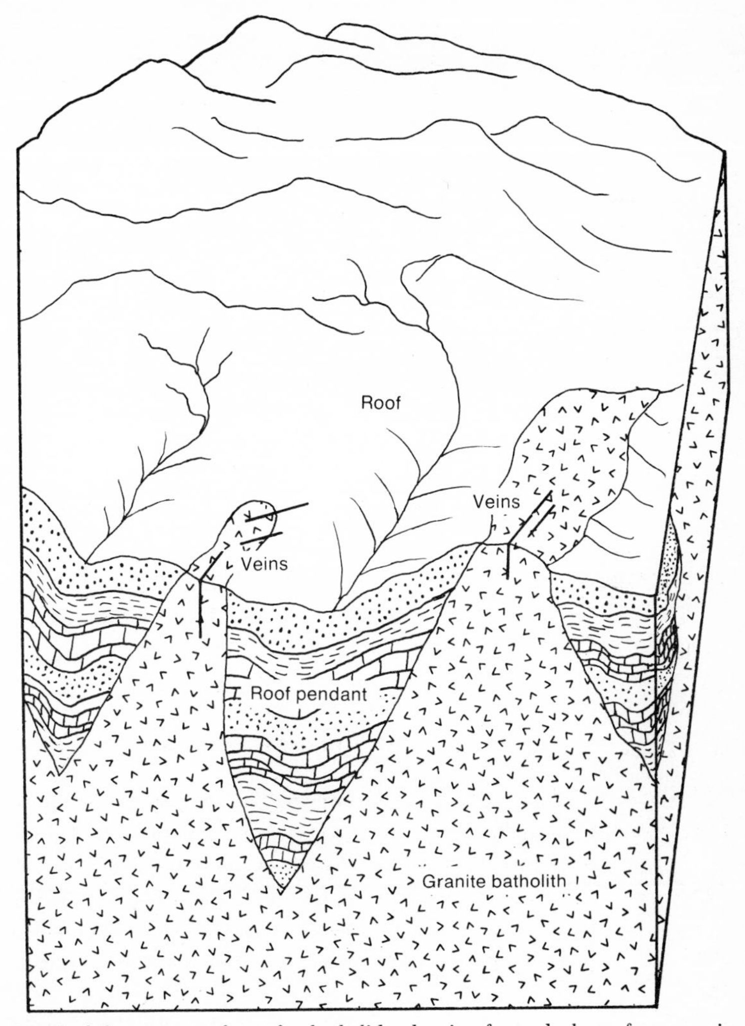

Detail of the upper surface of a batholith, showing favored places for ore veins. *(After Emmons, Bateman)*

can only be surmised, or determined by geophysical means, although the rock may be altered (perhaps to clay) and may show an abnormal concentration of useful metal. Where the volcanic rock has been eroded to sufficient depth, ores may be exposed if they are present.

Breaks in the surrounding rock may either be partly the cause of igneous intrusion or partly the result of it. The primary cause of cracks in rock is the cooling of molten matter (magma or lava) and the shrinkage of drying sediment. These yield joints in fairly regular patterns. Later cracks are produced in igneous, sedimentary, or metamorphic rock by earth movements, volcanism, or solution action. Individual cracks, or fractures, become fissures as they extend in depth, sometimes miles. All these openings are likely places for minerals to deposit as veins, although their distribution may not be easy to determine. Elliptical stocks are apt to be accompanied by parallel fissures, whereas circular ones are associated with radiating fissures.

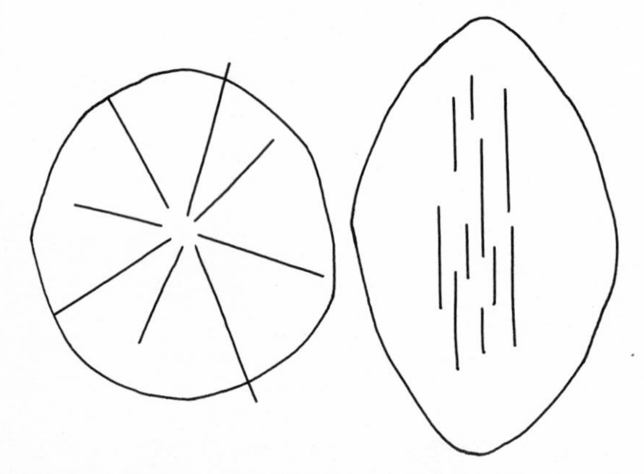

The arrangement of fissures and their ore veins may be determined by the shape of the igneous body. *(After Emmons, Bateman)*

The contact between a vein and the wall rock is often sharply defined but sometimes gradational. The veins may contain minerals that are different from those that have penetrated the wall rock and partly replaced it. Openings in veins are known as vugs. The nonmetallic minerals in a vein (quartz is the most common one) and also the generally worthless metallic ones (such as pyrite) are known as gangue. Weathering may drastically alter the appearance and composition of ore veins, which, furthermore, may widen and narrow, split, and disappear from place to place.

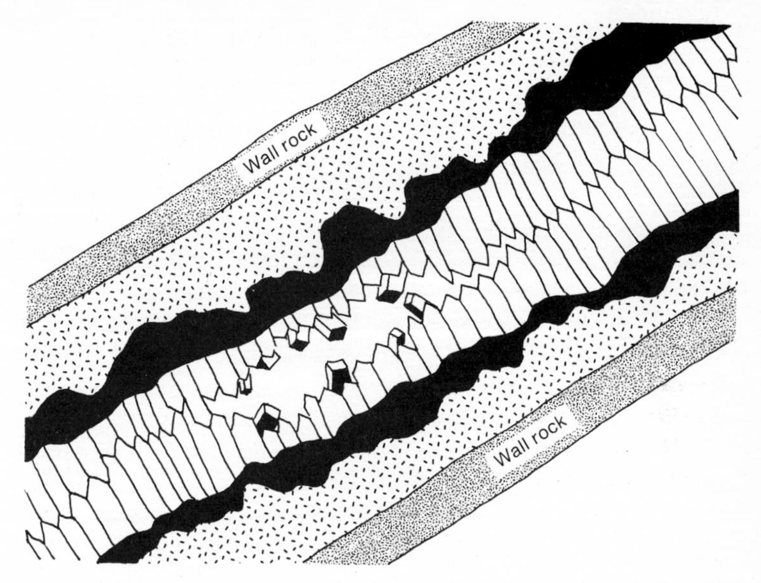

Comb structure containing crystal-lined vug. Dark pattern is typically sulfide mineralization. *(After Bateman)*

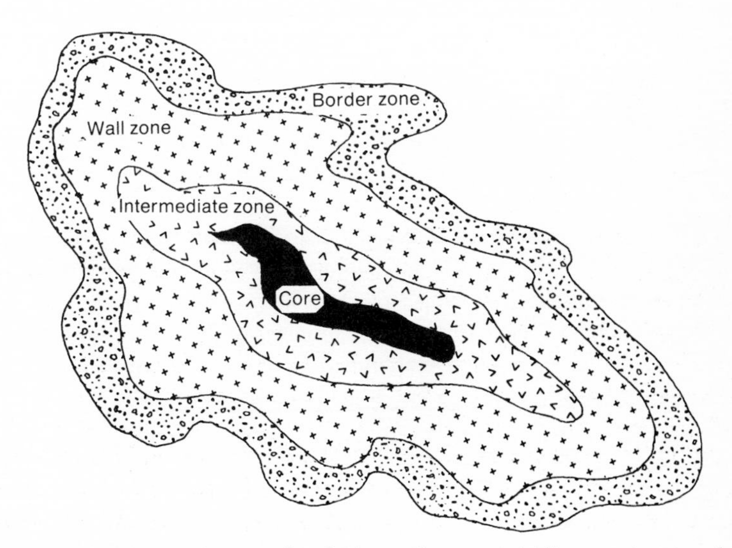

Pegmatite bodies are often zoned in fairly regular ways, yielding certain minerals from definite areas. *(After Cameron, Lamey)*

The cooling of a batholith leaves shrinkage openings in the shape of caverns as well as joints. Called miarolitic cavities, these are often lined or filled with pegmatite minerals. Those that form deep within a batholith are likely to contain little of value except feldspar, quartz, and mica; the rarer minerals of pegmatite occur in the higher levels and within the country rock itself.

Where the alignment of the surrounding rock is such that the more porous and permeable structures rise away from the intrusive body, they act favorably to carry the mineralizing solutions and hence to cause them to be deposited in a more or less direct course.

Movement along a fissure or joint produces a fault. Unless filled tightly with impervious matter (such as clay), faults may serve as passageways for mineral solutions and may become major sites of ore deposits. If faulting occurred after the rise of the mineral solutions, it may cause the deposits to be offset rather than providing a place for them to settle.

In mountainous regions, rocks are generally first folded (helping to produce fractures and faults), then intruded by igneous bodies, then faulted, and finally invaded by metal-bearing solutions. The geometry and nomenclature of folds and faults are quite complex.

The zonal arrangement of ore deposits around igneous bodies is so typical as to make the relationship obvious. The "pyrite halo" has been mentioned as the most commonly developed feature of most porphyry-copper deposits. The zoning is sometimes in a vertical direction, sometimes in both directions, mostly according to temperature. Sometimes, it involves ores of different metals; sometimes, of different minerals of the same metal. Ideal sequences have been described, but these may not be reliable indicators of actual deposits.

Certain other associations of geologic features and mineral deposits are important to understand. The margins of plateaus are sometimes marked by deposition of sediments, which may later be intruded by igneous matter. Basins of sedimentation, including beds of minerals that were deposited by evaporation, have rims that are often easy to map. Local variations in the size and shape of rock layers, and in their porosity and permeability, may control the accumulation of minerals. The presence of a caprock (such as shale) that has trapped mineral solutions may have prevented them from rising further. Descending solutions may be trapped by an impervious base rock. Minerals can likewise gather at the crests of folds or the troughs. The chemical nature of rock may influence mineral deposition;

thus, carbonate rocks (limestone, dolomite, marble) are especially favorable hosts. Countless examples are known that make interesting speculation for the prospector and miner.

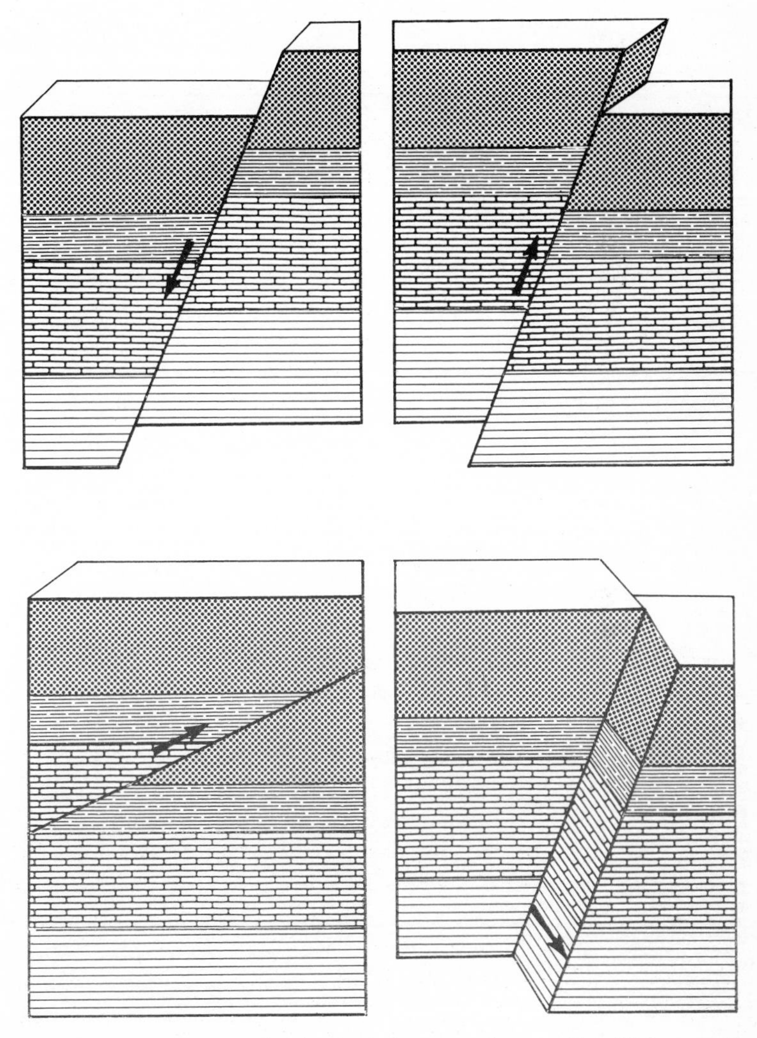

Several of the many kinds of faults: normal, reverse, thrust, strike. The upper side of each is the hanging wall; the lower one, the foot wall.

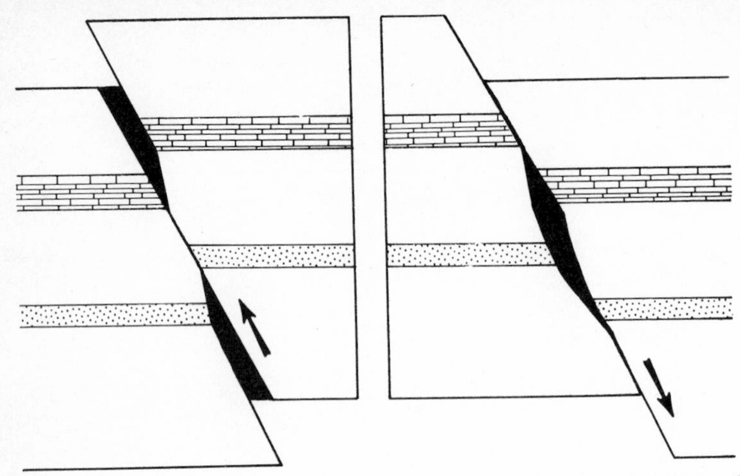

Veins likely to occur with reverse or normal faults. *(After Park and MacDiarmid)*

Sedimentary rock, especially, shows folding under intense pressure, as here in Montana. *(U.S. Geological Survey)*

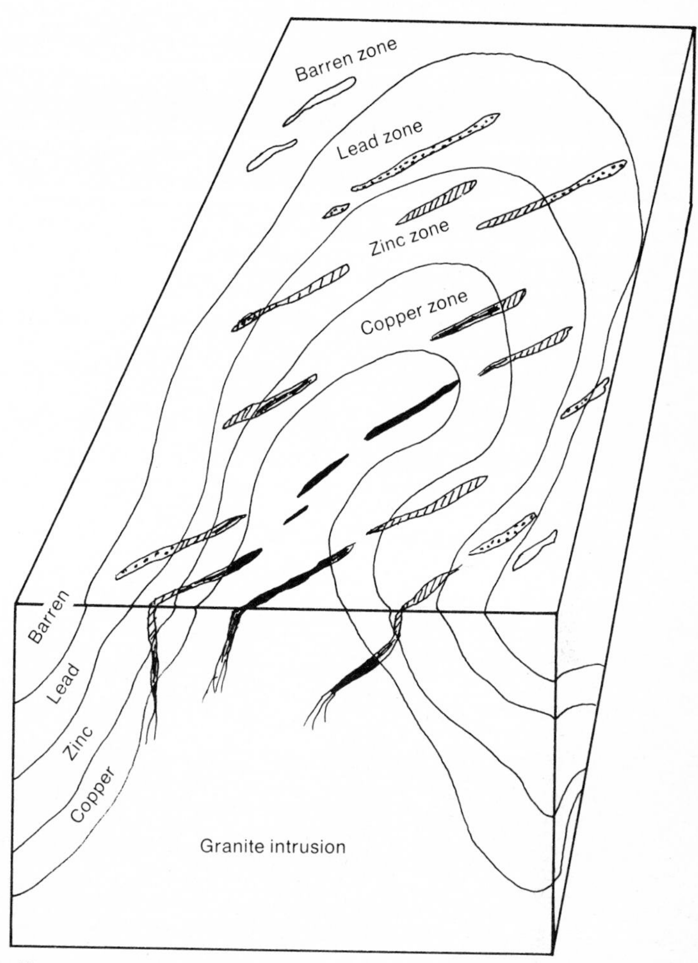

Metallic minerals are often deposited in concentric zones, as in this typical example. *(After Emmons)*

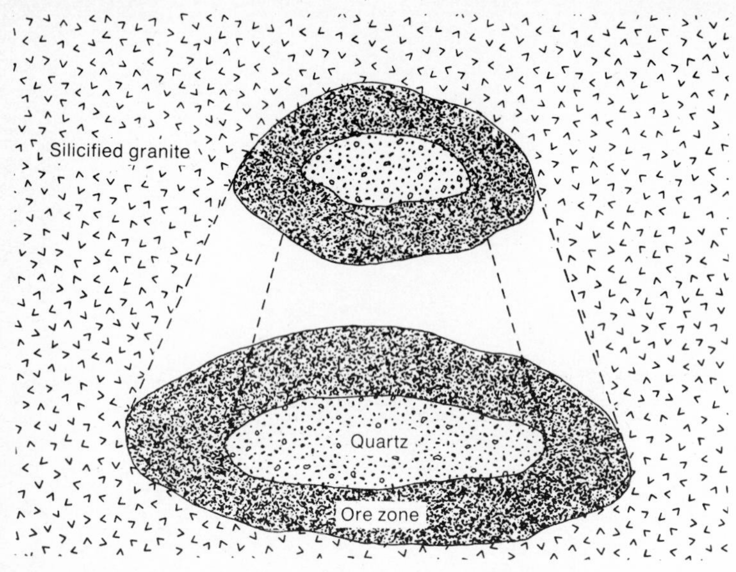

The cone of molybdenum ore at Climax, Colorado, is an unusual structure but related to other chimneys and pipes. *(After Butler and Vanderwilt, Emmons)*

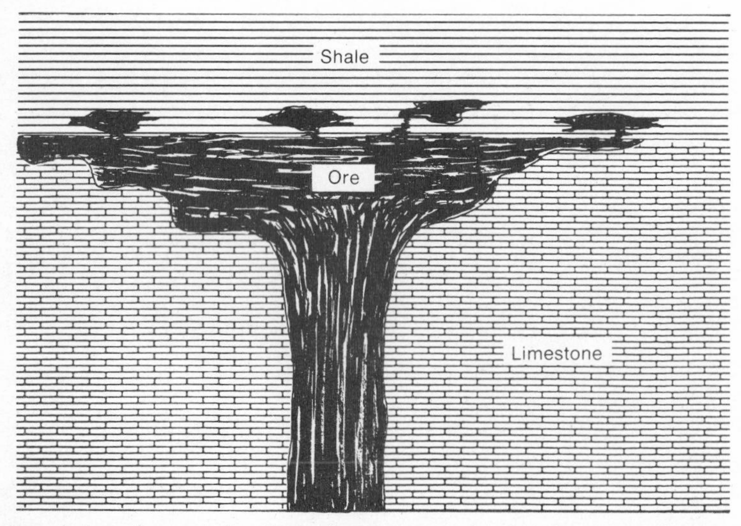

Ore solutions often spread out beneath nonporous rock, as in the Quartz Hill district, near Divide, Montana. This is one of the ways in which mineral deposits mushroom. *(After Taylor, Newhouse)*

The famed saddle reefs at Bendigo, Victoria, Australia, have ore at the crests of upfolded rocks. *(After Emmons)*

Secondary deposits of minerals form in a number of ways. Atmospheric weathering, together with related chemical changes that are due to groundwater, oxidizes certain minerals (especially pyrite, which should therefore be absent in the final product) and yields solvents that dissolve other minerals. These are carried downward by percolating water, and so porous or broken rock is a favorable condition. They may be deposited above the water table as oxidized ores (if sulfide minerals are scarce), or they may descend below the water table and be deposited as enriched sulfide ores. Copper, silver, and zinc are the metals most often involved in this process. The important subject of secondary enrichment is complex; here is formed the often colorful gossan described on page 115. The weathering effects of minerals are discussed in chapters 18 to 32.

Mechanical erosion of surface rock removes the durable minerals so that they can be transported by gravity and stream action, to be laid down as alluvial deposits, or placers.

A more or less individual kind of rock is termed a formation. It is usually a large and persistent stratum of sedimentary rock and is the unit of mapping. Groups of rock layers that were formed under rather similar conditions during a fairly limited span of time are also called formations. So also are units of igneous or metamorphic rock. Formations in the United States and Canada are named for places where they were first de-

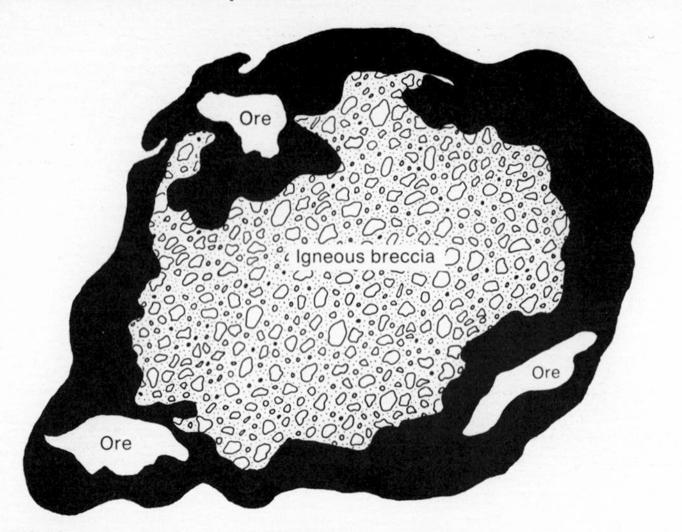

Ore cylinders, or pipes, are exemplified by Los Pilares near Nacozari, Sonora, Mexico. The rock was greatly shattered by explosion, but this is not a volcano. *(After Wandke, Emmons)*

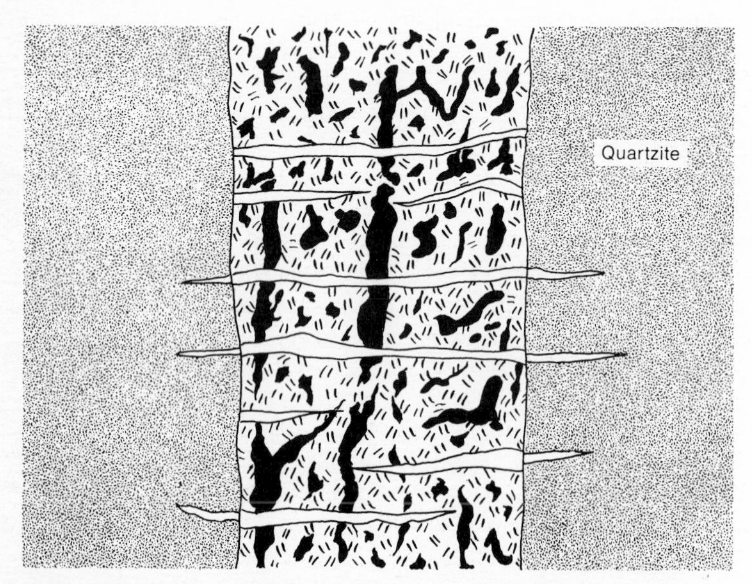

The interesting ladder vein in the Coeur d'Alene district of Idaho has barren quartz crossing lead and other sulfide ores. *(After Ransome, Emmons)*

The distinctive horsetail structure at Butte, Montana, illustrates the possible complexity of fissures and veins. *(After Sales, Emmons)*

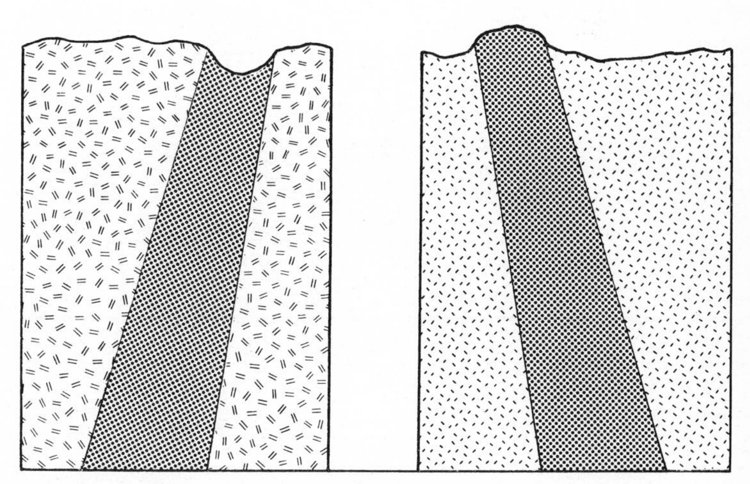

Outcrop less or more resistant than country rock. *(After Emmons)*

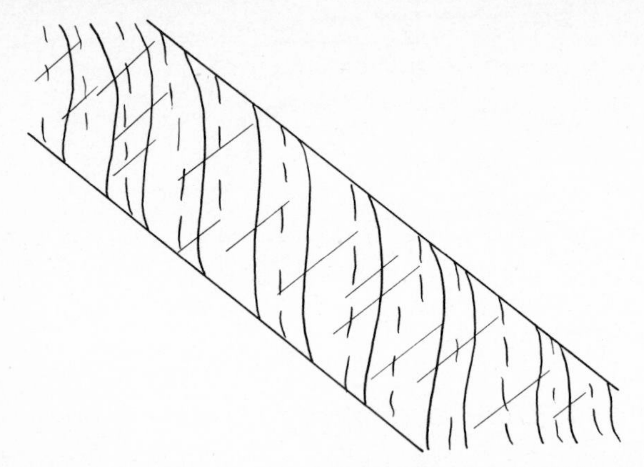

A shear zone contains numerous small veins closely spaced. *(After Emmons)*

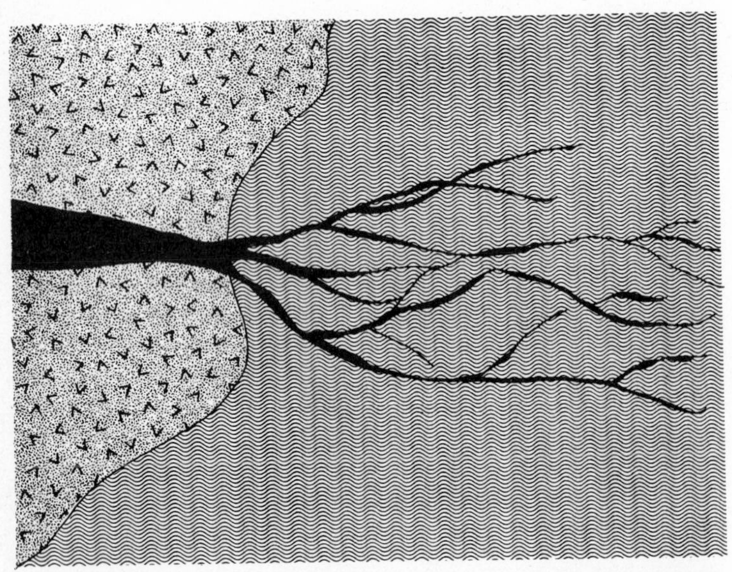

A thick vein may be split where it enters a weaker rock, as at Freiberg, Germany. *(After Beck, Bateman)*

A vein may widen and narrow, especially in schist. *(After Bateman)*

Sheeted veins or zones are usually mined as a unit, as this one at Cripple Creek, Colorado. *(After Bateman)*

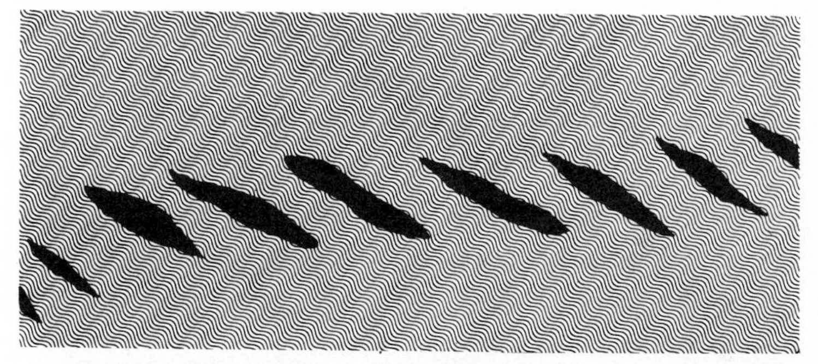

En echelon veins are offset. *(After Bateman)*

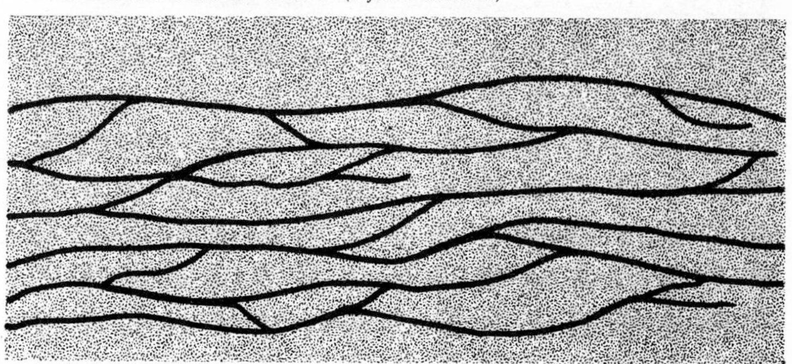

A linked vein is a sheeted vein with cross veinlets. *(After Bateman)*

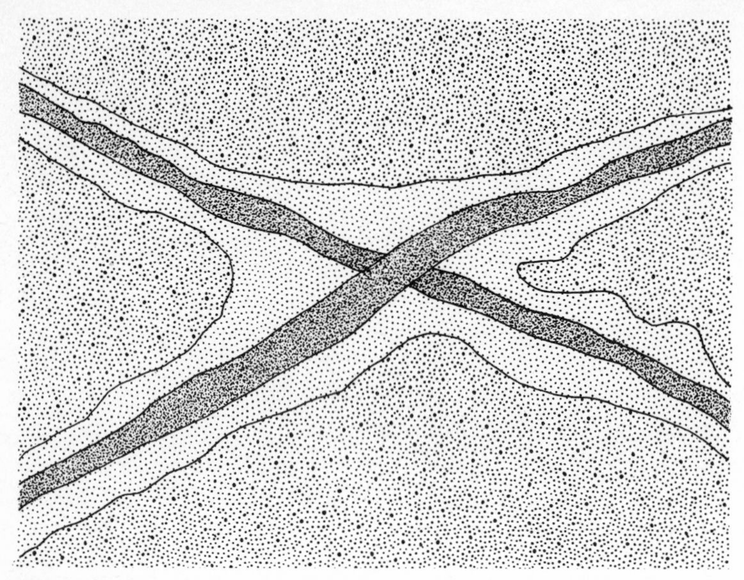

An orebody often enlarges vertically where two veins intersect. Replacement of the rock aids such increase in size. *(After Bateman)*

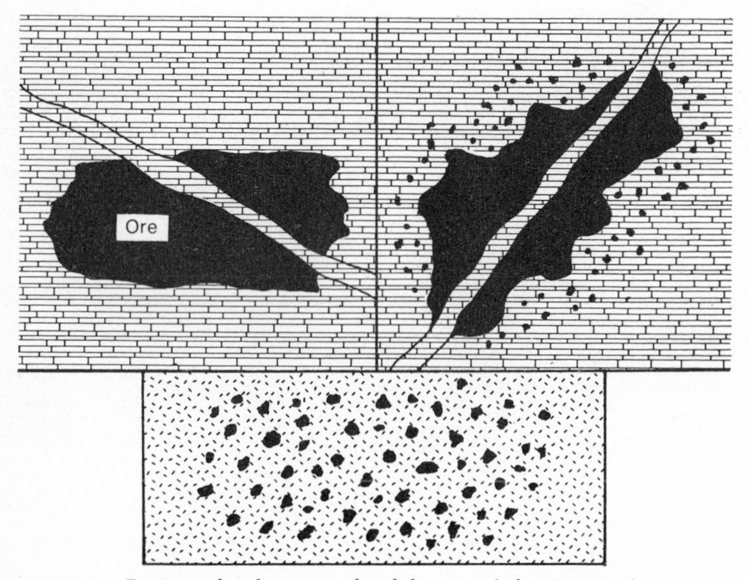

Patterns of replacement of rock by ore. *(After Bateman)*

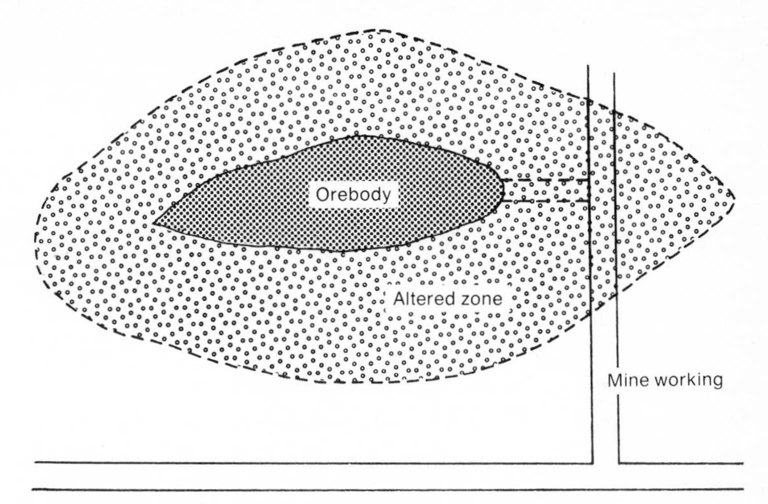

The alteration of wallrock next to an orebody strongly suggests its presence even where it has not yet been revealed. *(After McKinstry)*

Layering such as this, in the Arbuckle Mountains of Oklahoma, could be sedimentary (which it is), metamorphic (as slate), or even igneous (as lava). *(Standard Oil Company, New Jersey)*

scribed and typically occur. During the uranium boom, would-be prospectors commonly asked to be shown "Morrison"—the most productive formation in the Colorado Plateau—often without knowing whether it was a rock, a place, or a person. No magic resides in a formation (or any other) name, and success came also to those prospectors who broke the Morrison barrier.

Elementary geology cannot be discussed without reference to the time scale. It is given below, with the oldest time intervals at the bottom, just as the rock layers of the earth would appear if a complete sedimentary record were present (but it is not) in a single locality. The eras and periods are not of equal length; actual years can be assigned by means of radiometric dating.

G. Montague Butler, in a series of articles written a number of years ago and titled "Some Common Mining Fallacies," endeavored to correct

Geologic Time Scale

Era	Period	Epoch	Duration in millions of years	Began millions of years ago
Cenozoic	Quaternary	Recent	(Late archeologic and historic time)	
		Pleistocene	1–3	1–3
	Tertiary	Pliocene	12	13
		Miocene	12	25
		Oligocene	11	36
		Eocene	22	58
		Paleocene	5	63
Mesozoic	Cretaceous	(Early, middle, or late)	72	135
	Jurassic		46	181
	Triassic		49	230
Paleozoic	Permian	(Early, middle, or late)	50	280
	Carboniferous: Pennsylvanian		30	310
	and Mississippian		35	345
	Devonian		60	405
	Silurian		20	425
	Ordovician		75	500
	Cambrian		100	600
Pre-Cambrian eras			4,000	4,600

certain myths that had gathered around the subject of the formation and occurrence of ores. Brought more up to date, these corrections are stated below as briefly as possible. They are applicable worldwide, even though they were originally written mostly for Arizona prospectors and miners.

1. Ore deposits usually decrease in value with increasing depth. This observation excludes the instances of secondary (generally sulfide) enrichment and the finding of a deposit that did not outcrop. Reaching the zone of secondary enrichment after going through barren rock above it does not contradict the fact. This zone, moreover, is shallow and soon grades into the lean zone of primary ore below. Even somewhat richer ore might not offset the added cost of deep mining.

2. Ore does not change with depth in a way that is certain to make deeper mining more profitable. The zoning of mineral deposits—which depends mainly on temperature—does not follow any order of economic value. Some zones are transitional, some are absent, and of the sixteen zones named by William H. Emmons, not more than two or three have ever been seen in any single mine.

3. Leaching does not always enrich a deposit below. Veins and similar structures do not always contain valuable minerals; hence, nothing worth leaching may be present. Certain minerals (of gold, lead, tin, tungsten, molybdenum, chromium, iron, aluminum) are not normally subject to leaching. Dissolved matter may be carried sideward as well as downward, or it may be forced upward by rising flows of water. Limestone (or other carbonate) beds mostly neutralize the acid solutions and prevent the deposition of any metal except silver. Removal of worthless material by leaching may, however, serve to enrich deposits of the minerals that are left, or it may improve their workability and marketability.

4. The association of certain metals with certain rocks should not be overdone. Dr. Butler called this "the phonolite myth" after the close but entirely accidental relationship of gold and phonolite at Cripple Creek, Colorado. However, chromium, tin, platinum, and tungsten are metals that are most likely to be confined to distinct kinds of rocks. For wider diversity, sedimentary rock (especially limestone) and schist that are cut by moderate-sized intrusions of igneous rock are recommended as especially worthy of the prospector's effort. So also are large masses of recent volcanic rock, as brought out on page 258.

5. Valuable deposits are not always marked by prominent outcrops. This situation exists only when the weathered outcrop is harder than the surrounding rock. The reverse may even be true more often.

6. Veins do not always widen below their outcrops. Variations in width on the surface will probably be found similarly at depth.

7. No fixed relationship exists between length of outcrop and depth of profitable production. Some lodes are long, others wide, others deep.

8. Plants (see chapter 11) are not a very reliable indication of the existence of an outcrop. Except under local conditions, this seems generally to be so.

9. "Blowouts" do not exist. The resemblance of some occurrences (having wide outcrops) to volcanic eruptions can be explained as due to intersections of two veins or other structures, to fillings of chambers, and to other causes.

10. Mineralized structures do not always contain ore shoots. These favorable sites of higher grade ore within ordinary veins, or in other places, go under different names, but they are everywhere the ultimate hope of the prospector and miner. Ore shoots are formed by one or more conditions: the intersection of veins or other structures, the presence of certain wall rocks, the existence of crushed areas, the close approach of two separate veins, and swelling and pinching produced by faults. However, ore shoots are not always found even where conditions are favorable, and such conditions may not be equally favorable in each locality. No fixed relationship between the length and the depth of ore shoots can be depended on.

11. A large percentage of a given element does not make the mineral or rock valuable. Dr. Butler called this "the aluminum myth," based on the large amounts of this metal that occur in the earth's crust. And yet, bauxite is the only "mineral" (it is more nearly a rock) that is mined for aluminum. The caution here is not to take too seriously an assay report unless its true meaning is understood. Some elements (such as potassium and sodium) cannot be recovered profitably from many of their common minerals. Certain elements (such as arsenic and antimony) may even be penalized for; others (such as iron and zinc) are seldom present in paying quantities; others (such as platinum) are so rarely present that they are usually overlooked safely.

12. Divining rods are of no scientific value in prospecting. If you are superstitious, you might as well be so in this direction as in any other. But do not expect sympathetic treatment from engineers or geologists.

In addition, the "myth of minable 'leavings'" was exploded by Hugh E. McKinstry, who pointed out that because a mine was operated at a time when costs were so high that only rich ore could be mined profitably, it is not necessarily true that plenty of lower grade ore was left. Removing the best ore may leave only ore that is too poor ever to be mined; and the final days of a mining venture may see the extraction of anything that might help to pay the costs of closing the property.

Placer, or alluvial, deposits are the simplest kind to understand. They are nevertheless highly diversified and may present many interesting problems, both as to origin and as to means of recovery. The sort of minerals that are most apt to occur in placers, and the usual methods of extracting them—either for testing or for actual (though tedious) mining—were described in chapter 7, which is devoted to the techniques that the ordinary prospector could be expected to use. The present chapter explains how placers form and where they are found.

Let us emphasize that, in almost any placer, anywhere, the most abundant material will prove to be common quartz, magnetite, and other minerals that are seldom of commercial value. You must become accustomed to look through them without really seeing them, while you inspect the panning for signs of "color"—gold—and of unusual minerals, which alone are worth your while. When we refer to placer minerals, then, we mean the heavy, chemically resistant, and physically durable minerals of economic value. Magnetite can be included, however, for it can be concentrated on a 15-to-1 basis, becoming commercial ore. Chromite and ilmenite look like magnetite. Rutile and cassiterite are often rather similar to each other in appearance. Gold and platinum can scarcely be mistaken. Zircon, monazite, rutile, and certain gemstones may look much like one another. Cinnabar, native copper, wolframite, thorianite, and thorite are other placer minerals.

Alan M. Bateman has named five sources of placer deposits, as follows:

1. Commercial lode deposits. A major example is the erosion of gold from the Mother Lode gold veins in California, which came to rest in placers west of the Sierra Nevada.

2. Noncommercial lodes. Small veins (such as those in Malaysia containing cassiterite) may erode to accumulate as an economic deposit.

3. Disseminated ore minerals. Small amounts of ore minerals that are present in rock may collect in a workable deposit; the platinum of the Ural Mountains is an example.

4. Rock-forming minerals. These are like those in number 3 except that they are minerals that are ordinarily present in common rocks; the heavy-mineral beach sand of India, Australia, and Florida came about in this way.

5. Former placer deposits. These may be reworked over and over again, as were the latest gold placers of California.

John H. Wells has outlined placer deposits as follows, but other classifications have also been proposed:

1. Residual placers
2. Eluvial placers
3. Stream placers
   a. Gulch placers
   b. Creek placers
   c. River placers
   d. Gravel-plain deposits
4. Bench placers
5. Flood gold deposits
6. Desert placers
7. Tertiary gravels
8. Miscellaneous types
   a. Beach placers
   b. Glacial deposits
   c. Eolian placers

After being released by weathering, placer minerals either remain as residual materials or are carried downslope by gravity and running water. They may go no farther than the nearest hillside and become eluvial deposits. Tin (as cassiterite) in Malaysia is abundantly found under these conditions. So was gold in Australia, where large nuggets were picked up close to the underlying veins from which they had come. Wolframite

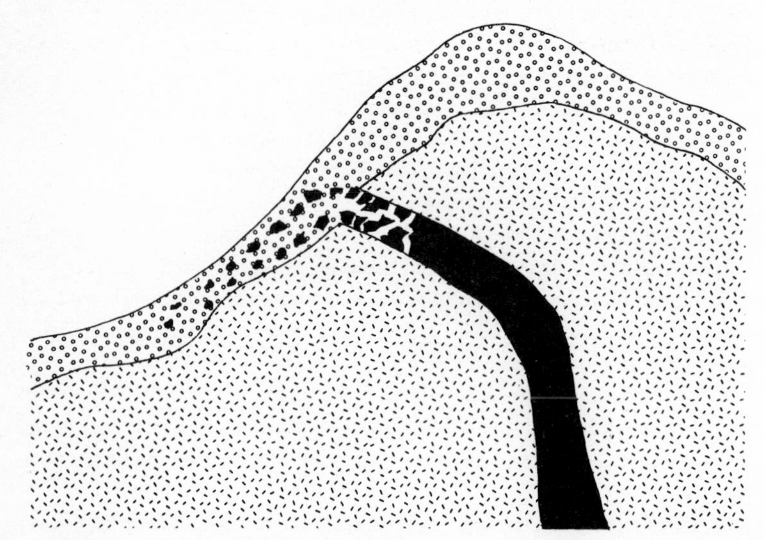

Eluvial or hill-slope deposit derived from veins. *(After Carter, Bateman)*

(tungsten), manganese, kyanite, and barite have also been obtained from eluvial placers.

Beach placers are formed by the action of waves and currents on minerals that are brought down to the seashore by streams, by erosion of rocky shores, and by coastal advance on former beach placers or stream placers. Among the minerals that are found in this type of occurrence are gold, ilmenite and rutile (titanium ores), magnetite and chromite, diamond, zircon, monazite, garnet, and formerly platinum. Nome, Alaska, is the classic locality.

In arid regions, wind may serve to concentrate so-called eolian placers. Gold may occasionally collect in this way and be recovered by simple means. So-called desert placers, however, are deposited by fast-rising streams after cloudbursts, and they are not dependable sources of minerals.

By far the most typical and most important of placer deposits are those that are formed by normal streams. These may later be left stranded above the stream as it erodes deeper or the land is uplifted; the result is a

Great Sand Dunes National Monument, Colorado, was long rumored to contain some eight billion dollars in gold. The rumor has been disproved! *(U.S. Department of the Interior)*

terrace placer, or bench placer, on one or both sides of the valley. Several stages may be represented by "low" and "high" benches, each standing where the valley floor used to be. Such placers may be buried under other sediment or lava; in California, extremely rich deposits of this sort—called high-level, or Tertiary, gravels—were exposed by new streams that flow in a quite different direction from the former ones. Some of this ground remains unexplored, awaiting future prospecting and mining.

Submerged placers, as in Alaska, are found in frozen ground. So-called cement gravels have been consolidated into conglomerate in a number of localities.

Glaciers as well as streams may have played a part in the accumulation of placers, but most such moving bodies of ice bury or disperse rather than gather useful minerals.

A few of the basic principles will help you to find stream placers of gold, platinum, tin, gems, and other heavy minerals. These are given below:

1. Placer minerals are concentrated where the carrying power of a stream is reduced. Slack water occurs where the stream widens (and in a pond or lake); on the inside of a bend (where back eddies form); on the far side of bottom projections (where quiet eddies form); when a dry season lessens the flow; where irregularities cause a jigging action that pushes the heavy minerals into the bottom gravels; where potholes and plunge pools serve as traps.

2. The middle reaches of a stream are the most favorable. The upper stretches provide too limited a source area, and the lower stretches are too sluggish. A gradient of about 30 feet per mile is perhaps the best for placer concentration. Most sediment is carried near the river bed, in mid-water, and along the sides of the channel rather than near the surface in the center of the stream, where the flow is greatest.

3. In a stream meander, the outside curve is a site of erosion, while the inside is a site of deposition—it is hoped—of placer minerals. Gravel bars accumulate at the junction of the inner and outer curves. The amount of flow should be taken into account; as William H. Twenhofel said, "a rising river erodes from its deeps and deposits on its shoals, while a falling river scours from the shoals to deposit in its deeps."

4. Meanders become more exaggerated and migrate downstream. The former position of a stream (when it may have carried placer minerals) can thus be determined by tracing the meanders to their earlier positions. Many buried placers of great value have been found, in California and elsewhere, by using this principle.

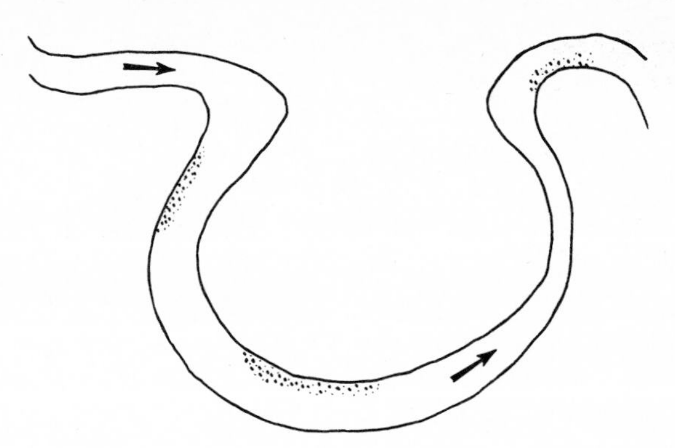

Placer deposits likely to occur in a meandering stream. Narrowing increases velocity, hence no gravel deposits. *(After Bateman)*

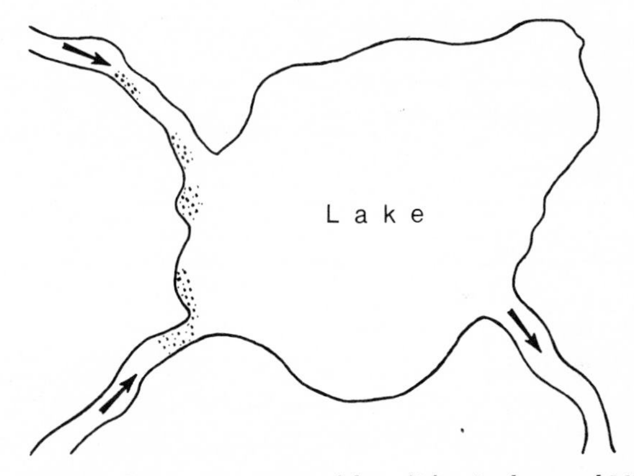

Placer deposits where streams enter a lake. *(After Raeburn and Milner)*

5. Steeply protruding layers of rock act as riffles in a stream bed. These help to trap placer minerals.

6. Placer minerals taper off downstream from a lode or vein that crosses the stream. Lodes should be looked for where this pattern exists.

7. Placer gravel tends to accumulate on the side of the main stream where a steeper tributary enters it, but some distance below the junction.

8. The solid rock beneath a stream bed is often richer than the placer itself. This is because native gold and the platinum metals, especially, are flattened and pounded into the bedrock by moving pebbles. A "false bedrock" of clay may overlie gold-bearing gravel. The deeper currents of water act, however, to lift particles against the force of gravity, which would otherwise cause an even earlier deposition.

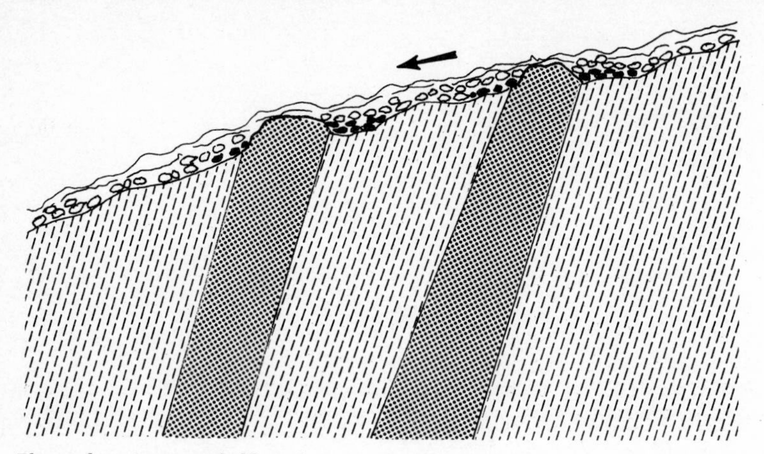

Placer deposits around dikes that intersect the stream, forming natural riffles. *(After Bateman)*

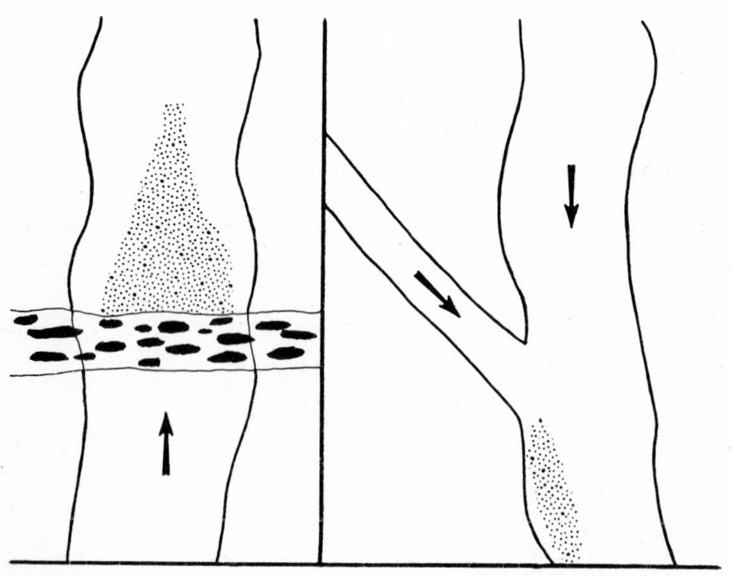

Placer deposit formed downstream from intersecting veins. *(After Bateman)*

Placer deposit where tributary enters the main stream. *(After Bateman)*

9. Placers are often marked by "white runs" (similar terms are also used) because of their high content of quartz, by black sand (mostly magnetite), or by yellow sand (monazite especially).

10. Placer minerals separate out according to their individual properties, of which specific gravity is only one.

BOOKS ON GEOLOGY

Besides the books on general prospecting that are listed at the end of chapter 1, the following publications emphasize placer operations in the United States and Canada:

Placer Mining in the Western United States, E. D. Gardner and C. H. Johnson, U.S. Bureau of Mines Information Circulars 6786, 6787, 6788, 1934–35.

"The Geology of Placer Deposits," W. E. Cockfield, *Transactions of the Canadian Institute of Mining and Metallurgy,* vol. 35, 1932, pp. 58–64.

The following books contain information that concerns aspects of geology that pertain to prospecting and mining. There are, of course, a great number of books on general geology; keys to 22 textbooks are given in *Geology,* 3d edition, by Richard M. Pearl (Barnes & Noble College Outline Series, 1963).

Elementary Geology Applied to Prospecting, J. F. Walker, British Columbia Department of Mines, Vancouver, 1953.

Geology Applied to Mining, J. E. Spurr, McGraw-Hill Book Company, New York, 1904.

Field Geology, 6th edition, Frederic H. Lahee, McGraw-Hill Book Company, New York, 1961

Principles of Field and Mining Geology, James Donald Forrester, John Wiley & Sons, Inc., New York, 1946.

Mining Geology, Hugh Exton McKinstry, Prentice-Hall, Inc., Englewood Cliffs, N.J., 1948.

In addition to the books on economic geology listed at the end of chapter 24 (for metallic minerals) and chapter 32 (for nonmetallic minerals and rocks), the following books discuss the geologic aspects of mineral deposits:

The Formation of Mineral Deposits, Alan M. Batemen, John Wiley & Sons, Inc., New York, 1951

Mining Geology, Hugh Exton McKinstry, Prentice-Hall, Inc., Englewood Cliffs, N.J., 1948.

Part 4

18

Precious Metals

Gold, silver, and platinum are termed the precious metals. Certain other metals, however—for example, gallium—bring higher prices. Platinum, furthermore, is one of six so-called platinum metals, which occur together and are used in more or less similar ways. Nevertheless, gold, silver, and platinum are thought of in relation to their uses in jewelry, which makes them precious. All three, however, have industrial uses for which they are more valuable and even indispensable.

Gold and silver have been known for thousands of years, as their frequent mention in the Bible makes clear. Silver was once more valuable than gold because it was less abundant. Either gold or copper was the first metal known to man. In spite of its occurrence in ancient times, silver did not become a common metal until the discovery of America, and it can fairly be said to be a New World metal, most of the production to date having come from the Cordilleras: the Rockies and Andes. Platinum (which means "little silver") certainly is an American metal, for it was not known—except a single grain found in a 7th century B.C. Egyptian ornamental case—until Spanish explorers acquired some platinum objects from South America. European technology brought platinum into use and separated the other platinum metals from it. In the 19th century, Russian coins were counterfeited by covering disks of platinum with gold, both being very heavy metals.

GOLD

Gold has been described as the forerunner of civilization—men followed gold and took society with them—and more prospectors have searched for it than for all other minerals put together. Gold is, or should be, the most readily recognized of minerals when found native, as it was when first seen in stream beds as long ago as any record exists—"when time began," as the saying has it.

Still, as we have already stated, gold offers today's prospector little incentive. Large mining companies are finding new reserves of the yellow metal in rock that yields only to mass production, but the lone prospector—with his mule or partner or wife (it makes little difference which!) is not apt to be well rewarded by looking for it. He will do better by seeking for a gravel pit or a vein of fluorspar or a bed of phosphate rock.

Yet, the demand for gold exists, whether or not the gold standard ever returns, for its industrial uses are forever expanding. And the story of gold is as exciting as ever, just as the rich beauty of this unique substance has as stong an appeal. Furthermore, the reader of a book such as this would have every right to feel cheated if gold were slighted. Hence, the subject is treated as conscientiously as possible. The last successful finder of a new gold mine has surely not passed from the earth: he may be you.

Gold is derived from three types of ore, as follows:

1. Placers
2. Ore in which gold is the main commercial metal
3. Base-metal ore that yields gold as a byproduct

Classified by origin, in the Bateman way (see page 258), the main gold deposits are the following:

1. Magmatic concentration
2. Contact metasomatism: as Ouray, Colorado
3. Replacement
   a. Massive: as Noranda (Quebec), Morro Vehlo (Brazil)
   b. Lode: as Kirkland Lake (Ontario), Homestake (South Dakota), Kolar (India)
   c. Disseminated: as Witwatersrand (South Africa)
4. Cavity filling—as Mother Lode (California), Cripple Creek (Colorado), Bendigo (Australia)
5. Residual concentration
6. Mechanical concentration: as placers in California, Alaska

From the mineralogic standpoint, gold may be said to exist in three ways, as follows:

1. As native gold: placer or vein. It is always naturally alloyed with silver; pure gold would be expressed as 1,000 fine, but vein gold runs from about 500 to 850 fine, whereas placer gold averages higher as some silver is dissolved out.
2. As a gold compound: the tellurides.
   a. Calaverite: $AuTe_2$, 39 to 44 percent gold
   b. Sylvanite: also called graphic tellurium; $(Au,Ag)Te_2$, 24 to 30 percent gold
   c. Krennerite: $AuTe_2$, 31 to 44 percent gold
   d. Petzite: Ag_3AuTe_2, 19 to 25 percent gold

   Other gold tellurides are doubtless recovered in association with these.
3. As byproduct gold: tiny bits of metallic gold in fine-grained pyrite and other minerals. Several of these minerals are called fool's gold because they somewhat resemble gold, not because of their small gold content, for they may have no gold at all; there are other fool's-gold minerals that never contain any gold. The best observation on this subject was by Mark Twain in *Roughing It:*

> By and by, in the bed of a shallow rivulet, I found a deposit of shining yellow scales, and my breath almost forsook me! A gold mine, and in my simplicity I had been content with vulgar silver! I was so excited that I half believed my overwrought imagination was deceiving me. Then a fear came upon me that people might be observing me and would guess my secret. Moved by this thought, I made a circuit of the place, and ascended a knoll to reconnoiter. Solitude. No creature was near. Then I returned to my mine, fortifying myself against possible disappointment, but my fears were groundless—the shining scales were still there. I set about scooping them out, and for an hour I toiled down the windings of the stream and robbed its bed. But at last the descending sun warned me to give up the quest, and I turned homeward laden with wealth. As I walked along I could not help smiling at the thought of my being so excited over my fragment of silver when a nobler metal was almost under my nose. In this little time the former had so fallen in my estimation that once or twice I was on the point of throwing it away.
>
> The boys were as hungry as usual, but I could eat nothing. Neither could I talk. I was full of dreams and far away. Their conversation interrupted the flow of my fancy somewhat, and annoyed me a little, too. I despised the sordid and commonplace things they talked about. But as they proceeded, it began to amuse me. It grew to be rare fun to hear them planning their poor little economies and sighing over possible privations and distresses when a gold mine, all our own, lay within sight of the cabin and I could point it out at any moment. Smothered hilarity began to oppress me, presently. It was

hard to resist the impulse to burst out with exultation and reveal everything; but I did resist. I said within myself that I would filter the great news through my lips calmly and be serene as a summer morning while I watched its effect in their faces. I said:

"Where have you all been?"

"Prospecting."

"What did you find?"

"Nothing."

"Nothing? What do you think of the country?"

"Can't tell, yet," said Mr. Ballou, who was an old gold miner, and had likewise had considerable experience among the silver mines.

"Well, haven't you formed any sort of opinion?"

"Yes, a sort of a one. It's fair enough here, maybe, but overrated. Seven thousand dollar ledges are scarce, though. That Sheba may be rich enough, but we don't own it; and besides, the rock is so full of base metals that all the science in the world can't work it. We'll not starve, here, but we we'll not get rich, I'm afraid."

"So you think the prospect is pretty poor?"

"No name for it!"

"Well, we'd better go back, hadn't we?"

"Oh, not yet—of course not. We'll try it a riffle, first."

"Suppose, now—this is merely a supposition, you know—suppose you could find a ledge that would yield, say, a hundred and fifty dollars a ton—would that satisfy you?"

"Try us once!" from the whole party.

"Or suppose—merely a supposition, of course—suppose you were to find a ledge that would yield two thousand dollars a ton—would *that* satisfy you?"

"Here—what do you mean? What are you coming at? Is there some mystery behind all this?"

"Never mind. I am not saying anything. You know perfectly well there are no rich mines here—of course you do. Because you have been around and examined for yourselves. Anybody would know that, that had been around. But just for the sake of argument, suppose—in a kind of general way—suppose some person were to tell you that two-thousand-dollar ledges were simply contemptible—contemptible, understand—and that right yonder in sight of this very cabin there were piles of pure gold and pure silver—oceans of it—enough to make you all rich in twenty-four hours! Come!"

"I should say he was as crazy as a loon!" said old Ballou, but wild with excitement, nevertheless.

"Gentlemen," said I, "I don't say anything—*I* haven't been around, you know, and of course don't know anything—but all I ask of you is to cast your eye on *that,* for instance and tell me what you think of it!" and I tossed my treasure before them.

There was an eager scramble for it, and a closing of heads together over it under the candle-light. Then old Ballou said:

"Think of it? I think it is nothing but a lot of granite rubbish and nasty glittering mica that isn't worth ten cents an acre!"

> So vanished my dream. So melted my wealth away. So toppled my airy castle to earth and left me stricken and forlorn.

Gold occurs in many kinds of rock, of all ages. The most productive deposits, however are geologically very old (Precambrian) or rather young (Tertiary). They are related to, and are found near, igneous bodies of intrusive origin and silicic composition—like granite. Yet, granite itself is not normally productive, although the extrusive (lava) and porphyry rocks that are related to it, and the dikes and other small bodies that are associated with it, are especially so. Dark and heavy rocks may or may not be encouraging, usually not. Metamorphic rock is neither favorable nor unfavorable, but its dikes may be worth examining. The richest deposits have been placers and small veins that contain quartz, but larger total amounts of gold have come from big deposits of lower grade. Other, more specific information about the conditions that assist the presence of gold is given below.

When vein gold is found, it is apt to seem richer toward the top. This is because gold is very resistant to weathering, and particles of native gold tend to accumulate in cracks and other open spaces, while other minerals are leached away. If sulfide minerals and other soluble minerals are few, the gold content of quartz veins may be fairly constant as they extend downward. It seldom gets better, but exceptions are known. More than in almost any other kind of prospecting, you are well advised to consider the mining history of the district, learning the clues that others have followed to their advantage. Gold tellurides present an entirely different problem, but these occur only in restricted areas.

The following list from Von Bernewitz gives well-known examples of the occurrence of gold worldwide:

1. Quartz in dacite
2. Quartz in schist
3. Quartz in breccia and granite
4. Quartz in schist near granite and gneiss
5. Quartz in slate near its contact with diorite
6. Quartz in slate or at the contact between slate and diorite
7. Porphyry dikes
8. Granite in gneiss
9. Shale and quartz porphyry
10. Quartzite in a region of granite or diabase
11. Faults or fractures and stockworks in schist or quartz monzonite
12. Areas near contacts of granite and diorite
13. Areas along contacts of granite and felsite

14. Calcite as a replacement in slate (pocket deposit)
15. Gravel lying on schist, slate, or decomposed volcanic matter
16. Feldspar-porphyry and quartz-porphyry intrusions in or near belts of conglomerate, slate, and greywacke
17. Primary hematite
18. Stringers of feldspar and adularia with little silica in shale
19. Barite

Placer gold can be found anywhere that conditions permit, even far from the rock types that gave rise to the veins from which it came.

The main characteristics of the gold-bearing minerals are given below; more complete descriptions of the properties should be sought for in books on mineralogy.

Native gold Pale to deep yellow, depending on the silver content; malleable, soft, scratches yellow, very heavy. The fool's-gold minerals are brittle (pyrite, chalcopyrite, pyrrhotite), or the flakes (of weathered biotite mica) can be chipped with a knife; they may be magnetic (pyrrhotite) or harder than a knife (pyrite, pyrrhotite).

Gold tellurides Calaverite, sylvanite, krennerite, hessite, petzite. These are silver colored to pale brass yellow and may be very bright; they give a yellow globule of gold, the color depending on the amount of silver.

Some minerals (such as arsenopyrite) are often given credit for containing gold in certain localities, but it seems to be a chemical impurity. To the prospector or miner, the nature of the atomic structure does not sound important, but the extent of the recovery in milling may depend very much on such details, which are the daily concern of the metallurgist.

In the newly opened Carlin and Cortez mines in Nevada, for example, the gold occurs in sub-micron-size particles.

The uses of gold need not be discussed at length. Everyone knows about its value in jewelry, decoration, dentistry, coinage, and as a monetary medium. Everyone should know that it has many industrial uses and that these are growing for coatings, electrical apparatus, scientific equipment, chemicals, and other needful purposes.

Although the literature on gold is enormous, this recent publication is worth mentioning for its information on gold-producing localities in the United States: *Principal Gold-producing Districts of the United States,* by A. H. Koschmann and M. H. Bergendahl, U.S. Geological Survey Professional Paper 610, Washington, 1968. A map, MR-24, is sold for 75 cents by the U.S. Geological Survey; Alaska is covered by maps MR-32 (lode gold and silver, 75 cents) and MR-38 (placer gold, 50 cents).

SILVER

Silver is the brightest of all known metals. Its industrial future is even brighter as it enters a larger number of essential products. Next only to gold—and often along with gold—silver has drawn the attention of prospectors throughout the centuries. Yet not until 1492 and Columbus did silver become a truly abundant material, as the western part of the Western Hemisphere, beginning in South America and spreading northward, began to pour its treasure of the white metal into the coffers of the Old World. Up to the discovery of gold in California in 1848, silver seemed to be the chief answer to the dreams of wealth that helped to bring the explorer and adventurer to America.

Silver continues to beckon the prospector today. In fact, it seems to offer him a greater opportunity than does gold, since the demand for it is rising faster than the supply and the price is going up with the demand.

The main kinds of silver deposits according to origin are the following:

1. Contact metasomatism
2. Cavity fillings: as Pachuca, San Francisco, and Fresnillo (Mexico), Sunshine (Idaho)
3. Replacement
   a. Massive: as Bingham, Tintic, and Park City (Utah), Kimberley (British Columbia), Leadville (Colorado), Santa Eulalia (Mexico)
   b. Lode: as Comstock Lode (Nevada), Potosí (Bolivia)
   c. Disseminated
4. Oxidation and enrichment: as Parral (Mexico)

Silver comes from the following types of mineralization:

1. Native silver: 100 percent silver.
2. Silver compounds, as follows:
   a. Sulfides:
      (1) Argentite: also called silver glance; Ag_2S, 87 percent silver
      (2) Stromeyerite: CuAgS, 49 to 53 percent silver
   b. Sulfosalts:
      (1) Polybasite: $(Ag,Cu)_{16}Sb_2S_{11}$, 61 to 74 percent silver
      (2) Pearceite: $Ag_{16}As_2S_{11}$, 57 to 77 percent silver
      (3) Pyrargyrite: also called dark ruby silver; Ag_3SbS_3, 60 to 61 percent silver
      (4) Proustite: also called light ruby silver; Ag_3AsS_3, 64 to 65 percent silver

(5) Tetrahedrite: also called gray copper ore, fahlore, freibergite; $(Cu,Fe,Ag)_{12}Sb_4S_{13}$, 0 to 17 percent silver

(6) Tennantite: $(Cu,Fe,Ag)_{12}As_4S_{13}$, 0 to 4 percent silver

(7) Stephanite: Ag_5SbS_4, 68 to 69 percent silver

c. Chlorides:

(1) Cerargyrite: also called horn silver; AgCl, 67 to 75 percent silver

d. Tellurides:

(1) Sylvanite: also called graphic tellurium, $(Ag,Au)Te_2$, 9 to 14 percent silver

(2) Hessite: Ag_2Te, 59 to 63 percent silver

(3) Petzite: Ag_3AuTe_2, 41 to 45 percent silver

Other silver tellurides occur in close association with these.

3. Other minerals in which silver is a chemical impurity, as tiny bits of silver compounds in galena and other minerals. Silver is commonly associated with base metals, mostly copper, lead, and zinc. The silver obtained from the large-scale, low-grade copper bodies that are called porphyry copper can be a valuable byproduct. The silver content may determine whether a deposit of base metal is worth mining at all: thus, in Missouri, where galena bears no appreciable silver, its price does not matter but in Idaho, the value of the silver means more than the lead content.

The relationship to gold is obvious (see page 289). Most gold deposits contain silver; native gold is always alloyed with more or less silver, and native silver may contain gold.

Silver ores occur in rocks of various geologic ages, of which the most productive have been Precambrian, Cretaceous, and Tertiary.

When veins of silver minerals weather and oxidize, some of the silver is dissolved away, while the rest of it remains even while the mineral that contains it (galena, for instance) alters to other minerals (such as anglesite or cerussite). Unlike gold, a lack of silver in the gossan at the top does not mean that silver is also absent below. Silver may, however—like gold—be spotty in this kind of ore. Cerargyrite is, for chemical reasons, richest close to the surface; a noted discovery was made at the Barrier lode in New South Wales, Australia, when a campfire melted this mineral.

In addition to the kinds of rock in which gold is also found, as previously described, silver often occurs with lead in limestone that is crossed by dikes or paralleled by sills of igneous rock. Where silver is known to exist in an area of lead and zinc ore, it is generally found closely associated with them. Either lead or zinc may be with silver but without the other metal.

Silver also occurs with manganese, but it is then difficult to treat. It occurs with manganese and zinc together; with tungsten (the mineral scheelite); and with tin.

Volcanic rock is especially suitable for the finding of silver: consider andesite, trachyte, rhyolite, and basalt; breccia (broken rock) and porphyry are favorable aspects of these rocks. Granite and diorite, as at Great Bear Lake, Canada, cannot be ignored. Within the past few years, the significance of black calcite as a host for silver mineralization has been recognized in the western United States; the color is due to dispersed grains of manganese oxide, which contain the silver.

Silver grains that are so tiny as to merit the term "no-see-'em" have been discovered in lake-bed sediments in California and Arizona and described in 1972. These beds overlie volcanic rock.

The Silver Snooper, recently devised by the U.S. Geological Survey, employs neutron generation as a means of finding silver by measuring the radiation that is produced. This instrument applies the methods that were already in use for detecting radioactive elements and beryllium.

Identifying native silver is easy, although it may be necessary to dissolve it in nitric acid and add hydrochloric acid to precipitate a white cloud of silver chloride. Native platinum does not react, nor does it tarnish in the familiar way of silver. Pyrargyrite and proustite are known as ruby silver, because they are reddish on a fresh surface, as crystals, or when rubbed on a streak plate. Cerargyrite is called horn silver on account of its waxy or hornlike appearance. The other silver minerals are metallic gray and probably require testing, for which information should be sought in a mineralogy textbook.

The uses of silver in jewelry, coinage, and tableware are well known. It is the basis of the photographic industry. The chemical and electrical industries use it extensively, and silver is consumed in dentistry, solder, batteries, and growingly in a huge variety of important products. Rising prices are offsetting some of the difficulties that have long bothered the miner in treating low-grade ores. In spite of substitutions, silver will be needed much more in the future. It is up to the prospector to help find it.

A recent technical summary on methods of prospecting for silver deposits (with emphasis on Canada) is Chapter V of Geological Survey of Canada Bulletin 160, *The Geochemistry of Silver and its Deposits,* by R. W. Boyle, Ottawa, 1968.

U.S. Bureau of Mines Information Circular 8427 is titled *Silver in the United States.* A map of the U.S. Geological Survey, MR-32, covers lode silver and gold in Alaska and sells for 75 cents.

PLATINUM

Exclusively a New World metal until found elsewhere much later, platinum is one of a group of six so-called platinum metals or platinoids. These are platinum, palladium, iridium, osmium, rhodium, and ruthenium. All except palladium are much more costly than gold; hence, indeed, these are precious metals. All occur together and are alloyed in varying proportions; analyzing them is one of the more expensive jobs of the assayer, and separating them is a tedious job for the chemist. The metals are then refined chemically or (if byproduct) electrolytically.

Platinum and its colleagues are found in three main types of ore, as follows:

1. Placers: always with gold
2. Segregations: in ultrabasic igneous rock
3. Disseminations: in basic and ultrabasic igneous rock

By origin, the main classifications of platinum deposits are as follows:

1. Magmatic concentration
   a. Early magmatic
      (1) Disseminated: as Urals (U.S.S.R.)
      (2) Segregated: as Merensky Reef (South Africa)
   b. Late magmatic
2. Contact metasomatism
3. Hydrothermal: as Sudbury (Ontario)
4. Mechanical concentration: as Urals (U.S.S.R.), Colombia

As minerals, the platinum group occurs as follows:

1. As native metal: alloyed with one another and with iron or copper
2. As platinum compounds: chiefly sperrylite but sometimes cooperite, braggite, stibiopalladinite. Sperrylite is platinum arsenide, $PtAs_2$, 55 to 57 percent platinum. Platinum and palladium may exist as chemical impurities in pyrrhotite at Sudbury, Ontario.

Very little platinum is mined for this group of metals alone, but mostly as a byproduct or in conjunction with other valuable metals. These are gold and silver, nickel or chromium, or copper.

The platinum metals occur in basic and ultrabasic rocks unless they have been eroded out and carried into placers. The original rock is dunite, pyroxenite, serpentine, peridotite, or other igneous rock of similar nature: heavy and dark, rich in iron and magnesium, and especially favorable for nickel, chromium, cobalt, and platinum.

A large platinum nugget has been preserved as a cast. *(Field Museum of Natural History)*

The prospector may have difficulty finding platinum, but he should have no trouble locating rocks of the right type. Thus, for example, the Sierra Nevada is hopeless, but part of the Coast Ranges offers good possibilities. Streams that drain large areas of basic or ultrabasic rock should always be panned for platinum; gold, at least, might be found. A geologic map or the advice of a local geologist—geology professors are easier to find than platinum—is worth consulting.

Mining platinum from an igneous body is expensive, but the price of the product is also high.

Native platinum and its alloys—all malleable and very heavy—look like silver. This is true even when they are coated with a dark film, as happens in some placer deposits. Platinum, however, is completely inert in nitric acid. The other platinum minerals are silvery to steel gray in color and would certainly require identification by an expert.

The use of platinum in jewelry is known, but few people are aware of other uses for these remarkable metals, uses that keep the market going and doubtless always will. Here is a brief summary: electrical industry, chemical industry, glass industry, jewelry and decoration, dental and medical uses, petroleum industry. Iridium is used more in jewelry than anywhere else; osmium, rhodium, and ruthenium have smaller usage, mainly to harden platinum and palladium. Rhodium makes a good plating metal.

19

Nonferrous Metals

These are often referred to as the base metals, meaning nonprecious or inferior in value to gold and silver (and now platinum) or more active chemically than these. As the term indicates, they must exclude iron. Generally, they are meant to include copper, lead, zinc, tin, and aluminum, but aluminum seems now better included among the light metals discussed in chapter 21. Nonferrous alloys also include alloys of magnesium and nickel. Nickel, however, is included among the ferroalloy metals, which are used mainly to alloy with iron to give steels having special properties. The terms *nonferrous metals* and *base metals* are therefore not synonymous. Copper, lead, zinc, and tin are represented in this chapter. Like iron and the ferroalloys, they require large-scale investment and operations for mining, treatment, and shipping. Copper can be a highly profitable discovery for the prospector, and tin is so little available in the parts of the world where it is utilized that any find promises wealth.

COPPER

Next to gold and silver, copper has been the object of more prospecting than any other metal in history. This beautiful and essential substance was either the first or second (after gold) metal known to man, even though it is not at all common—except in one locality, northern Michigan—in native form. Long after copper was used alone in primitive times—perhaps as early as 6000 B.C.—it was alloyed with other metals to become the basis of brass and bronze: the Bronze Age marks one of the great periods of human culture, its start dating earlier than 3500 B.C.

Nuggets of native copper have been found in a wide range of sizes. *(Filer)*

Copper comes from the two following types of ore:

1. Ore in which copper is the main commercial metal
2. Ore that yields copper as a byproduct

The classification as to origin gives the following main kinds of copper deposits:

1. Magmatic concentration
2. Contact metasomatism
3. Hydrothermal
   a. Cavity filling: as Butte (Montana), Nacozari (Mexico), Braden (Chile), Lake Superior
   b. Replacement
      (1) Massive: as Bisbee (Arizona), Bingham and Tintic (Utah), Noranda (Quebec), Flin Flon (Manitoba), Rio Tinto (Spain), Cerro de Pasco (Peru)
      (2) Lode: as Magma (Arizona)
      (3) Disseminated: as Bingham (Utah), Ely (Nevada), Ray, Miami, Ajo, and Clay at Morenci (Arizona), Santa Rita (New Mexico), Rhodesia (Africa), Braden and Chuquicamata (Chile)
4. Sedimentary
5. Oxidation (but not enrichment): as Chuquicamata (Chile), Katanga (Congo)

From the standpoint of its minerals, copper is derived from the following sources:

1. Native copper: 100 percent copper
2. Copper compounds, as follows:
   a. Sulfides:
      (1) Chalcopyrite: also called copper pyrites, yellow copper ore; $CuFeS_2$, 32 to 35 percent copper
      (2) Bornite: also called purple copper ore, peacock ore, horse-flesh ore; Cu_5FeS_4, 63 percent copper
      (3) Chalcocite: also called copper glance; Cu_2S, 80 percent copper
      (4) Covellite: CuS, 66 percent copper
   b. Sulfosalts:
      (1) Enargite: Cu_3AsS_4, 48 percent copper
      (2) Tetrahedrite: also called gray copper ore, fahlore; $(Cu,Fe)_{12}Sb_4S_{13}$, 30 to 46 percent copper
      (3) Tennantite: $(Cu,Fe)_{12}As_4S_{13}$, 36 to 53 percent copper
   c. Oxides:
      (1) Cuprite: also called ruby copper, red copper ore; Cu_2O, 89 percent copper
      (2) Tenorite: also called melaconite; CuO, 80 percent copper
   d. Carbonates:
      (1) Azurite: also called blue copper carbonate, chessylite copper; $Cu_3(OH)_2(CO_3)_2$, 55 percent copper
      (2) Malachite: also called green copper carbonate; $Cu_2(OH)_2(CO_3)$, 57 percent copper
   e. Chlorides:
      (1) Atacamite: $Cu_2(OH)_3Cl$, 15 percent copper
   f. Sulfates:
      (1) Antlerite: $Cu_3(SO_4)(OH)_4$, 54 percent copper
      (2) Brochantite: $Cu_4(SO_4)(OH)_6$, 56 percent copper
   g. Silicates:
      (1) Chrysocolla: $CuSiO_3 \cdot 2H_2O$, 36 percent copper
3. Other minerals in which copper is a chemical impurity. Pyrite is the chief of these, often containing gold also.

Copper ores are also classified as follows:

1. Native copper.
2. Sulfide ores: sulfides and sulfosalts. Some are primary, of igneous

origin; others are formed by secondary enrichment after weathering has taken place above.

3. Oxidized ores: oxides, chlorides, carbonates, sulfates, and silicates. These are formed by weathering.

4. Complex ores: mixtures of copper, lead, zinc, gold, and silver minerals. These are the most difficult to treat.

It is customary to refer to copper ore as high grade, intermediate, or low grade. These terms do not indicate the mineral itself, for native copper (which is 100 percent metal) may constitute the lowest grade ore, but it is easily treated. Sulfide ore may represent the highest grade (though possibly 30 percent) and is the most valuable.

Geologically, copper occurs in the following ways:

1. In veins and other open spaces.
2. In layered rocks.
3. Disseminations: the porphyry-copper deposits described below.

Besides the gold that may be recovered from copper-bearing pyrite, as mentioned above, the pyrite itself may prove of commercial use. So too might barite, another (but nonmetallic) gangue mineral found in similar deposits.

More than almost any other ore, copper is revealed by the bright green and blue minerals of copper that stain the surrounding rock. More color may be present than metal, however, so do not be misled. Pyrite and other sulfides of iron are typical associates of copper sulfides; a general rustiness (not too red) on a bleached rock is a good sign, as is the presence of cubic cavities in quartz that were once filled with pyrite. Outcrops and float of chalcopyrite or bornite are not to be considered reliable guides to commercial ore. The exceedingly low grade (tenor) of ore that is now the rule in many a multimillion-dollar property has made prospecting for copper a matter of geophysical prospecting—but the color of the ground may still furnish the first clue in future discoveries. If no color occurs in the gossan at the surface, copper is unlikely to be present at depth. But, if present at all, it may exist at shallow depths; limestone, however, may give an exaggerated effect. Copper is the most important metal for which secondary enrichment is significant, but limestone is not favorable for this kind of occurrence. This is altogether a difficult subject, which requires experience and careful observation. If you expect to specialize in copper prospecting, you cannot study gossans too much. In the long run, perhaps, the mining history of a given district may give the most useful information as to what to expect.

Igneous rock of intermediate composition (such as monzonite, quartz monzonite, diorite, granodiorite) is the best host for copper. Limestone beds may be closely associated, especially if altered to green or brown garnet or green epidote. Copper minerals are apt to be found with nearly as wide a variety of rock types as is gold or silver.

The porphyry-copper bodies are fairly much alike, although the country rock may be diverse. These tremendously important sources of copper (and molybdenum) are all low-grade and large-scale deposits, wider than they are deep, and are combinations of many small veins and a speckled arrangement (known as salt and pepper) of scattered minerals. These are termed disseminations. They are accompanied by replacements of the rock, both by ore minerals and by gangue. Later oxidization and enrichment have affected them all to some degree, and the porphyry coppers are more alike than they are different.

Features that are noticeable include a general bleached effect, giving a creamy color to the rock. A crackled appearance is typical, and fracturing is necessary. Visible quartz is usually prominent, perhaps with alunite or anhydrite. Look also for orthoclase feldspar and sericite mica, perhaps biotite mica or tourmaline. Pyrite, chalcopyrite, and bornite are the chief primary minerals. Where these sulfide minerals have been removed by weathering, their places are taken by voids that contain fillings of yellowish limonite in clay. Molybdenite is usually present but does not weather. An alteration halo of pyrite is the most common feature. Below the leached capping are (in most deposits) the enriched minerals, mainly chalcocite or covellite. The brassy or bronzy sulfide minerals in this lower zone may merely be coated with these gray or blue minerals rather than entirely replaced by them. Both the country rock and the intrusion are mineralized.

The copper minerals range greatly in appearance, because there are so many of them. Native copper—which gives a shiny scratch when cut with a knife—may tarnish to malachite, the green mineral that is so prominent in copper-bearing rock. Atacamite, antlerite, brochantite, and chrysocolla are also green. Chrysocolla may be blue, azurite always is, and covellite is dark blue. Cuprite is red and heavy. Chalcopyrite is brassy looking (one of the fool's golds). Bornite is usually seen with its iridescent tarnish instead of the coppery color that shows when fresh. The other minerals named are gray to black. Tests for copper minerals are not difficult to make.

The recent strength in the price of copper indicates its ability to compete with its substitutes. It is a chief metal in the electrical and light and power industries. In building construction, industrial equipment, and motor vehicles, copper finds a large market. Its uses run into the thousands, from coins to missiles, from jewelry to boats.

LEAD

An ancient and highly regarded metal, lead is no less essential in the economy of today. The prospector will do well to consider lead in his search, although you should keep in mind that geochemical and geophysical methods have become important in finding deeper deposits than those that have been of most significance in the past. Furthermore, lead (as well as zinc) may require good economic conditions if the deposit is small, for many such are known, but a large discovery is always of immediate value.

Lead is obtained from the three following types of ore:

1. Straight lead ore: simple mineralization, without byproducts or coproducts.
2. Lead-zinc ore.
3. Lead-silver ore: lead is often the byproduct and silver the principal metal of value. The silver occurs as small particles of silver compounds.

Following are the chief kinds, by origin, of lead deposits:

1. Cavity fillings: as Bawdwin (Burma)
2. Replacement
   a. Massive: as Leadville (Colorado), Bingham and Tintic (Utah), Sullivan (British Columbia), Cerro de Pasco (Peru), Santa Eulalia (Mexico)
   b. Lode: Park City (Utah), Coeur d'Alene (Idaho)
   c. Disseminated

Rarely seen as a native metal, lead is mined chiefly from three compounds, as follows:

1. Galena (lead sulfide), the primary mineral: also called galenite; PbS, 87 percent lead. Galena alters in the zone of oxidation to these two main secondary minerals:
2. Cerussite (lead carbonate): $PbCO_3$, 77 percent lead.
3. Anglesite (lead sulfate): $PbSO_4$, 68 percent lead. Gold and iron minerals also occur with lead in some places.

It is of interest to note that deposits of lead are usually sharply separated into those that are dominantly lead and those that occur with a rather abundant amount of zinc. For example, the lead deposits of southeastern Missouri, the greatest in the world, yield little except galena. In southwestern Missouri, however—extending into Oklahoma and Kansas—the famous Tristate district, the largest zinc district in the world, produces lead also. The mines in the Coeur d'Alene district of Idaho have a good deal of both lead and zinc, as do those in the western United States in general and those in most other countries too.

Again, deposits of lead can be divided into those that contain silver and those that do not. In this case, the price of silver may be of more concern to the miner than the price of lead. The Missouri ores yield no silver, whereas those of Idaho (and Utah, British Columbia, and elsewhere in western America) are rich silver ores as well.

Lead deposits mostly fill open spaces in rock or have replaced the rock as masses, veins, or scattered minerals. They are rare in igneous rock, but beds of limestone (and other rock) when cut by igneous dikes offer likely places to search.

Lead minerals are more or less resistant to weathering. You may therefore usually find them rather near the surface, if any reasonable amount of lead was present, unless oxidation has gone to extreme. Fortunately, certain of the rarer lead minerals of the oxidized zone—especially wulfenite, vanadinite, and pyromorphite—are brightly colored (which the common ones are not) and thus serve as a valuable clue to ore, even when they themselves have little if any commercial merit.

Galena is metallic gray, often bright, always heavy, and has a prominent steplike cleavage. So-called steel galena, however, which has the highest silver content, may not reveal this otherwise typical cleavage. Cerussite is white or gray and very bright. Anglesite is white or gray, either bright or dull. Besides these lead minerals, others are known, and several of the rarer ones are colorful in orange, red, yellow, brown, or green hues, which show conspicuously.

Our word *plumber* comes from the Latin word for lead, and this metal is still used for pipes, as it was in ancient Egypt. Storage batteries, however, consume the largest amount of lead, but most is recovered as scrap. Gasoline takes the next largest (but decreasing) amount, and this is all used up. Many are the miscellaneous uses for lead; paint, ammunition, solder, type metal, and so on and on.

ZINC

"Zinc is a metal that seldom comes to mind except perhaps as an alternative to zebra in giving alphabetical examples," says Andrew Fletcher. Though indeed less well known to the general public than the metals previously discussed, zinc is one of the base metals of value of our industrial civilization. It was a component of brass fully 2,000 years ago. Its present production actually exceeds that of lead, the metal with which it is most often associated. Both metals may require an economy of high prices if the deposit is small, inasmuch as such occurrences are fairly common. A large deposit will always be of interest to a mining company.

Zinc comes from the following two types of ore:

1. Straight zinc ore: interesting byproducts are often recovered commercially; but chemically, they must be considered impurities. These include cadium, germanium, gallium, indium, and thallium.
2. Zinc-lead ore; with galena or its alteration minerals.

By origin, zinc deposits fit into the following main classifications:

1. Contact metasomatism
2. Cavity fillings: as Jefferson City and Mascot (Tennessee)
3. Replacement
   - *a.* Massive: as Leadville (Colorado), Bingham and Tintic (Utah), Sullivan (British Columbia), Flin Flon (Manitoba), Cerro de Pasco (Peru), Santa Eulalia (Mexico)
   - *b.* Lode: Park City (Utah), Coeur d'Alene (Idaho), Ogdensburg (New Jersey)
   - *c.* Disseminated

The only zinc mineral of major economic value is sphalerite (zinc sulfide), the primary one. It is also called zinc blende, blackjack, ruby blende, ruby zinc, blende, rosin jack, resin jack. Its formula is ZnS, and it contains 57 to 67 percent zinc. Oxidized zinc minerals are mined in only a few places and at favorable times; these include smithsonite (zinc carbonate), hemimorphite (zinc silicate), and the remarkable association at Ogdensburg, New Jersey, of franklinite, zincite (both oxides), and willemite (a silicate). Iron, copper, gold, and silver minerals also occur with zinc in some places; the combination of zinc, silver, and copper promises to make the 1964 discovery at Timmins, Ontario, the largest producer of zinc and silver in the world.

Zinc occurs with lead, as in the Coeur d'Alene district of Idaho; at Kimberley, in British Columbia, Canada; and at Broken Hill, New South Wales, Australia. It also occurs with scarcely any lead or other metals, as in eastern Tennessee.

Deposits of zinc, like those of lead, usually fill cavities in rock or are replacements of the rock, or they may be both. Both irregularly layered masses and veins are represented.

In contrast to lead, zinc is quite soluble. Oxidized zinc minerals are therefore seldom seen at the surface. If lead appears in the gossan, however, even more zinc can probably be expected at some depth, especially in carbonate rock. The frequent presence of pyrite causes a limonite stain in the gossan, even though little or no iron may be incorporated in either the galena or sphalerite. The experience of miners in the district should be given attention.

Limestone that is cut by igneous dikes is especially conducive to zinc mineralization. Zinc ores that contain gold are associated closely with igneous rock.

Sphalerite is noted for its resinous luster; the color may be yellow to brown to red, even black or green. The other minerals are quite variable in appearance; smithsonite, for example, may even look like dried bone and be called dry-bone ore.

The main uses for zinc are for die castings and protective coatings on iron and steel: this is known as galvanizing. The making of brass, other zinc compounds, and zinc products that are wrought or rolled uses most of the rest.

TIN

More than any of the metals already discussed, tin has a restricted distribution. The 1,000-mile-long stretch of tin deposits in southeast Asia holds a monopoly of the world's output. It is also true that the Rand deposits of Africa fairly well monopolize the world's gold, but gold is nevertheless widely distributed. This is not so in regard to tin, which is a metal in seriously short supply not only in the United States but in most of the industrial nations on earth. Opportunities awaiting the prospector are good, but the odds are not encouraging.

The following two modes of occurrence yield tin:

1. Placers: cassiterite only. These deposits may be hillslope (eluvial), stream (alluvial), or marine (by submergence).

2. Veins and disseminated (scattered) deposits: cassiterite or (only in Bolivia) stannite.

Classified by origin, tin deposits fall into the following categories:

1. Hydrothermal
   a. Cavity filling: as Bolivia, Malaysia
2. Mechanical concentration: as Malaysia

The mineralization is limited to two species:

1. Cassiterite (tin oxide): also called tin stone, wood tin, stream tin; SnO_2, 79 percent tin.

Cassiterite, the chief ore of tin, is hard and heavy. This specimen is from Perak, Malaya. *(Malayan Tin Bureau)*

2. Stannite (tin sulfide): Cu_2FeSnS_4, 28 percent tin. Three sulfosalt minerals of tin—teallite, franckeite, and cylindrite—occur with stannite in Bolivia, but not significantly. Associated with the nonplacer tin deposits are minerals of numerous other elements, most of these minerals being sulfides.

The rocks in which tin occurs as veins and disseminated minerals are granite and related igneous rocks of silicic composition. These include pegmatite and range to granodiorite and quartz monzonite and the porphyries of these rocks. The other minerals that are found are typical of such rocks, especially pegmatite. Tin is the result of the action of hot, acidic gases, which cause alteration of the wall rock in a peculiar manner, producing a mixture of quartz, mica, and topaz. Learning to recognize this material, called greisen, is a great help in prospecting for tin; furthermore, wolframite (a tungsten mineral) may be present, and so may tourmaline, but this is an indicator mineral of small value. The so-called visor twins of cassiterite crystals are another means of identifying cassiterite. Little except cassiterite is found in the placers, but it may serve as a guide to covered deposits of gold in certain places.

Because of the several appearances it may assume, cassiterite is not always readily recognized, but the unique test previously mentioned is certain: boiled with zinc metal in hydrochloric acid, the tin oxide becomes coated with metallic tin, which brightens when rubbed. The other tin minerals are metallic gray and need to be tested for tin in other ways.

Used in ancient times in bronze, and later in pewter, tin enters industry as a protective coating (tinplate and other products), in solder, bronze, and brass, and in numerous other materials and substances.

20

Iron and Ferroalloy Metals

The most important metal of our industrial civilization is iron. Its alloy with carbon is steel. Special steels, desired for their useful properties, are made by adding the so-called ferroalloy metals. Of these, the major ones are manganese (which is necessary in all carbon steel), nickel, chromium, molybdenum, tungsten, vanadium, cobalt, and titanium. Except titanium, these are covered in this chapter. Titanium is included among the lightweight metals in chapter 21.

IRON

So much taken for granted is iron that its overwhelming importance to industry is seldom considered by the prospector. Iron is the second most abundant metal in the earth's crust (next to aluminum) and occurs in a huge number of minerals, including most of those that give color to the rocks we see. These are found everywhere, but a profitable discovery of iron ore would have to involve millions of tons. For this as much as any other reason, iron is not often the object of search by the lone prospector lacking geophysical equipment. New bodies are almost impossible to find otherwise in settled country. Remote regions, however, may still entice the prospector for iron; any discovery of value would doubtless be one of spectacular size.

The occurrence of iron deposits has been classified geologically in several ways, one of which follows:

1. In sedimentary basins. Combinations of processes have caused some of the complex deposits such as those of Lake Superior. Metamorphism produced banded and layered deposits of taconite (and itaberite and jaspilite), which are scarcely more than rock containing enough iron to work by modern means of treatment.

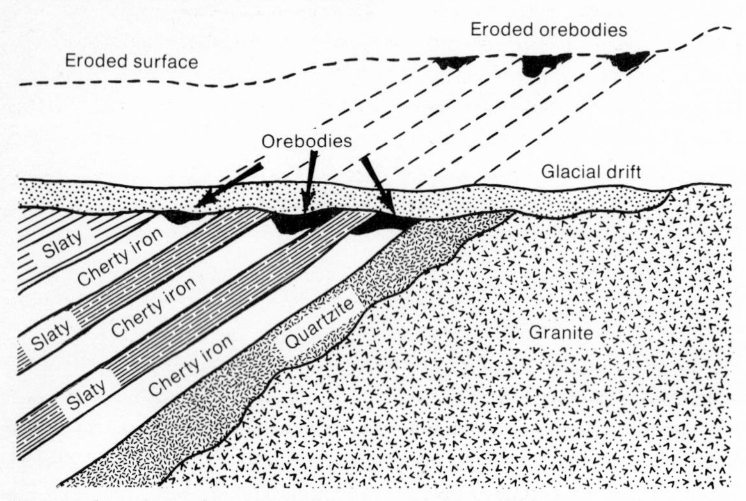

Stratigraphic relationships in iron deposits as in the Mesabi Range, Lake Superior region. The ore bodies erode downdip. *(After Newhouse, Bateman)*

2. With eroded igneous bodies of intrusive origin.

3. From tropical weathering: so-called residual deposits.

The types of iron deposits by the Bateman classification as to origin follow:

1. Magmatic: as Kiruna (Sweden)
2. Contact metasomatic: as Cornwall (Pennsylvania)
3. Replacement: as Lyon Mountain (New York)
4. Sedimentary: as Clinton ore at Birmingham (Alabama), Wabana (Newfoundland), Minette ore in Lorraine-Luxembourg

5. Residual concentration: as Lake Superior, Pao (Venezuela), Labrador-Quebec

6. Oxidation

The following classification of the ore itself is usually employed in the mining industry:

1. Hematite (red ore).
2. Magnetite (magnetic or black ore).
3. Limonite (brown ore). This is a group name, the material being mostly goethite (referred to as limonite) or a mixture of minerals.

Various other classifications of iron ore are used in the industry: iron ore pellet; natural ore versus dry ore; crude ore versus usable ore; bessemer ore, nonbessemer or low-phosphorous ore; high-phosphorous ore, manganiferous ore, and siliceous ore; lump ore, fine ore, high-sulfur ore, high-silica ore, titaniferous ore. These terms are not of interest to the prospector except to suggest the importance of the minor aspects of the composition of iron ore; buyers' specifications are rigidly adhered to, and otherwise similar deposits of iron may have quite unlike appeal to the prospective user, apart from their iron content.

Of the many iron minerals, the following are the commercial ones:

1. Hematite: the most abundant; also called red iron ore, specular iron, red ocher, specularite, micaceous iron ore, kidney ore; Fe_2O_3, 70 percent iron.

2. Magnetite: the richest; also called magnetic iron ore, lodestone; $FeFe_2O_4$, 72 percent iron.

3. Goethite: usually called limonite, which is a good field term for hydrous and mixed minerals; also called bog iron ore, brown hematite; $HFeO_2$, 59 to 63 percent iron.

4. Siderite: also called spathic iron, chalybite; clay ironstone, black-band ore; $FeCO_3$, 48 percent iron. Limonite and siderite are mined in other countries but are too low grade to have wide use in the United States. Several silicate minerals (such as chamosite) also occur in the taconite and related types of ore. Even pyrite and pyrrhotite (which are iron sulfide) have served as an ore of iron when needed, perhaps after being changed to iron oxide. Byproduct iron that is obtained with the mining of nickel and base metals brings a premium price when separated, roasted, and sintered.

The magnetic nature of iron deposits is the most important feature to the prospector; the stories about early-day discoveries in the Great Lakes region emphasize the compass. You would do well to review the history of exploration among the so-called iron ranges in Minnesota and Michigan before prospecting elsewhere; the same methods have been used successfully in Brazil, Labrador, and in other places. These involve the following steps, as given by Stanley A. Tyler and Hugh E. McKinstry:

1. Outline the extent of the iron body: make a magnetic map, using a dip needle. (Not all magnetic bodies represent iron ore, by any means.) The great Steep Rock Lake deposits in Ontario were opened in 1944 after Jules Cross, a "persistent prospector" (indeed!), had made a dip-needle survey of the lake from the ice. The Seine River had to be diverted and the lake pumped dry, in order to begin mining, but these were done.

2. Determine the favorable areas, eliminating the rest: select areas most likely to be oxidized and leached because of wide outcrops and the presence of fractures and faults.

3. Test the favorable areas by drilling.

The original rock in which primary iron minerals form is basic: dark and heavy and of course magnetic. Placer deposits of magnetite may be made commercial by beneficiation, but this is seldom done.

No matter what their appearance, hematite always gives a red streak, magnetite is always magnetic (the lodestone variety acts as a magnet itself), goethite and limonite always give a yellowish-brown streak, and siderite always fizzes in hydrochloric acid but only when hot. Other common minerals bear some resemblance to these, but so abundant are the iron ore minerals that they are recognizable even to beginners.

The uses of iron should require no explanation. It is used in paint and other products and alone as a metal. Steel is a refined alloy of iron and up to 1.7 percent carbon; alloy steel contains specified percentages of the metals that are discussed in the next chapters. It is imperative that the finder of an iron deposit investigate his exact potential market before spending much on developing a mine.

MANGANESE

This is the most important of the ferroalloy metals, because it is essential in the manufacture of all carbon steel. Additional amounts make special steel of wide usage.

Most commercial deposits of manganese are of necessity large-scale ones, of the size typically associated with iron deposits. Their origin, in fact, is often much alike, and they may occur together. Lacking the simple advantage of magnetic properties that iron has, manganese has less of an appeal to prospectors. Its minerals, moreover, are, with few exceptions, neither attractive to look at nor especially distinctive. The huge requirements of the steel industry for this metal, however, and the relative scarcity of it in North America, Europe (outside the Soviet Union), and Australia make manganese a strategic material worth keeping in mind.

The kind of deposit that offers the best possibilities to the prospector is the bog type, which occurs in basins, usually a few feet or yards in depth, or as small benches, or terraces, on gently sloping hillsides. Manganese minerals also are found in beds of limestone, where they branch into veins and pipes.

By origin, the following kinds of manganese deposits may be named:

1. Hydrothermal: as Butte and Philipsburg (Montana)
2. Sedimentary: as Russia
3. Residual concentration: as India, Ghana, Brazil
4. Metamorphic: as Postmasburg (South Africa), India

Manganese ore is classified in several ways. One difficulty with such arrangements is that the long-time usages of certain common names are no longer scientifically acceptable. Field terms do not correspond to modern laboratory identification.

The U.S. Bureau of Mines has three groups, as follows:

1. Manganese ore: 35 percent or more manganese
2. Ferruginous manganese ore: 10 to 35 percent manganese
3. Manganiferous iron ore: 5 to 10 percent manganese

The last two groups are included under manganiferous ore. So-called synthetic ore is manganese dioxide that is produced chemically or electrolytically and has the same uses as natural ore.

The United States national stockpile specifications call for three major grades of ore: (1) metallurgical, (2) battery, and (3) chemical. Only the first two of these are recognized by industry, the third being classed in statistics as miscellaneous.

The many minerals that contain manganese are often difficult to distinguish from one another. So intermixed are they that the term *wad* is applied in the field to bog deposits of hydrous manganese oxides in general or to soft, impure oxides in which those of manganese are dominant and in which sand, clay, and barite are the usual impurities. These kinds may also be classed as (1) psilomelane type (the hard, massive oxides) and (2) soft oxides. They include the following minerals:

1. Pyrolusite: MnO_2, 63 percent manganese

2. Manganite: MnO (OH), 62 percent manganese

3. Psilomelane: $(Ba, H_2O)_2Mn_5O_{10}$, 45 to 60 percent manganese; cryptomelane has potassium replacing barium

4. Hausmannite: $MnMn_2O_4$, 72 percent manganese

A carbonate, rhodochrosite ($MnCO_3$), is another important ore of manganese, as is a silicate, braunite $(MnSi)_2O_3$. Manganese may rarely be recovered from other minerals that are of greater value for other metals, such as heubnerite (for tungsten) and franklinite (for zinc). Other minerals, such as the silicate rhodonite, for example, may be mined for manganese when it is associated with the oxides.

Geologically, manganese deposits are varied and include veins and replacement bodies (igneous), sedimentary (including beds and disseminations), and metamorphic, as well as the residual deposits that are derived from these others by weathering. There seem to be no special techniques for prospecting for manganese. Its presence may indicate copper, silver, gold, or zinc, and should be investigated carefully, especially for copper. The rather abundant finds of manganese nodules on the sea floor are of no concern to the prospector and as yet have not proved commercial anyway.

The manganese oxides are black, with a somewhat bluish cast, and the soft ones may be sooty. Rhodochrosite is typically pink and fizzes in acid.

Manganese metal has, curiously, no use of its own as a metal. Yet, its use as a metallurgical material has no substitute. Such a situation is probably unique among all the materials known to man. In addition to its use in metallurgy in the form of ferromanganese, other ferroalloys, and manganese metal, this element is employed chiefly in making dry-cell batteries and chemicals. Different grades are used for different purposes, and so the finder of a mine needs to investigate his market before opening a mine.

NICKEL

Used abundantly in coins of a century ago—and many centuries ago, for that matter—nickel has come into its own as one of the most important of the ferroalloy metals of our time. It consequently is a prime target for prospectors, even though its distribution seems restricted to a few places in the world.

Three classes of nickel deposits exist, as follows:

1. Nickel-copper sulfides
2. Nickel silicates
3. Nickel-bearing serpentine and laterite

Classified by origin, deposits of nickel include the following:

1. Hydrothermal
   a. Replacement: as Sudbury (Ontario)
2. Residual concentration: as New Caledonia

The mineralogy of nickel is outlined as follows:

1. Pentlandite: $(Fe,Ni)_9S_8$, 34 to 35 percent nickel. So intimately is it associated with pyrrhotite that this last mineral has been wrongly said to be a nickel ore itself; the mixture, of course, is.

2. Garnierite: $H_2(Ni,Mg)SiO_4;nH_2O$, 1 to 4 percent nickel.

Nickel is a typical metal in basic rocks: those that are heavy and dark, often dark green, and high in iron and magnesium. It is therefore commonly found with chromium, platinum, and cobalt, which likewise favor this type of rock. The prospector should be aware of this natural combination of minerals. In addition, nickel ores occur with gold, silver, tellurium, and selenium in the Sudbury type of ore, which is found rather similarly at Petsamo (U.S.S.R.) and Lynn Lake (Manitoba). This kind of sulfide ore contains a large amount of chalcopyrite, which is described under copper.

The weathered type of nickel ore—most characteristic of New Caledonia but also of Cuba, the Philippines, and elsewhere—can be expected only in the upper levels of the peridotite (or dunite) or serpentine from which the metal has been derived. An apple-green stain, from the garnierite, is often a vital clue to nickel. It should be emphasized that most peridotite or serpentine does not contain nickel, and most pyrrhotite is not associated with pentlandite.

Pentlandite (as well as pyrrhotite) has a bronzy color and luster; it is not magnetic, but the pyrrhotite is, and so may the mixture be. Garnierite is apple green.

Apart from its dominant use as a ferroalloy, which puts it in this chapter, nickel is alloyed in larger amounts with other metals than iron. Electroplating consumes much of the metal, and there are actually hundreds of miscellaneous uses. A nickel mine must have facilities for smelting the ore, and these must be arranged for in advance.

Profile of nickel-bearing laterite, as in New Caledonia. *(After Chéte-lat, Park, and MacDiarmid)*

CHROMIUM

In a neon-lighted, gilt-edged, chromium-plated world, this metal can scarcely be ignored. More of it is used in ferroalloys than any of the others. Its restricted geology makes it possible for the prospector to concentrate on the few types of occurrence that are likely to contain it. These are much like those of nickel. Like manganese, chromium is found mostly in the nonindustrial countries, which have little use for it. This fact should stimulate the prospector in his search for it.

The following are the three types of chromium deposits:

1. Masses in peridotite or serpentine: these may be of either igneous or metamorphic origin.
2. Layered bodies in basic igneous rock.
3. Chromium-bearing laterite.

The origin of chromium deposits can be classified as follows:

1. Magmatic concentration
   a. Disseminated
   b. Segregated: as Rhodesia, South Africa, U.S.S.R., Turkey
2. Residual concentration
3. Mechanical concentration

Chromium mineralogy is exceedingly simple: one mineral, chromite, is the only ore; its formula is $FeCr_2O_4$; 36 to 46 percent chromium.

Basic rocks are the home of this metal, whether they are igneous or metamorphic in origin. These are peridotite, serpentine, pyroxenite, amphibolite. The deposits are in some places of enormous size, spread

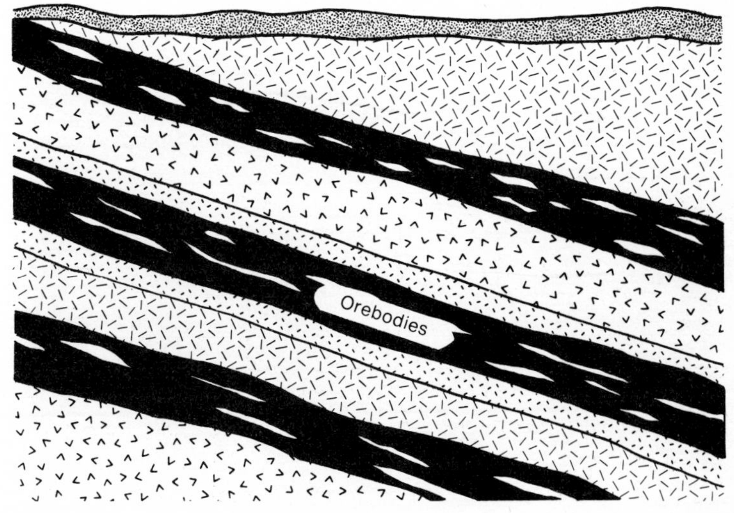

Chromium ore bands in basic igneous rock, as in the Bushvelt of Africa. *(After Zeschke)*

over tens of miles: the Great Dyke of Rhodesia, at the extreme, is 330 miles long and has bands of chromite in and near it for great distances. There is no use prospecting for chromium in any other geologic environment than this, or in places where such rocks have weathered, or in heavy stream sands eroded from serpentine. In turn, every area of peridotite and serpentine should be examined for chromium, as well as for platinum, nickel, and cobalt.

Chromite is black, but its streak is chocolate brown, and the mineral is sometimes weakly magnetic. No other similar-appearing mineral (magnetite, ilmenite, franklinite, hematite) will test for chromium.

Next to the use of chromium as an alloy metal (especially in stainless steel) is its use as a refractory, this also being in metallurgy. The chemical industry uses most of the rest that is produced. The price that is paid for chromium ore depends on the grade, for each of the three uses mentioned above requires a different grade, which is based on the ratio of chromium to iron in the raw material.

MOLYBDENUM

More a metal of recent significance than any of the others that have been discussed so far, molybdenum is also, more than any, including even silver, a metal of the Western Hemisphere. A new mine comparable to that at Climax, Colorado, which has given up more than half of the world's supply since 1925, has only lately been discovered in the Colorado Rockies at Urad, and it would seem that prospecting for molybdenum has by no means run its course.

Following are the two kinds of deposits:

1. Quartz veins
2. Porphyry-copper type: disseminated through the intrusive body and country rock, with small veins

The main kinds of molybdenum deposits according to their origin are the following:

1. Magmatic concentration
   a. Disseminated (in pipes)
   b. Residual liquid injection (in pegmatite)
2. Contact metasomatic
3. Hydrothermal
   a. Cavity filling

b. Disseminated replacement: as Climax (Colorado), Bingham (Utah), Santa Rita (New Mexico)

The mineralogy is limited almost to molybdenite, which (like the metal itself) is often referred to by miners as Molly or Molly-be-damned. Its formula is MoS_2; 60 percent molybdenum. A substitution of molybdenite for tungsten in scheelite grades toward powellite, which can yield both metals. Minor production comes incidentally from wulfenite, ferrimolybdite, and ilsemannite.

In distinct chemical contrast to nickel and chromium (just discussed) and to platinum—which are typical of basic rock—molybdenum is a characteristic metal in silicic rock. Tin and tungsten are associated with it in some places. Granite and quartz monzonite are thus the typical home of molybdenum ore, which penetrates the surrounding rock as well. The term *porphyry copper* that describes the large-scale, low-grade deposits of copper that are mined so widely on every continent also identifies the main kind of deposit in which molybdenum occurs. You ought to become familiar with it if prospecting for either metal. A yellowish to greenish buff color on rock suggests that molybdenite lies below; the shade tends to maroon when influenced by iron, as from pyrite. Powellite, when present, fluoresces in ultraviolet light. If lead is present, wulfenite forms, this mineral being brightly colored in a range of hues.

The mineral molybdenite is gray or black and soft-looking and marking paper like graphite but containing both sulfur and molybdenum.

By far the major output of molybdenum is used in alloys. The rest appears as plain metal and chemicals, allowing a small amount for paint and miscellaneous applications. Industry classes molybdenum concentrates according to their use.

TUNGSTEN

Either because of, or in spite of, its erratic record of demand and supply, tungsten (also called wolfram) has been a profitable item for prospectors. This metal comes into prominence whenever there is a surge of international trouble. A favorable aspect for the prospector is the fact that one of the major tungsten minerals, scheelite, is brightly fluorescent; its presence, furthermore, even in small amounts, usually indicates that other tungsten minerals are close by.

In terms of their economic status, two kinds of deposits are known, as follows:

1. Straight tungsten ore: other metals—especially tin, but also silver, copper, lead, zinc, antimony—may be obtained as byproducts.
2. Byproduct ore: with molybdenite especially in bodies of porphyry-copper type.

As to geologic origin, the following kinds of deposits yield tungsten:

1. Magmatic concentration
   a. Residual liquid injection (in pegmatite)
2. Contact metasomatic: as Mill City (Nevada), Pine Creek (California), Sandong (Korea)
3. Hydrothermal
   a. Cavity filling: as Kiangsi (China), Bolivia, Hamme (North Carolina)
   b. Replacement: as Yellow Pine (Idaho), Kiangsi (China), Malaysia
4. Mechanical concentration

Two groups of minerals, as follows, serve as tungsten ores:

1. Calcium tungstates:
   a. Scheelite: $CaWO_4$, 64 percent tungsten
   b. Powellite: $Ca(Mo,W)O_4$, 1 to 8 percent tungsten

Cuprotungstite alters from scheelite and is mined with it to a small extent; so may a few other tungsten minerals.

2. Iron-manganese tungstates (black ore): these are subdivided arbitrarily, but ferberite looks different from the other two.
   a. Ferberite: $FeWO_4$, 61 percent tungsten
   b. Wolframite: $(Fe,Mn)WO_4$, 61 percent tungsten
   c. Huebnerite: $MnWO_4$, 61 percent tungsten.

Very much like molybdenum and tin, tungsten is nearly always associated with siliceous rocks such as granite and prophyries. Its distribution is more irregular than that of most other metals. Whereas lode deposits—filling cavities as fissure veins—are the most important sources of tungsten on a worldwide basis, those in the United States are mainly of contact metamorphic (metasomatic) origin. The newly formed rock, called tactite, is found between bodies of granite and limestone. Prospecting for tungsten ore of this type has been very successfully done by locating intrusions of granite on geologic maps, where they have been noted as surface exposures. The tactite that contains tungsten may even form a halo that surrounds the granite and is almost always within 1 or 2 miles. Owing to its relative solubility, tungsten does not accumulate well in placers, but enough may collect nearby to justify panning in order to trace the source, as explained in chapter 7. Tactite is recognized by its complex silicate minerals, of which one may well be fluorescent scheelite. Prospecting for luminescent minerals is described in chapter 9.

Scheelite otherwise resembles greasy quartz but often with a yellowish or brownish tinge. Ferberite is black, sometimes occurring as shiny crystals, usually as thin veins. The rest of the wolframite series is yellowish brown to black; testing for tungsten may be necessary for these heavy minerals.

Tungsten has its largest use in cemented or cast carbides. Next is its employment in alloys, particularly high-speed steel. Tungsten metal uses most of the rest.

VANADIUM

Most of the world's supply of vanadium, and certainly most of the American supply, has come during the past two decades as a byproduct or coproduct of the mining of uranium. The ores of the Colorado Plateau characteristically contain both elements. A decline in uranium production has caused a proportional decrease in the output of vanadium, so that other sources are needed. This is especially true of high-grade ore of the sort that prospectors are most apt to be able to find without elaborate equipment. Low-grade deposits are, in fact, generally too expensive to bring into profitable operation except in time of war.

Vanadium comes from three main kinds of ore, as follows:

1. Straight vanadium ore: no byproducts
2. Byproduct ore: mined for uranium, lead, zinc, copper, iron
3. Sedimentary-rock byproduct: in clay, shale, phosphate rock, bauxite (aluminum rock), coal, petroleum

A classification by origin gives the following main types of vanadium deposits:

1. Magmatic concentration
2. Hydrothermal
   a. Cavity filling: as Colorado Plateau
   b. Replacement: as Colorado Plateau
3. Residual concentration: as Mina Ragra (Peru), Colorado Plateau

The vanadium minerals belong to the following groups:

1. Sulfides:
   a. Patronite: may be a simple sulfide, VS_4; or a complex sulfide; or a mixture with asphaltite. The large deposit in Peru was an unusual though not unique occurrence.
2. Micas:
   a. Roscoelite: $K_2(Al,V)_2(AlSi_3O_{10})(OH)_2$, 1 to 2 percent vanadium
3. Uranium minerals:
   a. Carnotite: $K_2(UO_2)_2(VO_4)_2 \cdot 3H_2O$, 11 to 12 percent vanadium
   b. Tyuyamunite: $Ca(UO_2)_2(VO_4)_2 \cdot nH_2O$, 10 to 12 percent vanadium
4. Lead vanadates:
   a. Vanadinite: $Pb_5(VO_4)_3Cl$, 10 to 11 percent vanadium
   b. Descloizite: $(Zn,Cu)Pb(VO_4)(OH)$, 10 to 13 percent vanadium

Deposits of uranium that also contain vanadium in sedimentary rock (no vanadium mineral contains uranium) are prospected for by the same methods that are used to discover radioactive materials. These are discussed in chapter 8. The sandstone in the Colorado Plateau has a tendency toward dark olive green, especially if wet, when vanadium is present. Other vanadium ores are looked for in connection with the other metals with which they are associated, especially lead, but also zinc, copper, and iron. The vanadium content of sedimentary rock will have to be determined by chemical analysis, and petroleum has little interest to most readers of this book.

Patronite is black and pitchy looking, found in red shale. Roscoelite is olive green and is seen in fine flakes. Carnotite and tyuyamunite are yellow and radioactive; these minerals are described further in chapter 22. Vanadinite is red to yellow, resembling pyromorphite and mimetite, and descloizite is brown to red to orange; both are bright minerals that are found with similar ones in the upper zones of metal deposits. Descloizite grades into mottramite, which may be mined with it.

Besides its use as an alloy in steel and other (nonferrous) metals, vanadium is important in several ways in the chemical industry.

COBALT

Although cobalt is a ferroalloy metal, it is not one that offers much opportunity for prospectors. Most cobalt of commerce is a byproduct of nickel, silver, or copper mining, for the cobalt minerals are seldom found in large enough amounts to be extracted alone. They occur in silver-colored arsenides and sulfides, and black oxides. Stains and coatings of red to pink (peach-blossom) colors, known as cobalt bloom—which is the mineral erythrite—serve as a clue to the metal and were partly responsible for the recognition of the enormously rich silver camp of Cobalt, Ontario. The discoverer of any metal deposit in which cobalt is likely to occur, especially nickel, copper, and silver, would of course inquire about marketing it.

TANTALUM

This metal, having its largest use in electronic components and chemical equipment (no longer as a steel alloy), comes mainly from pegmatite and always with columbium (niobium). Placer deposits that are derived from pegmatite or normal granite also may contain tantalum minerals, of which the following are examples: tantalite-columbite, microlite, formanite, stibiotantalite, tapiolite, euxenite, eschynite, samarskite, betafite, and rutile. These are all oxides, generally black and heavy. Tantalite, microlite, and euxenite are the chief ore minerals. Some of the occurrences of tantalum have resulted from the search for uranium, for certain radioactive minerals (among those named above) contain both elements.

COLUMBIUM

Invariably associated with tantalum, but having its main use still as a steel alloy, this metal is also called niobium. It occurs principally in pegmatite, in placer deposits that are derived from pegmatite or ordinary granite, and in a range of alkalic rocks. Some of the last mentioned are unusual ones, including the so-called carbonate-rich rocks of apparent igneous (magmatic) origin and occurring in regions of alkalic rocks. Following are examples of columbium minerals: columbite-tantalite, pyrochlore, fergusonite, stibiocolumbite, mossite, euxenite, eschynite, samarskite, betafite, and perovskite. These are all multiple oxides, generally black and heavy. Pyrochlore, which may be brown, is the leading source. Cassiterite (tin) and gold sometimes are obtained with columbium. Certain radioactive minerals, some of which are named above, contain this metal.

21

Lightweight Metals

Four exceptionally light metals, strong for their weight, have come into great prominence for structural use during the 20th century. These are aluminum, magnesium, beryllium, and titanium.

ALUMINUM

The oldest and best known of the light metals, aluminum is the most common metallic element in the earth's crust. However, only one material, called bauxite, yields it commercially, although numerous others have been tried without success to date. Bauxite also has other uses besides being the sole ore of aluminum. This rock—for so it is, rather than a true mineral—requires large-scale mining operations and enormously expensive plant facilities.

Bauxite is the result of residual weathering, which has taken place at or near the surface under special climatic conditions. These conditions include a humid climate of a tropical or subtropical nature; a groundwater circulation that permits the removal of dissolved matter; and, of course, the appropriate rock types on which this process can work. Therefore, bauxite deposits, though widespread in the tropics and temperate zones, are somewhat restricted in their occurrence. This explains why it would seem unprofitable to search for bauxite in any part of the world that is not now warm or that was not warm during the Tertiary Period just before the Ice Age. Other deposits have presumably been eroded away. Canada, for example, is the third-largest processor of aluminum metal in the world (because of its abundant water power) but does not mine a single pound of ore.

The deposits of bauxite, of remarkably similar origin, exist as nearly flat blankets of ore close to the surface, as layers or lenses between other strata, as irregular pockets in limestone or dolomite, and as transported (removed and redeposited) sedimentary beds.

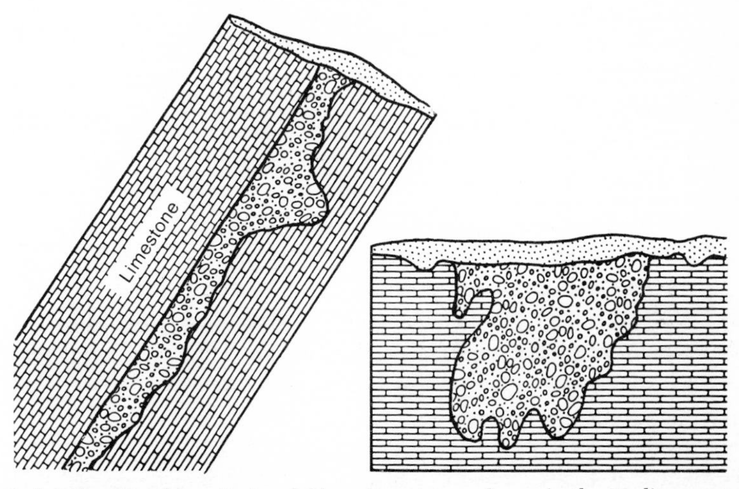

Pocket deposits of bauxite in soluble sedimentary rocks, as in the Mediterranean region. *(After de Lapparent, Bateman)*

As a rock, bauxite is a mixture of several hydrous aluminum oxide minerals—boehmite, $AlO(OH)$, diaspore, $HAlO_2$, gibbsite, $Al(OH)_3$—in varying proportions, which accordingly form the so-called Jamaican, Surinam, and European type of ore. A certain amount of iron oxide is present as the material grades into laterite, the characteristic residual soil of the tropics. Impurities include silica, silt, clay, and titanium oxides.

Three types of bauxite exist, as follows:

1. Oölitic or pisolitic
2. Sponge ore
3. Amorphous or clay ore

The first kind is the best known and, when it is found, is among the most readily recognized of all mineral substances. With its lumps of concretionary materials and pits where such nodules have fallen out, bauxite seems uniform enough to be regarded as a mineral, although the mixed composition has long been proved. *Bauxite* is at least a good field term, similar in application to *limonite* and *wad.* The color grades from white to red to brown with an increasing iron content.

The main uses of aluminum are familiar in wrought and cast products. The chemical and other industries also consume much aluminum. Bauxite, the raw material (which is turned into alumina, Al_2O_3), has subordinate uses in abrasives, refractories, chemicals, and for miscellaneous purposes, even for special cement and insulation.

MAGNESIUM

A metal that comes mostly from seawater scarcely seems like a fruitful object of search by the prospector. Magnesium, however, is also produced from the rock called dolomite or dolostone. This widespread sedimentary rock has other uses as well, and a prospector should know about it.

Although dolomite can be deposited by the evaporation of seawater under a particular set of conditions, most dolomite was probably once limestone or coral that has been replaced by the addition of magnesium from circulating groundwater. A sedimentary terrane is the only place where dolomite can be found.

Dolomite is usually white to gray, taking on other colors when impure. Powdered, it fizzes in warm acid. The formula is $Ca,Mg(CO_3)_2$; 13 percent magnesium.

In other parts of the world, outside of the United States, magnesium metal is obtained from other rocks. Even in the United States, magnesium compounds (but no longer the metal itself) are extracted from other minerals—brucite $(MgOH)_2$, magnesite ($MgCO_3$), olivine, or chrysolite $(Mg,Fe)_2SiO_4$—as well as from seawater, marine bittern (residual liquid upon evaporation), lake brine, well brine, and oyster shells.

Magnesium, as the lightest metal known, has grown into a major structural metal for all sorts of uses. The element also serves as a reducing agent for other metals, as a scavenger, as a deoxidizer, and as a protector of iron and steel.

Dolomite and the other magnesium minerals that are named above enter the metallurgical field in a number of ways and are used in construction and agriculture.

BERYLLIUM

One of the relative newcomers into the world of light metals, beryllium has been included in the atomic-energy field because of certain of its nuclear properties. It is, however, not a radioactive metal and should not be classed with those that are.

Nevertheless, a device closely related to the Geiger counter—one brand is called Berylometer—has been invented for prospecting for beryllium. This instrument, often portable, detects the element in any form. Radioactive antimony produces gamma rays, which convert common beryllium into a lighter isotope, releasing neutrons, which are counted by a scintillator.

This device, combined with some of the newer uses for beryllium, has brought the metal into view as one that is of nuclear interest. It has certainly made it of especial concern to the prospector, who, however, faces a situation that is as much a source of confusion to him as the Berylometer is a means of assistance. The problem referred to is this: after a long time during which only one mineral—beryl—often readily recognized when found, was practically the sole commercial ore, the element was discovered in nondescript mixtures that are difficult to identify and nearly impossible to name. Whereas beryl was known only from pegmatite, these new mixtures occur in ordinary granite, quartz monzonite, limestone, volcanic rock (altered to clay), and fluorspar. Their existence cannot be predicted in advance.

An occasional pegmatite is mined only for beryl. More often, beryl is a coproduct or byproduct of the mining of feldspar, mica, cassiterite (tin), columbite-tantalite, lithium minerals, or gemstones. Beryl itself is also a major gem mineral, which is described in chapter 25. In places, beryl is set aside for later shipment; in some of these places, rockhounds seem to arrive on collecting expeditions just before shipping day. Pegmatite mining is frequently a profitable operation on a small scale. Beryllium, however, presents health hazards as a toxic substance that causes lung disorders.

Beryl occurs in zoned pegmatites, the larger crystals being near the core, the smaller ones being near the outer zone. Sometimes, the zones are marked by certain shoots that are richer in beryl than others. Fractures may be filled with beryl, and some bodies represent later replacements of an earlier stage that did not contain beryl.

Characteristically, beryl is found in six-sided crystals, which may at times weigh hundreds of pounds. A pale-greenish cast is typical, but as the mineral is leached by weathering, it turns white and chalky, losing part of its beryllium content, which may reach as much as 5 percent. Apatite crystals look like beryl but are softer; massive beryl closely resembles quartz. Beryl is a beryllium aluminum silicate, $Be_3Al_2(SiO_3)_6$.

The other, usually indistinguishable minerals of beryllium include bertrandite, phenacite (phenakite), barylite, and chrysoberyl.

As a light metal, beryllium can hardly compete with aluminum, but its industrial uses are expanding with continued research. Its alloys, especially with copper—in which its earliest triumph was scored—furnish the chief market, for here is a light metal with various remarkable properties that rival or exceed those of steel. Beryllium chemicals have some uses too.

TITANIUM

This is—like iron and manganese, for example—a metal that is largely obtained by large-scale methods of recovery. The discovery of a titanium deposit may therefore lead to a big operation. Most mining of titanium is rather well integrated, so that a company that uses it is apt to own the full range of extraction, processing, and production facilities. The lone prospector has little opportunity in this business unless he sells out to a well-established firm. This does happen, however.

Titanium deposits are classified according to their geologic form as follows:

1. Rock deposits
2. Sand deposits

Likewise, two different minerals are the commercial source of titanium, as follows:

1. Ilmenite: also called titanic iron ore, menaccanite, $FeTiO_3$, 29 to 35 percent titanium.
2. Rutile: TiO_2, 60 percent titanium

Each of the above minerals has an entirely different use, as given below. Furthermore, each of them has its own characteristic type of occurrence except when they are mined together in sand deposits (occasionally in firm rock). Otherwise, ilmenite is taken from solid rock, in which it may be found alone or with iron ore (magnetite, hematite). Rutile, on the other hand, is recovered almost entirely from placer sand, although it occurs in many rocks of all types.

Other titanium minerals are associated with ilmenite and rutile and often get into the concentrates, but they are not mined deliberately. These minerals include anatase and brookite (oxides) and leucoxene (an alteration). Still other titanium minerals—including sphene, perovskite, pyrophanite—are found nearby, but have not been used. With titanium in certain places, monazite (thorium), cassiterite (tin), zircon, and staurolite are recovered at the same time.

When occurring with iron, titanium belongs with basic rock, especially anorthosite. Ilmenite when alone (without iron ore) is found in schist and in a dominantly ilmenite-bearing rock called nelsonite. Rutile needles in blue quartz have served as a clue for the prospector in Virginia.

Ilmenite is black, more or less metallic; a chemical test may be needed to tell it from franklinite (manganese) or chromite (chromium). Magnetite is strongly magnetic, and hematite has a red (or redder) streak. Rutile is red to reddish brown to black, more or less metallic, and bright; a titanium test distinguishes it from garnet and cassiterite.

The main uses of titanium are few in number. Its light weight as a metal puts it in the present chapter. This use consumes about half the total amount of rutile. The other half is mostly for welding-rod coatings, a peculiarly specific use. Nearly all the output of ilmenite goes into titanium dioxide pigments, which are greatly superior to most other white coloring matter.

22

Radioactive Metals

Uranium and thorium are metallic elements, but their metallic nature is of little concern to the prospector. For uranium, it is of no concern at all, for uranium has no present or likely future use apart from atomic fuel. Its former uses as a source of radium, a steel alloy, and a coloring agent in glass and ceramics are of historic interest only.

The general aspects of prospecting for radioactive materials were covered in chapter 8, which should be read closely with this chapter. As explained before, many natural elements have radioactive isotopes, but only uranium and thorium are involved in the production of nuclear energy. Thorium is not now employed in the United States, and its future as a nuclear fuel is not certain, even though it is being used to some extent in a few other countries. It should be considered on a long-time basis at least.

Here, then, we deal with the mineralogy and geology of uranium deposits particularly, and the techniques that are especially applicable to uranium prospecting. Emphasis is placed on the principal kinds of productive deposits in the western United States, where most American production has taken place. The minerals and geology of Canada and Australia are also discussed, more briefly.

Of the more than 100 minerals that contain uranium (about 185 are uranium or thorium minerals), only a few have commercial significance. A considerable amount of uranium is doubtless obtained in forms that are not yet correctly identified by name; and large reserves of uranium exist as chemical impurities, in uncertain forms, in rocks that offer large-scale opportunities for extraction in the future. These include phosphate rock (in Florida), black shale (the Chattanooga Formation), lignite coal, copper ore (in southwestern United States), and—already being used as a source—the gold-bearing rock of the Witwatersrand, Republic of South Africa. The condition in which uranium is present in these deposits is by no means fully understood; the uraniferous lignite of North Dakota, South Dakota, and Montana, for example, has uranium in association with asphaltic hydrocarbon and other carbonaceous materials.

The definite uranium minerals may be classified in several ways. There are primary minerals and secondary minerals derived from them by oxidation. There are oxides, carbonates, sulfates, phosphates, vanadates, and silicates. Some minerals are compounds of uranium; others contain only small amounts of uranium as chemical impurities. Certain of the minerals are pretty well distributed worldwide, whereas others seem to be confined to particular countries.

The primary uranium minerals are as follows:

1. Uraninite: uranium oxide, UO_2, usually with U_3O_8, variable, 46 to 88 percent uranium.

2. Pitchblende: the massive, colloform variety of uraninite; impure uranium oxide, $UO_2 + U_3O_8$.

3. Coffinite: uranium silicate, $U(SiO_4)_{1-x}(OH)_{4x}$, 41 to 60 percent uranium.

4. Brannerite: uranium oxide, UTi_2O_6 and variable, 26 to 44 percent uranium. This largely Canadian mineral may be at least partly a very fine mixture of two simpler uranium and titanium minerals.

5. Davidite: iron oxide, $FeTi_3O_7$, variable, 0 to 4 percent uranium.

The secondary uranium minerals are as follows:

1. Carnotite: potassium uranyl vanadate, $K_2(UO_2)_2(VO_4)_3 \cdot 3H_2O$, 53 to 55 percent uranium.

2. Tyuyamunite: calcium uranyl vanadate, $Ca(UO_2)_2(VO_4)_2 \cdot 5\text{-}8H_2O$, 49 to 54 percent uranium. Metatyuyamunite is similar.

3. Torbernite: cupric uranyl phosphate, $Cu(UO_2)_2(PO_4)_2 \cdot 12H_2O$, 47 percent uranium. Metatorbernite (with $8H_2O$) is similar.

4. Autunite: calcium uranyl phosphate, $Ca(UO_2)_2(PO_4)_2 \cdot 10\text{-}12H_2O$, 48 to 50 percent uranium. Meta-autunite (with $2\text{-}6H_2O$) is similar.

5. Uranophane: also called uranotite; uranium silicate, $Ca(UO_2)_2(SiO_3)_2OH_2 \cdot 5H_2$, 56 percent uranium. Beta-uranophane is similar.

6. Gummite: a mixture of uranium oxides and other minerals, most suitable as a field term for mixtures of yellow, orange, red, and other colors.

Many minerals (about 55) contain minor amounts of uranium. Some of these, such as zircon and allanite, are common but are not at present amenable to treatment. Thucolite is a familiar black mixture of uraninite and hydrocarbons.

Uraninite and pitchblende are usually referred to by either name; they are black with tinges of brown, gray, or green. Carnotite and tyuyamunite are canary yellow when pure, greenish yellow when impure, and look alike. They can scarcely be told from other secondary uranium minerals of the same color, including uranophane and less important ones. Removing the filter from an ultraviolet lamp causes carnotite (but not tyuyamunite) to appear more vivid. Autunite and meta-autunite also are lemon yellow to pale green and look alike. They are more or less strongly fluorescent in yellowish green. Torbernite and metatorbernite look alike and are bright green, but are not fluorescent. Uranium minerals do not phosphoresce. Coffinite looks like other black minerals of the Colorado Plateau and can be identified only by X-ray. Brannerite is black to yellowish and is nearly restricted to Canada; davidite is black and found mostly in Australia.

Few of the uranium minerals can be recognized for certain by sight. Uranium carbonates occur in hexagonal plates, blades, or needlelike (acicular) crystals. Uranium carbonate (and sometimes schoepite) effervesce in acid, as do all other carbonates. Uranium sulfates are typically needlelike or spherical. Uranium arsenates and phosphates tend to occur as rectangular plates and tablets. Uranium vanadates are found in scaly, platy, or spherical aggregates; a spot test will prove vanadium. Uranium silicates are needlelike, usually in tufts or rounded clusters. Uranium-copper minerals are mainly deep green or bluish green and are less fluorescent than minerals without copper.

Radioactive minerals reveal themselves by causing photographic film (or plate) to be exposed when they are wrapped light-tight and placed on it, with a flat metal object (such as a key or coin) between them. The outline of the object will show on the film after it is developed.

To distinguish uranium minerals (except refractory ones such as multiple oxides) from other radioactive minerals, the contact-print test can be used. First, thoroughly moisten (with 1:5 hydrochloric acid) gelatin-coated paper (obtained by "fixing" glossy photographic print paper, then washing and drying it). Then, hold the gelatin surface firmly (without slipping) against the smooth rock surface for about 20 seconds. Develop the print by moistening it with a saturated solution of potassium ferrocyanide. The presence and location of uranium minerals show by brown spots. Iron shows blue, and copper shows light brownish violet. Tests might be made first on known samples, for strong iron may mask uranium or copper.

Only two types of uranium occurrence are economically valuable in the United States: flat-lying deposits in sandstone and conglomerate (on the Colorado Plateau and in basins in Wyoming) account for about 96 percent, and vein deposits (mainly in Colorado and Washington) account for the rest. In order of importance, the ages of the rocks are Jurassic (53 percent), Early Tertiary (27 percent), and Triassic (13.5 percent). Owing to the instrumental techniques that are used in exploring for uranium, the advice already given, to prospect already proved areas first, should perhaps be less strongly emphasized here. Surface prospecting for radioactive minerals might better favor entirely new areas that have geologic environments similar to those of past productive areas. Nevertheless, some minor districts may be worth reevaluating in the light of new knowledge about uranium geology and the tendencies of ore. Improved drilling methods, lower drilling costs, and better bore-hole logging may aid in reappraising older, shallow deposits.

The following suggestions for prospecting have been given by Elton A. Youngberg:

1. Reevaluate areas where uranium minerals or anomalous amounts of uranium have already been found.
2. Look for areas of unexplained radioactivity.
3. "Investigate geologically favorable areas for the presence of elements commonly found in varying amounts with uranium, such as vanadium, copper, molybdenum, selenium, and arsenic deposits in sedimentary host rocks; or silver, lead, cobalt, nickel, and copper in vein deposits." Aluminum, titanium, and ferric iron are especially lacking, and magnetic properties are typically negative.
4. In known uranium-bearing areas, examine extensive sandstone outcrops that are bleached or iron-(limonite-) stained. Drilling may be warranted if geologic conditions are favorable after radioactivity has been detected, even if weakly.
5. Follow known trends rather than testing at random. Large deposits

often cluster in elongate trends, which offer better-than-average places to drill, for they may mark the direction in which the ore-forming solutions moved.

6. Investigate the marginal zones of base-metal deposits, which are better places to prospect than the districts themselves.

Inasmuch as 96 percent of the uranium deposits of commercial ore in the United States are in sediments that are continental (formed on land) and clastic (fragmental, broken), the following prospecting suggestions that are specific for these deposits are worth emphasizing:

1. Select a uranium province: this refers to the Rocky Mountains, in which nearly all the uranium of the United States has been mined to date.
2. Find, by regional reconnaissance, an ancient or modern basin, 1,000 to 10,000 square miles in area, that was fed by granite wash not too far from its source and that contains tuff (from volcanic ash) and sandstone that was laid down by streams. Large oil and coal resources are negative, although scattered ones need not be. A positive reading (anomaly), especially of total gamma radiation, recorded from the air is an added sign of encouragement, and each area has its own set of air guides, which should be listed.
3. Look for the following features:

   a. Sandstone or conglomerate:

   (1) Thick but not too thick

   (2) Interbedded with mudstone beds, lenses, or zones

   (3) Carbonaceous: medium content of organic matter—"dirty"

   (4) Poorly sorted: of different sizes

   (5) Pastel colors: bleached gray, green, buff sandstone, caused by a favorable reducing environment in contrast to red and brown sandstone in an oxidizing environment, which is generally less favorable; the gray versus red color contrast is mostly limited to the Colorado Plateau, except that pink is a good indicator for some deposits of moderate size

   (6) Moderately fractured: permeable enough to have admitted the original solutions but not so permeable as to allow too much leaching away of uranium

   In summary: "continental, fluvioclastic, porous, unconsolidated, carboniferous, arkosic sandstone and conglomerate," "dirty, cherty, trashy, arkosic, mud-seamed sands."

   b. Flattening of dipping beds, in the transitional zone between fast-water (sand) and slow-water (silt) deposition (especially in Wyoming-basin type)

c. Halo of uranium anomaly: chemical, radiometric, or mineralogic
d. Yellow or black uranium minerals and pyrite

4. Trace to their sources any geochemical patterns in stream water.
5. Map in detail and drill locally to outline a favorable zone of 10 to 100 square miles in area. A grid pattern is used for broad outcrops of gently dipping formations and a linear pattern for rim exposures, steeply dipping hogbacks, and structural alignments.

A new kind of uranium occurrence, called roll type, has become prominent in the Tertiary basins of Wyoming. The ore deposits occur at the curved bondary (C- or S-shaped or more intricate) of the roll with unproductive sandstone. The ore-containing sandstone is gray, whereas the barren sandstone is grayish-white, green, light yellow green, or buff.

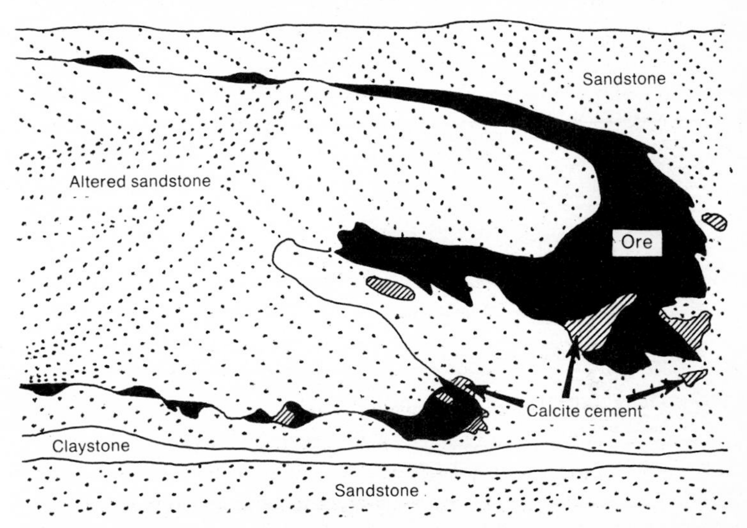

Roll-type uranium deposit in tongue of altered sandstone. *(After Harshman)*

In Canada, all the known commercial deposits of uranium occur in ancient (Precambrian) rock of the Canadian Shield. These deposits are of three distinct types, as follows:

1. Conglomeratic: layered
2. Pegmatitic
3. Veined

With the termination of the United States government's procurement program at the end of 1970, only private markets remain. Prices are subject to negotiation, as, in fact, they were with the government as customer, but they are no longer based on a flexible arrangement that was designed as much as possible to keep the producer in business. Price schedules of the sort that have appeared in older books are not now applicable. At present, 0.21 percent U_3O_8 is about the minimum level of commercial ore, but this can be made up by balancing higher and lower grades. The formula U_3O_8 does not necessarily mean U_3O_8 chemically but expresses the equivalent composition, as is done with other mineral products.

Below is a directory of uranium ore processing mills in the United States in 1972. Inquiry should be made first to the one nearest your mine or prospect, and preferably before active mining is undertaken. The amenability of an ore—that is, the ease with which the uranium in it can be extracted and refined in actual mill practice—must be determined in order to see if an excessive consumption of chemical reagents, special circuit improvisations, and other items of high cost can be avoided.

U.S. URANIUM ORE PROCESSING MILLS

Milling Company	*Mill Location*
Anaconda Company	Grants, New Mexico
Atlas Corporation	Moab, Utah
Conoco & Pioneer Nuclear, Inc.	Falls City, Texas
Cotter Corporation	Canon City, Colorado
Dawn Mining Company	Ford, Washington
Federal-American Partners	Gas Hills, Wyoming
Humble Oil and Refining Company	Powder River Basin, Wyoming
Kerr-McGee Corporation	Grants, New Mexico
Mines Development, Inc.	Edgemont, South Dakota
Petrotomics Company	Shirley Basin, Wyoming
Rio Algom Corporation	La Sal, Utah
Susquehanna-Western, Inc.	Falls City, Texas
Susquehanna-Western, Inc.	Ray Point, Texas
Union Carbide Corporation	Rifle, Colorado
Union Carbide Corporation	Uravan, Colorado
Union Carbide Corporation	Natrona County, Wyoming
United Nuclear-Homestake Partners	Grants, New Mexico
Utah International, Inc.	Gas Hills, Wyoming
Utah International, Inc.	Shirley Basin, Wyoming
Western Nuclear, Inc.	Jeffrey City, Wyoming

The byproducts of uranium production in the United States include vanadium (in the Colorado Plateau, as already discussed); molybdenum (especially in the Dakotas and Montana), which in turn may yield rhenium; and copper (in Arizona and Utah), from which silver may be recovered.

Inasmuch as thorium has its principal uses outside the nuclear-energy program, a few words about them are in order. Magnesium alloys consume the largest amount. Thorium chemicals are much employed in making incandescent-gas mantles and have some uses in other products.

Monazite has always been the only important thorium-bearing mineral, although it is a cerium phosphate, $(Ca,La,Nd)PO_4$, which contains up to 11 percent thorium. Thorianite (oxide) and thorite (silicate) are true thorium compounds but are much less abundant. Monazite is a heavy mineral, reddish brown to yellow; thorianite and thorite are dark gray to black or reddish brown. Thorium minerals, though radioactive, do not fluoresce or phosphoresce.

Radium continues to be used in medical technology.

BOOKS ON RADIOACTIVE MINERALS

Mineralogy and Geology of Radioactive Raw Materials, E. W. Heinrich, McGraw-Hill Book Company, New York, 1958.

Mineralogy of Uranium and Thorium Bearing Minerals, D'Arcy George, U.S. Atomic Energy Commission RMO-563, 1949.

Glossary of Uranium- and Thorium-Bearing Minerals, Judith W. Frondel, Michael Fleischer, and Robert S. Jones, U.S. Geological Survey Bulletin 1250, 1967.

23

Other Major Metals

RARE-EARTH METALS

Neither rare nor earths (in the usual sense), the 15 elements that make up this peculiar family offer opportunities to the prospector provided that he learns something of their occurrence but does not try to tell them apart. Even the analytic chemist has trouble doing so. These curious elements, which represent nearly one-sixth of all the 92 natural elements but occupy only a single position in the periodic table of the elements, are fairly abundant metals. Cerium is the most common of them, actually more plentiful in the earth's crust than lead or tin; even the least abundant, thulium, is more plenteous than iodine or mercury. The full list, with their chemical symbols, follows:

Lanthanum . . . . . . La
Cerium . . . . Ce
Praseodymium . . . . Pr
Neodymium. . . . . . Nd
Promethium. . . . . . Pm—not natural

Samarium . . . . Sm
Europium . . . . Eu
Gadolinium . . . . . . Gd
Terbium . . . . Tb
Dysprosium . . . . . . Dy

Holmium . . . . Ho
Erbium . . . . Er
Thulium . . . . Tm
Ytterbium . . . . Yb
Lutetium . . . . La

Two other elements—yttrium and scandium—are sometimes included by industry among the rare earths. The name lanthanides, or lanthanons, is sometimes used by chemists. The above elements are often divided into the cerium (or light) and yttrium (or heavy) subgroups, or the cerium, terbium, and yttrium subgroups. Because certain of the rare-earth minerals also contain uranium or thorium, these have been confused with the radioactive raw materials, but there is no necessary connection otherwise. The actinide rare earths are not related to these elements.

Igneous rocks are of course the primary home of the rare earths, but they are more abundant in the acidic (silicic) rocks like granite than in the basic rocks. The pegmatite variety of granite is a good source. A unique vein deposit in the Republic of South Africa was formerly of very great importance. A peculiar type of rock, only recently come into prominence, is carbonatite, a carbonate-silicate rock, which is perhaps formed as a result of an alkalic intrusion of igneous rock. This has become, at Mountain Pass, California, the world's largest mass of rare-earth mineralization, in the form of bastnaesite. The erosion of these and other bodies produces placer deposits, which occur as stream sand and, especially on a large scale, as beach sand, both ancient and modern, in several parts of the world (Florida, India, Australia, Brazil).

The chief rare-earth minerals are the following:

Monazite: Cerium phosphate, $(Ce,La,Nd)PO_4$
Xenotime: Yttrium phosphate, YPO_4
Bastnaesite: Cerium fluo-carbonate, $(Ce,La)FCO_3$
Allanite: also called orthite. Calcium silicate, $(Ca,Ce,Th)_2(Al,Fe,Mg)_3Si_3O_{12}(OH)$
Gadolinite: Yttrium silicate, $Be_2FeY_2Si_2O_{10}$

Among the common minerals that may contain significant amounts of rare-earth elements (or yttrium) are apatite, feldspar, biotite mica, amphiboles, pyroxenes, garnet. Several multiple-oxide minerals also contain these elements: the most important are euxenite (a commercial source), samarskite, and fergusonite.

The above multiple-oxide minerals also contain columbium, tantalum, titanium, and sometimes uranium. Monazite is the major ore mineral of thorium and is recovered as a byproduct of titanium, zirconium, and tin mining. It can be radioactive, owing to a high content of uranium, which may in fact be due to an intimate mixture of uranothorite.

The presence of uranium or thorium with the rare-earth elements in many of the above minerals would make them easier to recognize, but do not forget that they may not be radioactive to any appreciable degree. Monazite is heavy and reddish brown to yellow; xenotime, though lighter in weight, is similar; so is bastnaesite. Allanite is black to dark brown, and gadolinite is usually black and shiny. Euxenite, samarskite, and fergusonite are grouped with the so-called radioactive blacks, which describes their typical color when fresh. Green rare-earth minerals are never fluorescent. Removing the filter from a shortwave ultraviolet lamp causes the transparent grains of placer monazite (especially) to show as specks of dull emerald green, but this is not a fluorescent effect.

Owing to the complex chemistry of the rare earths and the various byproducts that are associated with their extraction, the discoverer of any such deposit needs expert advice. The processors of rare-earth chemicals should be consulted before property is opened or disposed of. Furthermore, the value of monazite is seldom in its rare-earth content, and so a coproduct is usually needed: gold, cassiterite (tin), ilmenite (titanium), or phosphate rock, for example.

By far the largest use for the rare-earth metals is in petroleum refining, followed by various applications in the glass industry. Certain of these miscellaneous uses are of remarkable interest. Misch metal is a curious alloy, mostly cerium, used in lighter flints and sparking metal.

MERCURY

One of the most interesting of all natural substances because it is a liquid metal at ordinary temperatures, mercury has been of major importance as a principal reason for the successful extraction in large quantities of native gold. California gold production, for example, would not have been possible on so successful a scale except for the availability of mercury close at hand.

Quicksilver ("living silver") is one of the easiest metals to prospect for, inasmuch as practically all of it comes from the mineral cinnabar, which is bright red when reasonably pure. This conspicuous color, in fact, is a drawback to prospecting for mercury, for only a little of it colors rock so intensely that the grade (tenor) can be widely overestimated, more so than with copper or iron colors. But at least you cannot overlook a potential mercury mine. Mining mercury can be done cheaply (but not always safely) on even low-grade ore, because retorting is simple though dangerous.

Mercury is deposited close to the surface in areas of recent volcanic and hot-spring activity. Wherever almost any kind of rock is fractured enough to permit the passage of rising solutions into this sort of terrane, mercury may form, held in by impervious rock that acts as a trap. Nearness to silicic (acidic) igneous rock is considered helpful. Mercury may replace the rock extensively, as it did in the world's largest deposit, at Almaden, Spain. Cinnabar may also be rather disseminated (scattered) through rock, grading into richer accumulations elsewhere.

Cinnabar is mercury sulfide, HgS, 86 percent mercury. Solid pieces are noticeably heavy. A small amount of other mercury mineralization is mined commercially, including native mercury, metacinnabar (sulfide), livingstonite (sulfosalt), and calomel (chloride). These are mostly associated with the dominant cinnabar.

The main single use of mercury is in electrical apparatus, followed by use in the manufacture of chemicals. Many industrial products require this amazing metal.

24

Minor Metals

"Minor metals" they may perhaps be to industry, but none would be so regarded by the prospector who finds a profitable deposit of one of them, nor by the company whose chief business it is to use it as the raw material in production. The designation of these as minor metals is largely arbitrary. Some are peculiar, some are fairly ordinary—all are of interest and importance to someone. Those that are likely to be found and extracted by the prospector are given attention here. A few metals that are recovered only as byproducts of smelting or other operations are merely mentioned in passing; still, the prospector owes it to himself to determine the full potential of any mineral deposit that he may find. Byproducts have been known to become coproducts, and these may eventually turn to dominant ones as markets change, prices improve, or technology advances.

Antimony

The typical antimony mine is in a small deposit of stibnite (the primary sulfide) and the minerals that have been oxidized from it. This simple type of operation usually recovers some other mineral products as well.

In turn, antimony is recovered from complex ores in which it is of equal value to other metals, or perhaps much less. Lead is the main coproduct. Most American antimony comes from this sort of occurrence, whereas foreign antimony is chiefly from stibnite deposits.

Antimony mineralization seems to favor the silicic (acidic) and intermediate igneous rocks—granite, granodiorite, monzonite—although any rock in the path of the rising solutions should receive them and be replaced by them. Near-surface conditions, even hot springs, are the best geologic environment.

Stibnite is Sb_2S_3, having 72 percent antimony. It is metallic gray, bright unless tarnished; enargite and bismuthmite, which resemble it, are chemically different. The oxides (and similar minerals) of antimony include cervantite, stibiconite, valentinite, and senarmontite, all of which are white to pale yellow.

One of the oldest metals in continuous use, antimony is now employed mostly in alloys, especially with lead in storage batteries, and in various chemicals. Antimony is regarded as a semimetal, or metalloid, for its metallic appearance is offset by its nonmetallic chemical properties, its compounds acting as either acids or bases.

Arsenic

This metal comes mostly as arsenic trioxide, called white arsenic, as a byproduct of copper mining and to a lesser extent from lead, gold, and silver ores. It is a nuisance element to many smelters or roasting mills, and so it is not sought but rather avoided. It is sometimes recovered only in order to prevent dangerous fumes from escaping. Nevertheless, arsenic is a useful substance, particularly as an insecticide and wood preservative; the metal itself is alloyed mostly with lead and copper. Red realgar and yellow orpiment, both sulfides, make beautiful mineral specimens, but these compounds are more likely to be made synthetically. Having nonmetallic chemical properties—its compounds acting as either acids or bases—arsenic is considered a metalloid, or semimetal.

Bismuth

Mainly a byproduct of lead refining and ores of copper, tin, molybdenum, tungsten, silver, or gold, this metal has also been produced from a few deposits in which it is the sole ore. Bismuthinite, the sulfide (Bi_2S_3), is metallic gray, looking like enargite, stibnite, and jamesonite but containing bismuth. Native bismuth and bismite (oxide) are mined in some places. Considering the relative rarity of bismuth, its frequent appearance in minerals is surprising, making it a reasonable object of the prospector's search.

Alloys and chemicals consume most bismuth, a metal that is easily fusible and that expands when solidifying.

Cadmium

A plating and alloy metal, and (as with all the elements) a constituent of chemical compounds, cadmium comes almost exclusively as a byproduct of the processing of zinc ore. So there is little reason for the prospector to look for it, although you may see it from time to time as a coating of yellow sulfide (the mineral greenockite) on specimens of sphalerite.

Cesium

Useful mostly in research, cesium is a metal that is best known for its ability to produce electricity from light. Infrared and photocell devices can utilize it, but other elements have largely replaced cesium. The uncommon mineral pollucite, which resembles quartz, is the only commercial source.

Gallium

Recovered solely as a byproduct of bauxite (aluminum) and sphalerite (zinc), this metal offers no opportunity for direct prospecting. Consumption is small, the use of gallium being mostly for semiconductors and electronic devices.

Germanium

A principal metal in semiconductors, germanium comes as a smelter byproduct of ores that chiefly yield zinc, lead, and copper; coal is a small source but potentially a considerable one.

Indium

This metal comes as a byproduct of zinc (sphalerite) mostly, but ores of copper, lead, molybdenum, and vanadium contain it. On this account, the many minerals in which indium has been detected are not apt to be of concern to the prospector. The largest use for indium is in electronic and communication equipment.

Rhenium

A metal of growing interest, rhenium is in short supply, but it seems likely to be recovered only from molybdenum ore in which molybdenite occurs in the porphyry-copper type of deposit. Find the copper body first, and then worry about rhenium. This curious metal hardens platinum and softens chromium, tungsten, and molybdenum.

Selenium

Copper refining supplies all the selenium that is yet needed; gold, silver, lead, and tellurium are obtained from the same electrolytic slime. Lead smelting also leaves selenium. Prospectors should therefore look elsewhere for something to discover, although selenium is frequently associated with uranium, and this relationship should be kept in mind. The main uses of selenium are in the glass, electrical, and duplicating-machine (Xerox) industries. Selenium is semimetallic, a metalloid, behaving either as a metal or as a nonmetal.

Tellurium

Tellurium is semimetallic, a metalloid, acting chemically as either a metal or a nonmetal. Like selenium, enough tellurium is furnished by copper and lead refining to meet the needs of the market, which is fairly diversified. It includes the metal, rubber and plastic, and chemical industries. Besides occurring in base-metal minerals, tellurium is found in many minerals. Of these, the tellurides of gold or silver (or both) are of great value though somewhat limited distribution. The tellurium prospector is really looking for gold and is referred here to chapters 7 and 17.

Zirconium and Hafnium

A rather new metal in industry, zirconium comes from zircon (a silicate) and baddeleyite (an oxide). Zircon is obtained at times from a number of rocks but especially when granitic rock has been weathered and eroded to form heavy-mineral sand along streams and beaches. Zircon-bearing beach sand is found around the world, especially on the coasts of Florida, India, and Australia. Baddeleyite comes only from Brazil. The metallic element hafnium always occurs with zirconium, but in lesser amounts (averaging about 1 in 50), and is extracted as a chemical impurity from zircon. Recovered with zircon are ilmenite and rutile (titanium), monazite (thorium), cassiterite (tin), and native gold.

Zircon is used mostly as the complete mineral. Its use as a metal and in alloys ranks next. Chemical compounds have a growing market. Hafnium is employed in nuclear reactors.

BOOKS ON METALLIC MINERAL DEPOSITS

Introductory Economic Geology, W. A. Tarr, McGraw-Hill Book Company, New York, 1930.

Mineral Facts and Problems, 4th edition, U.S. Bureau of Mines Bulletin 650, Washington, 1970 (1971).

Economic Geology, 5th edition, H. Ries, John Wiley & Sons, Inc., New York, 1925.

Economic Mineral Deposits, 2d edition, Alan M. Bateman, John Wiley & Sons, Inc., 1950

The Geology of the Metalliferous Deposits, R. H. Rastall, Cambridge University Press, London, 1923.

Ore Deposits, 2d edition, Charles F. Park, Jr., and Roy A. MacDiarmid, W. H. Freeman and Company, San Francisco, 1970.

General Economic Geology. A Textbook, William Harvey Emmons, McGraw-Hill Book Company, New York, 1922.

Metallic and Industrial Mineral Deposits, Carl A. Lamey, McGraw-Hill Book Company, New York, 1966.

Economic Geology of Mineral Deposits, Ernest R. Lilley, Holt, Rinehart and Winston, New York, 1936.

The Principles of Economic Geology, 2d edition, William Harvey Emmons, McGraw-Hill Book Company, New York, 1940.

Our Mineral Resources, Charles M. Riley, John Wiley & Sons, Inc., New York, 1959.

Mineral Deposits, 4th edition, Waldemar Lindgren, McGraw-Hill Book Company, New York, 1933.

Part 5

25

Gemstones

Beauty, durability, and rarity are the qualities sought for in gem and ornamental stones. Beauty may be represented by color, brilliancy ("fire") when cut properly, or special optical effects of a considerable variety. Durability requires that the substance be reasonably hard, in order to resist being scratched by the silica-laden grit in the air; but certain nonmineral gems are rather soft. Rarity is a psychological necessity; nevertheless, man-made gems have a wide popularity though a low value. Fashion is a factor that influences price and causes it to change from time to time. Most gems must be skillfully cut to bring out their full beauty. Although some substances of organic origin—amber, jet, coral, pearl—are gems, and artificial substances are called gems, only the mineral and rock gems are described in this chapter, as being those of interest to the prospector and mineral collector.

A fair amount has been written about looking for gems in the field. Not much new of general application can probably be said except that each specific locality doubtless has its special rock and mineral associations, a knowledge of which may be helpful in prospecting. Certain geologic principles of likely value to the prospector are given in this chapter. Different gems occur under different conditions. It was the discovery of diamonds, rubies, sapphires, and emeralds together that—as much as anything else—led to the exposure of the notorious Great Diamond Hoax in the American West a century ago. This information is condensed here, for another book on gem collecting is not attempted in these pages. Only the main gems are included. For adequate identification of gems, especially when cut and polished, books on gemology should be consulted. The following gems are described alphabetically.

Beryl

Treated in chapter 21 as the principal ore of beryllium, beryl is also a major gem mineral. The green, chromium-bearing variety of this silicate mineral, $Be_3Al_2(SiO_3)_6$, is called emerald and is one of the most highly regarded of all gems. The blue to bluish-green variety is aquamarine. Other colors of beryl occasionally appear as gems: morganite is a pretty pink; golden beryl is magnificent.

Whereas the origin of the best (Colombian) emerald is disputed—calcite veins in black shale—aquamarine and other kinds of beryl are characteristically found only in pegmatite. Emerald comes also from metamorphic rock (schist, marble). Inasmuch as emerald is extremely rare, it is advisable to check any complex pegmatite closely at least for aquamarine.

Corundum

Aluminum oxide, Al_2O_3, is corundum. The red variety is called ruby; all other colors, but especially blue, are termed sapphire. Thus, we have pink sapphire, golden sapphire, the poorly named "Oriental topaz," and a host of others. Star ruby and star sapphire reveal asterism—a star-shaped pattern—when properly cut. Even though these are, above all the gems, most effectively manufactured synthetically (and in enormous quantities), the natural gems, particularly ruby, bring high prices for good quality.

Ruby is typically a product of the contact metamorphism of limestone. Placer deposits acquire it secondarily. Sapphire has a similar occurrence but is more widely distributed. Dikes of certain rocks contain sapphire, but careful examination of the contents of a gold pan offers the best hope of finding any of the hues of corundum.

Chrysoberyl

An oxide of beryllium and aluminum, $BeAl_2O_4$, this unusual hard mineral (number 8 in the Mohs scale) exists in three gem forms. True cat's-eye

Corundum crystals are frequently shaped like these.

is chrysoberyl. Alexandrite is a variety that changes color in different light: at best it is green by daylight and raspberry red by artificial light. Yellowish-green chrysoberyl goes largely under its own name. These gems may bring high prices. Pegmatite and the adjacent, intruded rock constitute the original source of chrysoberyl. Placers may be looked into also.

Diamond

The only primary home of the king of gems is kimberlite, an altered (highly serpentinized), porphyritic, and broken (brecciated) variety of peridotite. This is a basic igneous rock, high in magnetic properties and low in radioactivity; on this basis, it can be prospected for successfully. Not all masses of peridotite are diamond bearing by any means, but those that are tend to occur as funnel-shaped pipes, which are modified downward into tabular bodies (dikes, sills). As the kimberlite alters downward from the surface, its color changes, causing it to be referred to as yellow ground (above) and blue ground (below). The fresher kimberlite is greenish blue. The diamond crystals are disseminated (scattered) very sparsely through the rock and inside inclusions of a garnet-pyroxene rock known as eclogite. Pieces of the country rocks also are found in the kimberlite.

The African pipes are classic, but kimberlite is known in rather similar fashion on other continents. Associated minerals include olivine, chromium diopside, enstatite, pyrope garnet, phlogopite (brown) mica, zircon, chromite, and magnesium ilmenite. Some of these are considered good clues for prospecting. Recent studies show the close association of kimberlite with carbonatite, a carbonate-rich rock of apparent igneous (magmatic) origin. This, together with alnoite (a dark rock in dikes) or alkaline complexes, seems to serve as a clue to areas that are suitable for prospecting for diamonds.

Eroded from kimberlite, diamond becomes an ideal placer mineral and ends in stream and beach gravel, which is apt to contain much garnet, diopside, and ilmenite. The nature varies according to local conditions, and a profitable operation today employs a vacuum device to suck diamonds offshore from South-West Africa (Nambia).

The isolated (but fairly numerous) diamond crystals that have been noted in the Great Lakes region doubtless came via glacier from a yet undiscovered source in Canada. Those in California are an even greater mystery. The true African-type pipes in Arkansas may not be the only ones there. Many, many diamonds have been found in the Yakutia and Aldan regions of Siberia, in the Soviet Union, since a systematic search got underway. Pyrope garnet is the best pathfinder, or *sputnik* mineral, followed by magnesium ilmenite and chromium diopside. Few diamonds have ever been seen in the native rock, so dispersed are they; only by picking up a loose specimen on the ground can they be expected to be found. Until such a time as gem-quality diamond is made synthetically, this pure-carbon mineral is a reasonable candidate for successful prospecting. In North America, David P. Gold, of the Pennsylvania State University, has noted favorable considerations that include zones of crustal warping, arches, and hinge zones that are marginal to basins; areas that contain faults of the rift-valley type; block-faulted areas; and areas of flood basalt.

Feldspar

The complicated group of aluminum silicate minerals that feldspar represents includes several gems of relatively low value but some interest. Orthoclase includes true moonstone, as also does some albite. Oligoclase (of the plagioclase series) includes sunstone. These gems appear much as their names suggest. Amazonstone is blue to green microcline. Labradorite shows a sheet of blue or green color across the surface. Feldspar is a typically igneous mineral, surviving into sedimentary and metamorphic rock.

Garnet

This group of silicate minerals includes a number of attractive gems. Their chemical compositions and geologic origins are rather diverse. Few specimens bring a high price. Pyrope garnet (magnesium-aluminum) is mostly of igneous and metamorphic occurrence. Almandite (iron-aluminum) is chiefly metamorphic. Spessartite (manganese-aluminum) is igneous and metamorphic. These kinds are typically garnet red to brown in color. Grossularite (calcium-aluminum) has pink, golden, orange, and green hues as well; it is metamorphic. Andradite (calcium-iron) is usually black to brown, but its yellow and especially green colors are fine; the bright-green variety is called demantoid. Andradite is metamorphic, but all the garnets make suitable placer minerals and are usually found first in this geologic environment.

Jade

Two different silicate minerals are properly called jade. One is jadeite, a kind of pyroxene; the other is nephrite, an amphibole. They are much more alike than different; together they run the gamut of opaque and translucent colors from black to white, including, of course, jade green. Both jades are mostly of metamorphic origin and are found in the parent rock as well as in boulders. Nephrite is somewhat widespread (as in Wyoming, California, Alaska, New Zealand, China), but jadeite is not at all, being confined mostly to Burma. Only a serpentine terrane seems to be worth searching for jade; loose material guides the way to the solid deposits.

Lapis Lazuli

Only one important, though still not common, gem is a rock rather than a mineral. This is lapis lazuli. It can be recognized as a mixture of several intensely blue minerals (mostly hauynite and lazurite) of the feldspathoid group, in which are white streaks of calcite and brassy flecks of pyrite. The whole is a product of metamorphism and has a restricted origin. Few gems would seem to attract the prospector more temptingly.

Opal

Consisting of alternating layers of spherical cristobalite (a silica mineral, SiO_2) and hydrated silica glass, opal is found chiefly in arid regions as a near-surface deposit. No other gem has so strong an appeal to the connoisseur, and the prospector should keep it clearly in mind as he prowls the deserts of the world. Furthermore, common opal, as in opalized wood, is an abundant mineral. Only the special structure that causes the glorious hues of gem opal is lacking in this material—and you may be the next person to discover another deposit like those in Australia and Nevada.

Peridot

The bottle-green, gem variety of olivine is known as peridot. (The rock peridotite gets its name thus.) Olivine is a very common mineral in basic igneous rocks, and it is here that peridot must be looked for. Readily weathered, olivine may accumulate for a time in arid-land gravel and sand. Gem olivine is a silicate of magnesium and lesser iron, $(Mg,Fe)_2SiO_4$.

Quartz

So abundant a mineral is quartz that virtually every conceivable kind of rock may contain it: igneous, sedimentary, metamorphic. It is least common in basic igneous rock but not unknown even there. So many names have been given to gem varieties of quartz that the beginner may well be confused. Some of these names are very ancient and well known. Altogether, however, they may be divided into two groups: quartz proper (which usually, though not always, occurs as crystals) and chalcedony (which is finer grained, usually opaque or translucent, and cannot develop as crystals). The individual varieties are named according to their color and pattern; a few structural peculiarities have also resulted in separate names. Below is merely a listing of the more familiar varieties, but detailed descriptions can be found in books on gemology.

Quartz proper Rock crystal, amethyst, citrine, aventurine, rainbow (iris) quartz, milky quartz, gold quartz, sagenite, quartz cat's-eye, tiger's-eye, hawk's-eye, rose quartz, smoky quartz, cairngorm, morion.

Chalcedony quartz Carnelian, sard, prase, chrysoprase, plasma, bloodstone (heliotrope), jasper, onyx, sardonyx, agate, moss agate, petrified (silicified, agatized, jasperized) wood. Many other names are used for both these types of quartz, some having only local acceptance.

Spinel

Found with corundum in stream gravel, and mistaken for it, spinel is a multiple oxide of magnesium and aluminum, $MgAl_2O_4$. Its original rock is contact-metamorphosed limestone, but common spinel is formed in other high-temperature rocks. The most familiar of the gem varieties is rubicelle, which is orange red. Gem spinel is not a likely find for the prospector.

Spodumene

This mineral is known in enormous crystals, weighing up to 90 tons, but its gem varieties are uncommon and fragile. Besides its yellow and yellowish-green hues, spodumene is lilac colored (called kunzite) and green (hid-

Huge tree trunks turned to lively colors of jasper lie in Petrified Forest National Park and on surrounding land. *(U.S. National Park Service)*

The banding of gem-quality agate is beautifully shown in this polished section. *(Courtesy H. C. Dake)*

denite). This member of the pyroxene group is a silicate of lithium and aluminum, $LiAl(SiO_3)_2$. It is confined to pegmatite, enough of which occurs throughout the world to make the gem worth looking for.

Topaz

A major gem, topaz does not always have what passes for topaz color. In fact, most jewelry-store topaz is actually citrine, the yellow quartz. True topaz may be blue or pale green, and brownish stones are often heated to give them rose or pink tints. This is a pegmatite mineral, an aluminum fluosilicate, $Al_2(F,OH)_2SiO_4$. Very large crystals are not rare. Its cleavage does not permit topaz to travel far as a placer mineral. Prospectors may do well to keep topaz in mind as they look about.

Tourmaline

An important pegmatite mineral of complex chemistry, tourmaline comes in almost every color. Many crystals are multicolored. The names are in surplus: achroite (colorless), rubellite (pink, red), indicolite (dark blue), and others, including the usual assortment of misleading names. Tourmaline is a favorite of the rockhound-collector.

Turquoise

The opaque blue to green color of turquoise is well known, although other minerals can be mistaken for it. This hydrous copper aluminum phosphate, $Cu(Al,Fe)_6(PO_4)_4(OH)_8 \cdot 4H_2O$, is found mostly in arid regions near copper deposits. It is in demand enough to justify a careful search for it under these conditions.

Zircon

Discussed also in chapter 24 as an ore of zirconium metal, this silicate mineral, $ZrSiO_4$, is a popular gem. Seldom seen in any of its natural colors, zircon is heat-treated to bring out the commercially desired hues: colorless, blue, green, golden. Most gem zircon is picked up in placers and seems to be of little interest to prospectors.

26

Fertilizers

To restore depleted soil to its original state of productivity, to furnish nutrients that may never have been present, to encourage the growth of certain organisms and discourage others, and to manipulate the chemistry of agriculture in many ways—fertilizer minerals are used in large amounts all over the world. These are nonmetallic minerals that are not related except in use; some, of course, are more important than others. In form, they exist as rocks rather than minerals.

Many a rock deposit that has special qualities is more valuable for fertilizer chemicals than for any other product. Beds of limestone, gypsum, and phosphate rock should be analyzed carefully before committing their sale to a less productive use. The chief fertilizer materials are potash, nitrates, phosphates, gypsum, lime, and sulfur. A few sentences about each are given below. Lesser ones include greensand, dolomite and magnesite, borax, and epsomite; they are not apt to be prospected for to use as fertilizers.

Sulfur

Sulfur is used on alkaline soil to neutralize it. Its nonagricultural uses somewhat exceed its use in fertilizer, and so sulfur is also included among the chemical minerals. Sulfur is recovered as native sulfur, from sedimentary beds (mainly as sulfates), from pyrite and other sulfide minerals, from oil resources, and as an industrial byproduct. The average prospector, however, has little expectation of profiting from a discovery of sulfur.

Lime

Much less important as an agricultural raw material than for chemical and metallurgical uses, lime is employed on acid soil to neutralize it. The so-called lime comes from limestone, other calcium carbonate rock, and oyster shells. Like gypsum, limestone is abundant and widespread, the most familiar of all the sedimentary rocks that are formed by evaporation and organic activity.

Gypsum

In its use as a fertilizer material, gypsum is much less important than as a constructional material, but gypsum adds sulfur (as sulfate) to soil and is a soil conditioner. The term "land plaster" for gypsum as a soil conditioner shows the relationship of this rock to its major use as the basis of the plaster industry. Large deposits of this marine evaporite occur widely, making deliberate prospecting for it unlikely.

Phosphate

Most phosphate production goes into fertilizers. The primary phosphate mineral is apatite, which is mined in some places. Phosphate rock, mostly of sedimentary origin, forms in a number of ways and goes under diverse names such as nodular land rock, hard rock (Florida), soft rock, brown rock (Tennessee), blue rock (Tennessee), white rock (Tennessee), land pebble (Florida), river pebble. Phosphate rock is usually not distinctive in appearance, and you will do well to make a simple test that can be done in the field: roughen the rock, put a drop of nitric acid on it, and add a crystal of ammonium molybdate or a drop of ammonium molybdate solution; phosphates turn yellow. Phosphate rock can be worked on federal land only by lease.

Sedimentary beds of phosphates (black), as in northern Africa. *(After Cayeux, Bateman)*

Nitrates

Chile is the only country that supplies nitrogen in the form of mineral
nitrate has come into production—as did the cave deposit in 1957 in
element, but it must be admitted that an occasional deposit of organic
nitrate has come into production—as did the cave deposit in 1957 in
the Grand Canyon, Arizona. Cave explorers can probably keep in mind the discovery of bat guano in large quantities. The chief consumption of nitrogen is in the making of ammonia, an enormously important industrial chemical in the fertilizer, chemical, and explosives industries. Nitrate rock is available for mining on federal land only under lease.

Potash

Because potassium is leached so readily from soil or removed in other ways, it must be added. Potash salts accumulated by evaporation in salt lakes, playas (intermittent lakes in arid regions), and marine basins. Of the numerous potassium minerals, the most important include: sylvite, carnallite, kainite, langbeinite, and niter. These are all soluble salts in brine or beds. The exact mineral composition is of little significance, for the deposits are extracted in bulk, and other minerals are used at times and may be in the future. Potassium gives a violet flame when seen through blue glass. Potassium is also used for other chemicals besides fertilizers.

27

Abrasives

Modern industry requires materials that are hard or have high melting points, as these properties are useful for abrasive purposes, including grinding, cutting, polishing, and other uses. These minerals and rocks are often classified as (1) high-grade (meaning harder than quartz). (2) siliceous (composed of quartz and other forms of silica), and (3) miscellaneous, or soft. They may also be classified as: (1) used as naturally occurring, (2) shaped, (3) crushed, and (4) formed. The minerals and rocks in this chapter are otherwise unrelated to one another. Prospecting for abrasives does not ordinarily seem justified as a sole objective, although it may be combined with a search for sand and gravel, and it could prove profitable if a large deposit were encountered.

Industrial Diamond

In spite of the synthesis of diamond and the availability of the artificial product, natural diamond still has a market as the premier abrasive. This is especially true of the larger sizes, which have not yet been made in the factory—although they are certain to be. Far less in the public mind than gem diamond, the industrial varieties are more essential, if less exciting. A detailed analysis of their structure is not necessary here, but the several types include bort, ballas, carbonado (or carbon), which comes only from Brazil. The uses of these extremely hard materials run the gamut: cutting glass and other diamonds, truing grinding wheels, turning machine tools, drawing wire, sawing rock, and drilling holes in rock, as well as doing a myriad of other chores for modern industry. No prospector would look only for this kind of diamond, but he would surely not pass it by because it is too discolored or flawed to be cut for jewelry. The prospecting clues that are given on pages 351–352 apply here too.

Corundum

The second-hardest mineral is used today in small amounts, being largely supplanted by Carborundum, a manmade product (silicon carbide) that was named after it. Pegmatite that is associated with basic rock—most pegmatite is related to acidic (silicic, granitic) rock—may contain crystals of corundum. These are typically blue or brown, six-sided, and may be very large. Other rock types also yield corundum occasionally. Emery, described below, is related.

Emery

This is a natural mixture of several minerals and is hard enough to be a valuable abrasive. One kind consists of corundum and magnetite. Another is spinel and magnetite, with or without corundum. Another is spinel, hematite, and perhaps corundum. The proportions vary, emery not being a true mineral. The hardness is intensified by an increase in the percentage of corundum, and three commercial types are recognized: Greek, Turkish, and American.

Garnet

Described in chapter 25 as gemstone, garnet has a relatively small but steady use as an abrasive. Almandite, the iron-aluminum garnet, is the subspecies that is most used, but it is not the only one. Almandite occurs in metamorphic rock, although a little placer garnet is recovered in some places.

Siliceous Abrasives

These are divided into natural silica and natural silica stone abrasives. They represent a wide range of geologic products, some properly minerals (such as crushed quartz), others properly rocks (such as pumice). Some are pure silica, SiO_2 (such as chert); others are silicates (such as pumice).

Garnet crystals are often easy to recognize. *(Westinghouse Electric Corporation)*

Geologically, they include igneous, sedimentary, and metamorphic materials. Many of them are used for other purposes as well. The prospector may keep in mind that common substances such as these, ordinarily overlooked, can make a simple mining operation profitable, although he may have to inquire extensively in order to find a suitable outlet. Railroads, state bureaus of mines, chambers of commerce, and others are often able to put the prospector in touch with buyers.

The following outline of names is taken from Alan M. Bateman:

Material	Abrasive	Chief uses
Sandstone	Grindstones	Grinding saws, knives, metal etc.
	Pulpstones	Grinding wood for paper pulp
	Sharpening stones	Hand sharpening
	Oilstones	Fine sharpening of steel
Quartz, flint	Burrstones	Grinding flour, pigments, etc.
	Pebbles	Grinding ore in mills
	Slabs	Fine hand sharpening
	Crushed	Soft wood (sandpaper)
Sand.	Grains	Sand blasting, glass grinding
	Paper and cloth	Wood, metal
Pumice	Blocks	Rubbing paint and varnish
	Grains	Glass, scouring powders
Diatomite	Powder	Metal polish, dental powder
Tripoli	Powder	Metal buffing
Volcanic dust.	Grains	Scouring powder, cleansers
Rottenstone.	Powder	Scouring powders

Soft Abrasives

These are even more diversified than the so-called siliceous abrasives, representing all sorts of minerals and rocks, all softer than quartz. Only an outline seems appropriate, as follows:

Material	Abrasive	Chief uses
Fuller's earth	Powder	Polishing metal
Feldspar	Powder	Scouring and cleaning
Clay (china clay, pipe clay)	Powder	Buffing metal
Bauxite	Powder	Buffing metal
Chalk (whiting)	Powder	Buffing silverware and metal
Black rouge (magnetite)	Powder	Buffing metal and mineral surfaces
Red rouge (iron oxide)	Powder	Buffing metal, optical glass, and mineral surfaces
Green rouge (chromium oxide)	Powder	Buffing hard metals and mineral surfaces
Tin oxide	Powder	Buffing metal and mineral surfaces
Talc	Powder	Polishing rice, leather, metal
Crocus (iron oxide)	Paste	Buffing metal
Magnesite	Powder	Making emery wheels
Calcite	Powder	Polishing metal
Manganese dioxide	Powder	Polishing metal

28

Ceramic Materials

Common, low-priced minerals and rocks constitute the foundation of the ceramic industry, but clay is the basis of it. Numerous other materials are used, some in fairly large quantities, and special ceramics, like special steels, draw upon an array of interesting mineral substances. Ceramic products include familiar materials that go under such names as brick, porcelain, china, enamel, whiteware, pottery, tile, stoneware, and terra cotta. Refractories and other kinds of ceramics also have a large use.

Clay

Besides being an indication of particle size in sediments (smaller than silt), the term *clay* means sedimentary rock that consists of one or more so-called clay minerals; impurities, such as silica or iron oxide, are usually present to some extent. Clay also refers to a natural material that has plastic properties. Not all true clay, however, is suitably plastic when wet and before it is fired. The clay minerals are hydrous aluminum and magnesium silicates; they can be identified only by X-rays, the electron microscope, and other complex means. The names mean little to the prospector except insofar as they can be relied on to indicate the origin. The most familiar names include kaolin (kaolinite), halloysite, dickite, nacrite, allophane, montmorillonite, nontronite, saponite, illite, sepiolite.

Solid pieces of clay look and break like this. *(Filer)*

The exact properties of any clay deposit may make a great deal of difference in its commercial value. The presence of rock fragments, hydrous oxides, and colloidal matter (in size between dissolved and suspended matter) may make or break a deposit, at least for a particular use—for, obviously, the manufacturers of art ware and sewer pipes have unlike specifications. The properties that are generally considered most vital are plasticity, strength, shrinkage, and fusibility. For your information, the industrial classification of clay by the U.S. Bureau of Mines is given below:

- Kaolin (china clay)
  - Halloysite
- Ball clay
- Fire clay
  - Plastic, semiplastic, semiflint, flint, nodular flint
- Bentonite
  - Swelling (foundry sand, drilling mud), nonswelling
- Fuller's earth
- Miscellaneous clay
  - Common, brick, sewer pipe, tile

Ceramic clay is not the only kind, for clay is a major material in the construction industry and in miscellaneous manufacturing, which are considered in other chapters.

Geologically, all clay results from the decomposition of aluminum-rich rock. The deposits may form by weathering in place, producing residual clay. Landslide masses produce colloidal clay. Transported clay may be sedimentary (deposited in river floodplains, in lakes, on deltas, in bays, and in the seas) or of glacial or wind origin. Prospectors probably do not search for clay; rather, they come across a likely deposit and then try to find a buyer for that kind of clay. Large and well-situated deposits of bentonite should always be kept in mind by prospectors in the western United States and Canada.

Feldspar

Less important in ceramics than in the glass and enamel industry (see chapter 31), feldspar nevertheless belongs among the ceramic raw materials. Feldspar is consumed in large amounts in making pottery, both for the body (as clay is used) and in the glaze. The potassium feldspar (orthoclase, microcline) is most in demand. Pegmatite is the only source, and a nearness to transportation is often a critical factor.

Other Ceramic Materials

A fairly large number of rocks and minerals are utilized, but in small amounts, in the ceramic industry. Most of them have more important uses in other ways. The following list is not complete, for new substances find their way into ceramics from time to time:

- Barite
- Bauxite
- Borax
- Cornish stone (china stone)
- Diaspore
- Diatomite (diatomaceous earth)
- Fluorite (fluorspar)
- Lithium minerals: lepidolite, spodumene
- Magnesite
- Potash minerals
- Pyrophyllite
- Sillimanite minerals: andalusite, dumortierite, kyanite, sillimanite, topaz
- Talc
- Zirconia

29

Metallurgical Materials

The rocks and minerals in this group, not being otherwise related to one another, are used to make steel and other products of metallurgy, including the heat-resistant (refractory) materials needed in the furnaces.

Fluorspar

This is the name given to the mineral fluorite, which is of major importance in steelmaking. Over the long run, fluorspar would seem to represent a very worthwhile target for prospecting. Occurring in veins and replacing beds of limestone, fluorite is usually purple; the color is frequently banded. It is calcium fluoride, CaF_2, recognized by its low hardness (number 4 in the Mohs scale) and its usual (but not invariable) fluorescence, which was named after it. Fluorspar is also the source of hydrofluoric acid and is much used in the glass and enamel industry. Metallurgical, ceramic, and acid grades of fluorspar are marketed.

Large masses of fluorspar may occur as cubes of the mineral fluorite. *(U.S. Geological Survey)*

Graphite

The foundry is the chief user of natural graphite. This carbon mineral, also known as plumbago and black lead, occurs in three types:

1. Lump: from vein deposits
2. So-called amorphous: from metamorphosed coal beds
3. Crystalline flake: from layered metamorphic rock

Graphite is black and soft; other minerals of similar appearance (molybdenite especially) give simple chemical tests, but graphite is inert. A lot of what is listed as graphite from South Korea is actually anthracite coal.

Sillimanite Group

Five minerals, called the sillimanite group, are used as high-aluminum refractories, having outstanding properties for tough, shock-resistant porcelain, which is suitable for spark-plug cores and other products. Sillimanite, kyanite, and andalusite have exactly the same chemical composition, Al_2SiO_5, and all yield silica and a material called mullite when heated to different temperatures. Two other minerals, dumortierite and topaz, react the same way. The entire group is often listed these days, as by the U.S. Bureau of Mines, under kyanite.

Sillimanite, kyanite, and andalusite are all of metamorphic origin; the first two are recovered from weathered boulders in India. Sillimanite, occurring in quartzite and schist, is white and fibrous or prismatic. Kyanite, also originating in and amidst quartzite and schist, is typically streaked in blue and white or green and white and is remarkable for being the only mineral that has a distinct difference in hardness in different directions: it can be scratched by a knife along the "grain" but not across it. Andalusite is usually brown and occurs in square-ended crystals in slate and other rocks. Dumortierite is usually also metamorphic and is seen as blue, violet, or pink masses that are columnar or fibrous. Massive topaz looks like white quartz and is found in pegmatite and quartz veins. The prospector can hope to discover deposits of these minerals only by learning to recognize them in the field. They are conspicuous because of their tendency to outcrop as low ridges and hills.

Magnesia

Several minerals of magnesium content serve as important refractory materials for so-called basic processes. Magnesite, the carbonate, $MgCO_3$, is a high-grade refractory when "dead burned" and made into bricks or granules; unburned brick is even better. Dolomite, the double carbonate of calcium and magnesium, $CaMg(CO_3)_2$, is even more abundant and therefore cheaper; it is likewise dead burned for use. Magnesite is formed in a number of ways, but dolomite is usually an altered limestone or coral. Both fizz in warm acid. Brucite, the hydroxide, $Mg(OH)_2$, has a similar utility; it occurs in veins. These three minerals (or rocks) have other uses besides the making of refractories.

Other raw materials of considerable value for refractories include fire clay (the most important), silica (as quartzite—called ganister—sandstone, crushed quartz, sand, diatomite), bauxite, diaspore, chromite, zircon and baddeleyite, and a miscellany of substances. These have other uses as well, some being given attention elsewhere in this book.

Additional metallurgical materials include foundry sand, furnace sand, limestone (and lime made from it or from dolomite), dolomite, phosphate rock, strontium minerals, borax, and bauxite. Their origin, occurrence, and appearance are diverse.

30

Chemicals

These minerals and rocks, as well as certain compounds taken from water and air and even a few that are made synthetically, are used in the great chemical industries, which are a hallmark of our industrial civilization. The prospector is interested only in the natural deposits that he is likely to find. Some of these substances are little known to the general public, but have potentially a good deal to offer the seeker after mineral wealth. Except that a number of them are formed by marine or lake evaporation, these substances are not necessarily related.

Sulfur

Probably the most important of the chemical minerals, sulfur is among the less likely object of the prospector's search, although it is not beyond possibility that a new mass of pyrite may be discovered. Such a deposit, however, is apt to be more valuable as an ore of copper or gold. Bedded sulfur deposits are in a similar category, except that they yield no byproducts. Sulfur in limestone, however, may be commercial if it is abundant enough. The large bodies of native sulfur that are found in the caprocks of salt domes can be located only by geophysical means. The extraction of native sulfur from volcanoes is not a job for the prospector either.

Salt

The general term *salt* refers to many chemical compounds that have certain characteristics or compositions. Common, or table salt—sodium chloride, $NaCl$—is a mineral of much popular interest and is exceedingly important as a raw material in the vast chemical industry. Any salt deposit should be checked by a chemist. So abundant is it that prospectors scarcely spend their time looking for it in any of its five types of occurrence, as follows:

1. Seawater: strictly a commercial proposition
2. Brines: having various degrees of saturation, including other dissolved salts; taken from water bodies or wells
3. Playas: intermittent dry lakes
4. Sedimentary beds: with gypsum, anhydrite, potash minerals, clastic (nonevaporite) strata
5. Salt domes: with gypsum and anhydrite

Salt has almost more commercial uses than can be enumerated. Besides water, it is the indispensible mineral.

Other Sodium Compounds

Sodium carbonate and sodium sulfate are two other compounds of sodium that are essential to the chemical industry. They make up what is referred to as the alkali industry. Both of them are used in many products. They are taken from basins of evaporation in which they originally accumulate as precipitates in lakes. Alkali lakes contain mostly sodium carbonate, although other chemicals are present, whereas bitter lakes contain mainly (but not solely) sodium sulfate. Such water bodies may be buried beneath later sediments after they have dried up. Isolated deposits of this sort have made recent mining history in Wyoming.

Borates

Boron compounds are employed in a long list of chemical commodities. Borax and other boron minerals occur in four different ways:

1. Hot springs, fumaroles (gas vents): probably the original source of other deposits
2. Brines: of salt lakes and marshes

3. Playas and lake encrustations
4. Sedimentary beds: playas buried under clay

The boron minerals of commercial importance are the following:

1. Borax: also called tincal, $Na_2B_4O_7 \cdot 16H_2$
2. Kernite: also called rasorite, $Na_2B_4O_7 \cdot 4H_2O$, only from the Kramer district, Kern County, California
3. Ulexite: also called boronatrocalcite, $NaCaB_5O_9 \cdot 8H_2O$
4. Colemanite: also called borocalcite, $Ca_2B_6O_{11} \cdot 5H_2O$
5. Priceite: also called pandermite, $Ca_4B_{10}O_{19} \cdot 7H_2O$
6. Boracite: also called stassfurtite, $Mg_7Cl_2B_{16}O_{30}$
7. Sassolite: natural boric acid, H_3BO_3

These are all colorless or white minerals, some readily turning chalky upon exposure to air. The borate minerals tend to change composition, being converted into one another. A green flame (see page 227) may mean boron. The first three minerals above, being sodium borates, cannot be claimed under the federal mining laws in the United States but are subject to leasing. Other sodium, potassium, bromine, and lithium compounds are obtained as byproducts or coproducts of borate extraction, which is as much a well-recovery operation as a straight mining one.

Lithium Minerals

This lightest of all metals has its principal uses in various chemical compounds rather than as a metallic substance. It is perhaps best known in multipurpose greases. Lithium has the following commercial sources:

1. Brine: at Searles Lake, California
2. Minerals:
   a. Spodumene: $LiAlSi_2O_6$; also a gem mineral (see chapter 25), 4 to 7 percent lithium
   b. Amblygonite: $(Li,Na)Al(PO_4)(F,OH)$, 8 to 9 percent lithium
   c. Lepidolite mica: $KLi_2Al(Si_4O_{10})(OH)_2$, 3 to 4 percent lithium
   d. Petalite: $LiAlSi_4O_{10}$, 4 percent lithium
   e. Eucryptite: $Li(Al,Si)_2O_4$

The above minerals occur typically (and some solely) in pegmatite. Spodumene is platy and prismatic, white to gray. Amblygonite looks like white feldspar. Lepidolite is a flaky mica. Petalite is gray to white. Eucryptite usually fluoresces red to rose in shortwave ultraviolet light. A crimson flame test for lithium may be needed for identification of these minerals.

Spodumene crystals may resemble feldspar, which is much more common. *(Foote Mineral Company)*

Strontium Minerals

Compounds of strontium have a number of chemical uses, especially for pyrotechnics and signals. This element is derived mostly from the following two minerals:

1. Celestite: $SrSO_4$
2. Strontianite: $SrCO_3$

These minerals often occur together. Celestite grows in tabular crystals, thin or thick, and colorless or white or bluish. Strontianite tends to be in elongated crystals and is usually colorless to gray.

Calcium chloride, magnesium chloride, bromine, and iodine are much used in the chemical industries, but they are taken from brine and therefore not prospected for in the usual sense. Nitrogen compounds come mostly from artificial sources. A number of other minerals are likewise of value for chemical purposes, but it does not seem feasible to discuss them in detail.

31

Building Materials

The construction industry, with its diversity of operations, utilizes a vast variety of natural products, most of them being minerals and rocks, many of them used in large bulk. Some, such as metals, are discussed elsewhere in this book, and nearly all the commercial mineral resources do, of course, interact with one another to a certain extent. The structural and building materials of this chapter include many rocks and minerals that are so familiar that they are taken for granted by the average person. The prospector, however, should not regard them lightly. As has been suggested earlier in this book, a gravel pit near a city may be much more valuable than a gold mine in the hills. The future of these materials is virtually unlimited.

Stone

This word does not have a technical standing, for solid earth material is either rock or mineral; but economically useful rock may properly be referred to as stone. Thus, we have building stones, gemstones, and numerous others. Broken rock—termed dimension stone—and crushed rock, or crushed stone, have been used as structural and building materials for 12,000 years or more. Their history would be the story of civilization. Rock may also be crushed, ground, and treated for use in many industrial products, such as cement, plaster, and glass.

A detailed discussion as to the commercial characteristics of building stone is out of place here; the factors are not only complex, but they change with changing fashions in architecture and changing needs of the market. Strength is seldom a mattter of concern. Neither are hardness and the resulting workability, but an adequate spacing of bedding planes, joints, and fissures is helpful in quarrying rock and in its usability. Texture and color are mostly matters of taste, but porosity may affect the durability. The nearness to a finishing plant can have a critical influence on the market value of a deposit of building stone that is to be utilized as dimension stone. You must take into consideration whether a given body of rock can be claimed under the federal mining law.

Broken and crushed stone is a major product in today's economy. Do not overlook its possibilities or those of ordinary sand and gravel. The requirements for special sand are suggested by these names: molding sand, engine sand, abrasive sand, filter sand, fire sand, furnace sand, glass sand. Road metal is rock employed for surfacing roads and for the foundations of roadways.

The prospector who is interested in sand and gravel either looks for a visible surface deposit (where someone else has probably looked already) or else applies his knowledge of sedimentary geology, which tells him how and where streams and lakes have carried and deposited their burden in the past. Sampling can be done by driving a 2-inch (or larger) pipe into the material and withdrawing it filled. The depth of the deposit can be determined by hammering in a narrow metal bar. Engineering laboratories will make the tests that are needed to decide whether a sample is of commercial value. Or a potential buyer, such as a local building contractor, might be glad to test the samples at his own expense.

Cement

The basis of cement is limestone; natural cement has given way to portland cement. Other rock materials (clay, gypsum) are added for specific reasons, but limestone is by far the main one. Adding crushed rock and sand to cement makes concrete. Limestone or dolomite when heated becomes quicklime, which flakes with water and becomes mortar when it is mixed with sand. These sedimentary rocks occur in abundance, and so the convenient location of a deposit of "cement rock" is essential. It must fulfill the commercial requirements, which are fairly rigid.

Gypsum

As limestone is the basis of the cement industry, so gypsum supports the plaster industry. When impure, gypsum contains substances that may aid its utilization, as, for example, retarding the rate of setting. Heating drives off three-fourths of the water, yielding plaster of paris; this powder solidifies when water is again added. There are five varieties of gypsum as follows:

1. Rock gypsum: a sedimentary rock
2. Gypsite: impure, earthy
3. Alabaster: compact, used ornamentally
4. Satin spar: fibrous
5. Selenite: clear, often in crystals

Gypsum—calcium sulfate dihydrate, $CaSO_4 \cdot 2H_2O$—originates almost entirely as a marine evaporite, associated with and related to anhydrite, $CaSO_4$. Both sedimentary rocks occur in beds of vast extent, many of which are known, and so it seems likely that these will be mined before new ones are. Moreover, the prospector should be sure that he can tell gypsum and anhydrite apart; gypsum yields water when heated in a glass tube.

Magnesite

Magnesite (or dolomite as a cheaper substitute) is used in flooring, stucco, and wall board as well as serving as an ore of magnesium and being used for ceramics, abrasives, chemicals, and other industrial purposes that are discussed in other chapters. Magnesite has three occurrences:

1. Veins and pockets: in basic rock
2. Sedimentary beds: much rarer than dolomite strata
3. Replacement of limestone or dolomite beds

Other compounds of magnesium have different occurences as discussed elsewhere.

Pigments

Natural mineral pigments are used in paint and in various other products in order to provide color, opacity, or body. Common deposits of familiar minerals may contain enough material of the right quality to justify mining them for the pigment industry. These minerals include hematite, limonite, and magnetite, with or without manganese oxides and clay. The substances are called ocher, umber, and sienna, which go under rather old names. The geologic origin is chiefly by weathering. In addition, some natural pigments are made from green, white, or black minerals and rocks. Most pigment, however, is derived from the treatment of natural minerals or the manufacture of paint, using, of course, minerals and rocks as raw material.

Insulation

Mineral wool, or rock wool (or other names), is a valuable insulating material. Various natural materials are utilized, and the processing changes with new technology, so that even formerly waste substances, such as blast-furnace slag and mine dumps, have proved satisfactory. Manufactured material, such as glass, is also used. Except perhaps wollastonite—a calcium silicate, $CaSiO_3$, of contact-metamorphic origin—no one mineral is deliberately sought for this purpose, but the shrewd prospector may be able to find and sell a suitable deposit of rock if it is conveniently situated with respect to the plant or the market. Impure rock, of little use for anything else, may be just what the insulation industry wants.

Other insulating minerals and rocks include mica, vermiculite (an altered mica), diatomite (or diatomaceous earth), gypsum, asbestos, dolomite, pumice, and perlite.

Bitumens

Asphalt, a hydrocarbon, occurs naturally, as well as being made by the distillation of petroleum. In addition to roofing, asphalt is used for paving. Several solid substances are of like composition, gilsonite being the best known.

Other Building Materials

Clay and mica are employed as structural and building materials, but their chief use is for other purposes, as discussed elsewhere.

32

Manufacturing Materials

Apart from the uses that are covered in other chapters, the manufacturing industries utilize a large number of minerals and rocks of entirely unrelated nature. Certain of these—such as the asbestos minerals and the iceland-spar variety of calcite—are among the most intriguing of earth resources. Some of them are of little interest to the prospector and so are passed over quickly here.

Asbestos

Not a single mineral but any that is fibrous, asbestos includes some minerals that are commercial and others that are not. The length of fiber determines the price, as spinning fiber (for spinning and weaving) is most in demand. Textiles are therefore the chief product, but more than 2,000 uses are known. Cross fiber and slip fiber can be spun; mass fiber cannot be. The useful kinds of asbestos are the following silicates:

1. Serpentine: chrysotile asbestos, the best variety, $Mg_6Si_4O_{10}(OH)_8$
2. Amphibole group:
   a. Anthophyllite: including amosite, $(Mg,Fe)_7Si_8O_{22}\ (OH)_2$

Asbestos occurs as veins in basic rock, especially dark-green serpentine. *(Johns-Mansville)*

b. Riebeckite: crocidolite, blue asbestos, which may be converted naturally into hawk's-eye and tiger's-eye gems (see chapter 25), $Na_2Fe_5Si_8O_{22}(OH)_2$

c. Tremolite: $Ca_2Mg_5Si_8O_{22}(OH)_2$

These minerals occur in basic igneous rock (peridotite or dunite) that has altered to serpentine and in metamorphic rock (including serpentine). A little asbestos is found in dolomite that has changed to serpentine. There is no use looking elsewhere for asbestos. A large deposit is usually needed for a profitable operation, and the material must be the right grade.

Mica

The electrical industry is the chief user of mica, especially muscovite and phlogopite. The physical peculiarities of these kinds of mica (elastic, cleavable, nonconducting) are unique. Sheet mica and scrap (including flake) mica have entirely unlike markets. Only pieces of muscovite or phlogopite that are more than 6 inches in size would seem to have much commercial appeal at present. Other mica and micalike minerals have other uses: lepidolite is a source of lithium compounds (see chapter 30), roscoelite is an ore of vanadium (see chapter 20), and vermiculite is an insulating medium (see chapter 31). Muscovite is also called white mica (although thick pieces may appear dark) and potassium mica: $KAl_2AlSi_3O_{10}(OH)_2$. Phlogopite is brown mica, or magnesium mica: $KMg_3(AlSi_3$-

$O_{10})(OH)_2$. These occur only in pegmatite: muscovite in granitic (especially zoned) pegmatite; phlogopite in basic pegmatite (especially in pyroxenite dikes). Mica is in general a declining product, owing largely to competition from synthetic material, but mica is only one of the minerals for which pegmatite is a source.

Talc

Soapstone and pyrophyllite are grouped with talc, for they have similar industrial uses. Talc is the softest mineral (number 1) in the Mohs scale: $Mg_3Si_4O_{10}(OH)_2$. The massive form is called steatite, and soapstone is much the same but perhaps more impure and a rock rather than a mineral. Pyrophyllite is actually a different mineral, having the composition $Al_2Si_4O_{10}(OH)_2$. Wonderstone is black pyrophyllite from the Republic of South Africa, but this name is familiar to American rockhounds as a picturesque volcanic rock. The familiar uses of these materials are in cosmetics (talcum powder) and tailor's chalk, but the quantity uses are much more prosaic, and the properties are so variable that different deposits may have quite different applications in industry. These include ceramics, paint, roofing, insecticides, and fillers. Talc comes from basic igneous rock and metamorphic carbonate rock that is rich in magnesium (dolomite). Pyrophyllite is metamorphic.

Barite

This common mineral—barium sulfate, $BaSO_4$—is used to make a heavy "mud" that is extremely useful in drilling oil and gas wells. Barite has other industrial uses of less importance, mostly in the glass, paint, and rubber industries. Witherite, which is barium carbonate, $BaCO_3$, shares some of the industrial uses, but it is much less abundant. As a commercial source, barite occurs mainly as veins and as replacement deposits in sedimentary rock and as residual deposits that have weathered out of such beds as heavy lumps of barite embedded in clay. Because of its weight, barite is most profitably mined when it is close to transportation.

Fillers

Many minerals and rocks, as well as products that are manufactured from them, are used when finely ground to put body, weight, opacity, or wearing qualities into almost every article of consumers' goods, from asphalt paving to candy. The list is too long to be helpful, but it serves to emphasize the possible sale of ordinary mineral substances to industries that otherwise might have no use for them.

Filters

Diverse minerals and rocks are employed to filter and purify all sorts of substances. Sand for filtering water supplies is perhaps too commonly used to merit much attention. Another important natural filter is diatomite, also known as diatomaceous earth (and by other names); it is the accumulation of vast numbers of tiny shells (made of silica) of fossil algae called diatoms. Geologic conditions do not favor deposits in the eastern and central parts of the United States, where they are wanted.

Fuller's earth is bleaching clay; its strongly absorptive power, naturally active, makes it of value. Bentonite may be either swelling or nonswelling. Both fuller's earth and bentonite are derived from volcanic ash and are found as layered rocks. Bauxite and alunite are other natural filters.

Optical Crystals

Clear calcite, known as iceland spar, is used for optical purposes. It is typically found in gas cavities in lava rock. Fluorite, quartz (rock crystal), tourmaline, muscovite mica, and gypsum (selenite) are likewise wanted for optical instruments.

Other Manufacturing Materials

Industry has uses for phosphates, nitrates, bauxite, magnesite, limestone, and still other earth substances besides those applications that have previously been mentioned. Alan M. Bateman includes sepiolite, better known as meerschaum, a mineral that is mostly carved into pipe bowls. Since being found in New Mexico, meerschaum might be expected to occur wherever else serpentine or magnesite has been altered.

A final word is in order: Do not overlook any deposit of earth material without first considering whether it can serve a useful purpose of some sort.

BOOKS ON NONMETALLIC MINERAL DEPOSITS

Mineral Facts and Problems, 4th edition, U.S. Bureau of Mines Bulletin 650, Washington, 1970 (1971).

Geology of the Industrial Minerals and Rocks, Robert L. Bates, Dover Publications, Inc., New York, 1969.

Industrial Minerals and Rocks (Nonmetallics Other than Fuels), 3d edition, Joseph L. Gillson (editor), American Institute of Mining, Metallurgical, and Petroleum Engineers, New York, 1960.

Economic Mineral Deposits, 2d edition, Alan M. Bateman, John Wiley & Sons, Inc., New York, 1950.

Nonmetallic Minerals, 2d edition, Raymond B. Ladoo and W. M. Myers, Mc-Graw-Hill Book Company, New York, 1953.

Metallic and Industrial Mineral Deposits, Carl A. Lamay, McGraw-Hill Book Company, New York, 1966.

Our Mineral Resources, Charles M. Riley, John Wiley & Sons, Inc., New York, 1959.

The Principles of Economic Geology, 2d edition, William Harvey Emmons, McGraw-Hill Book Company, New York, 1940.

Economic Geology, 5th edition, H. Ries, John Wiley & Sons, Inc., New York, 1925.

Introductory Economic Geology, W. A. Tarr, McGraw-Hill Book Company, New York, 1930.

Mineral Deposits, 4th edition, Waldemar Lindgren, McGraw-Hill Book Company, New York, 1933.

General Economic Geology: A Texbook, William Harvey Emmons, McGraw-Hill Book Company, New York, 1922.

Economic Geology of Mineral Deposits, Ernest R. Lilley, Holt, Rinehart and Winston, New York, 1936.

Minerals for the Chemical and Allied Industries, Sydney J. Johnstone and Margery G. Johnstone, John Wiley & Sons, Inc., New York, 1961.

Part 6

33

Opening a Mine

"There is no substitute for mining experience," says Harry E. Klumlauf. "It is improbable that a person with little or no mining experience could select a suitable mining method or develop a mine for production. Inexperienced people generally mine by 'gophering' or 'gouging,' and unless the ore is high grade, the operation is apt to fail." "Knowledge of mining is only storybook lore," says H. Byron Mock; few understand this complicated business. Different kinds of deposits may need to be tested to a greater extent than others in order to determine their value.

Development of a mineral deposit falls in between the period of exploration (when the size, grade, shape, and position of the mineral body are determined) and the systematic mining that continues until either the mineral body or the capital gives out. Development includes blocking out the productive body by driving the needed openings; preparing haulage ways, access openings, air shafts, ore chutes, and manways; planning the methods of mining and starting stopes; installing mining, ventilation, and haulage equipment; and building ore bins, mills, access haulage roads, and surface structures preparatory to the actual mining.

Development takes into account the type of deposit, the mining method to be used, the desired rate of extraction, the waste-disposal plan, and the physical environment of the deposit, including climate, topography, and subsurface water conditions.

The purpose of this chapter is not so much to give instruction in mining as it is to outline the main aspects of the subject so that you will know what is involved and to what extent you may wish to proceed. A better way may prove to be selling or leasing the property and letting someone else do the work. Those with some experience will find that this summary is useful as a reminder of the techniques of small-scale mining. The references at the end of the chapter will provide much more detailed information. Placer mining is discussed after lode mining.

Dr. John D. Ridge, advisor to the publisher of this book, reports that "of the 1,372 metal mines in the United States in 1967, 820 produced less than 10,000 tons of ore per year and certainly qualify as small mines: another 194 produced less than 100,000 tons of ore per year and probably would fit the small-mine category. Of the 6,912 nonmetal mines, 1,839 produced less than 10,000 tons a year, and another 2,449 produced less than 100,000 tons." Similar figures are available for a few other countries; elsewhere, they are not so reliable.

After sampling, testing, and assaying (as discussed in chapter 12) have showed that a likely geologic structure, rock outcrop, or mineralized body has been found and one or more mining claims have been staked in accord with the existing laws (as explained in chapter 13), the prospector is in a position to take the more serious (more expensive) step of opening his mine. These operations include the following stages:

1. A thorough sampling of the deposit to determine the grade and tonnage
2. Benefication or smelting tests
3. A marketing survey
4. A study of engineering considerations: mining methods, concentration or treatment plants, water and power, estimates of costs

Even then, you are well advised to hire the consulting services of a mining engineer or mining geologist, or both. Such a person may be registered; almost all engineers and many geologists (in California, for example) are required to be. Or else he should be recommended favorably, as, for example, by a state or provincial school of mines, where, in fact, he may perhaps be engaged. The mining directories that are listed on page 407 of this book give names and addresses of mining consultants;

also, advertisements appear in the mining periodicals that are listed on page 405. If capital is needed, an outside professional opinion may be required before it can be obtained. Both adequate financing and sound management must be available for the success of any mining venture.

Unless it is advised that the property should be abandoned, owing to a poor showing of commercial minerals, either further trenching or additional test pitting should be done on the outcrop or else drilling should be tried at depth: both these steps are taken for the purpose of learning more about the structure or the extent of mineralization.

Drilling to 2,000 feet in an Arizona copper prospect as done with a Mountaineer skid-mounted drill. *(Acker Drill Company, Inc.)*

A limited amount of trenching can be done with a pick and shovel, perhaps to a width of 3 feet and a depth of 7 feet. This spells work! A larger amount of trenching can be done (more quickly but also more expensively) with a bulldozer. Where the overburden is too thick for trenching, test pits are sunk, typically up to 4 by 5 feet in area and 40 feet in depth. Diamond drilling goes to greater depths, and wagon drilling is often used for shallow depths in dry rock.

Koehler S. Stout has enumerated the factors that affect the cost of mining, as follows. Present-day inflation makes it unsuitable to give here any estimates of price.

1. Acquiring and prospecting the ore body
2. Studying the possibilities of the venture
3. Developing and preparing the ore body for production
4. Erecting the plant and buying or renting the equipment
5. Mining the rock and bringing it to the surface
6. Taxes, insurance, and other business expenses
7. Milling, smelting, or otherwise preparing the mineral product for sale, including transportation from mine to mill or smelter
8. Selling the mineral product, including transportation to market

The expenses of underground exploration, listed by Harry E. Krumlauf, involve the following considerations:

1. Cost of equipment and supplies
2. Rent of equipment
3. Cost of labor: foreman, miner, mucker, topman, hoistman
4. Cost of roads
5. Cost of surface plant: headframe (including shaft collar), buildings (for hoist, changeroom, shop, storage), compressor shed, powder house, outhouse, leveling ground, water tank, air receiver, ore bin
6. Cost of sinking shaft
7. Cost of drifting, including laying of track and pipe
8. Cost of raising
9. Cost of sampling and assaying

The three major operations that present difficult problems in underground mining are the following:

1. Breaking the rock. This is less a problem in a placer mine!
2. Supporting the excavation. This is a less a problem in a shallow open-pit mine or a placer mine.
3. Loading and transporting the rock.

Other important problems include the following:

1. Supplying materials and tools to working areas
2. Supplying compressed air, water, fuel, electricity
3. Providing maintenance and machine service
4. Ventilation
5. Pumping of water

MINING METHODS

The mining method that is to be followed should be both suitable and economical under the conditions that prevail. The most important factors that are involved include the following:

1. Physical nature and strength of the rock: ability to support itself
2. Physical nature of the mineral body
3. Size and attitude of the mineral deposit: length, width, dip (can gravity be an aid?)
4. Value of the mineral body: costs must be related to possible recovery
5. Experience available
6. Capital available

Each kind of deposit can usually be mined only by one or two methods. Sand and gravel bars, for example, may perhaps be mined profitably only by simple hand methods.

Open-pit mining, or strip mining, removes the rock by surface excavation. It is usually cheaper than underground mining, but only after

Open-pit mining is cheapest in most iron deposits, as at Cornwall, Pennsylvania. *(Bethlehem Steel Corporation)*

stripping has got under way and the higher capital investment has been offset. In general, the deposits that are best adapted to open-pit mining are shallower, larger, and more continuous, having relatively unconsolidated overburden. This especially describes many uranium deposits of the Rocky Mountain region.

Since the turn of the 20th century, mining has completely changed over from underground to surface mining methods. These have the advantage of earlier operation, lower cost, greater productivity and total recovery, and less skilled labor, as summarized by Eugene P. Pfleider, of the University of Minnesota. Various other authorities believe, however, that future conditions—rising standards for metals, environmental problems, new geophysical techniques—will reverse the trend. Surface mining may be undertaken as an early stage, which may later continue underground by one of the following methods.

This typical stone quarry at Amherst, Ohio, produces Buckeye sandstone. *(The Cleveland Quarries Company)*

An adit, which is a tunnel that opens from the surface, is preferred to a shaft, for it is cheaper to open and it drains itself. Drifts, which follow the ore, should be driven at frequent intervals in order to intersect as much of the rock body as possible. Crosscut tunnels, which intersect the ore, are usually too expensive to be justified. Shafts, where necessary, should follow the ore until the property can be reclassified from the prospect stage to that of a producing mine. Underground work, in general, should be done as quickly and cheaply as possible, even if some is wasted after the mine is proved.

Open stoping is divided into four categories, as follows, but many variations are known under other names:

1. Gophering, or gouging. Small, irregular, high-grade ore deposits are followed closely regardless of their course.
2. Glory holing. Like open-pit mining, but removal of ore is underground.
3. Stull stoping. An open stope that is timbered to support the wall or provide a platform for mining operations.
4. Pillar (or room and pillar) stoping. The rock itself is left to support the wall.

Cut-and-fill stoping is divided into three categories, as follows:

1. Resuing, or stripping. Narrow, high-grade veins are mined by removing the ore and waste separately.
2. Horizontal cut and fill. The ore moves horizontally from the stope, while the waste moves into the stope by slushers (loaders) or ore cars.
3. Inclined cut and fill, or rill. The broken ore moves by gravity into a center raise, while the waste fill moves by gravity from the end raises.

Shrinkage stoping is employed in a vein of hard and firm rock. The broken ore provides a platform from which to work.

Square-set timber stoping advances the workings by successive stages. The square timber forms a continuous series of floors and compartments, or sets.

Underground-caving methods of mining are suitable only for large bodies of ore. They must be planned and conducted by experts.

The following outline is a checklist for placer investigations, as suggested by the U.S. Bureau of Land Management. Using it as a field guide will help the prospector to avoid overlooking any important factor, especially certain physical details that may prove to be more significant than the mineral content itself.

1. Date of examination
2. Name of claim(s) or property
3. State, county, district
4. Township, range, section(s)
5. Reason for examination
6. Examined by
7. Assisted by
8. Others present
9. Number of claims or acres

10. Names of locators and present owner
11. Owner's address
12. Type of deposit (stream, bench, desert, etc.)
13. Terrain
14. Gradient of deposit: Less than 5 percent ( ); more than 5 percent ( ). Remarks
15. Is the deposit dissected by deep washes or old workings? Yes( ); no( ). Remarks
16. Type and extent of overburden
17. Depth to permanent water table
18. Depth to bedrock
19. Kind of bedrock (rock type)
20. Hardness of bedrock
21. Bedrock slope or contour to be expected
22. Are high bedrock pinnacles or reefs in evidence? Yes( ); no( ). Remarks
23. Gravel is: well-rounded ( ); subrounded or subangular ( ); angular ( ). Remarks
24. Does gravel contain rocks over 10-inch ring size? Yes( ); no( ). Remarks
25. Boulders (maximum size, number, distribution, etc.)
26. Rock types noted in gravel
27. Predominant rock type (if any)
28. Sand (kind, amount, distribution, etc.)
29. Sorting or bedding patterns (if apparent)
30. Sticky clay? Yes( ); no( ). Remarks
31. Cemented gravel? Yes( ); no( ). Remarks
32. Caliche? Yes( ); no( ). Remarks
33. Permafrost? Yes( ); no( ). Remarks.
34. Buried timber? Yes( ); no( ). Remarks
35. Hard or abrasive digging conditions? Yes( ); no( ). Remarks
36. Character of gold: coarse ( ); flaky ( ); fine ( ); rough ( ); shotty ( ); smooth ( ); bright ( ); stained or coated ( ). Remarks
37. Can good recovery be expected by use of riffles or jigs? Yes( ); no( ). Remarks
38. Is recovery said to depend on secret process or special equipment? Yes ( ); no( ). Remarks
39. Are black sands said to contain locked gold values? Yes( ); no( ). Remarks
40. Have black sands been checked for valuable minerals other than gold? Yes( ); no( ). Remarks
41. Distribution of values in deposit (if known)
42. Record or evidence of previous sampling
43. Results of prior sampling (if known)
44. Are old workings in evidence? Yes( ); no( ). Remarks
45. Past production (if known)
46. Date of last production or work

47. Reason for quitting
48. Present work (if any)
49. Applicable mining method
50. Possible cost to bring property into production
51. Possible mining cost
52. Dimensions of (physically) minable ground
53. Possible extensions
54. Maximum yardage indicated to date
55. Mining equipment on ground
56. Accessory equipment or improvements on ground
57. Water supply
58. Power supply
59. Does property have adequate tailings dump room? Yes( ); no( ). Remarks
60. Would mining in this area come under county, state, or federal water quality-control regulations? Yes( ); no( ). Remarks
61. Fish and game regulations? Yes( ); no( ). Remarks
62. Can settling ponds be built to effectively retain or clarify the muddy water? Yes( ); no( ). Remarks
63. Is property subject to resoiling or other surface restoration regulations? Yes( ); no( ). Remarks
64. Elevation of property
65. Climate
66. Working season
67. Season governed by
68. Surface cover and its effect on mining
69. Merchantable timber or other surface values
70. Nearest town
71. Access
72. Reference maps
73. Aerial photos (USGS, Forest Service, etc.)
74. Reference literature
75. Previous examinations or reports
76. Other reference sources
77. Sampling (describe or attach notes)
78. Additional information and remarks
79. Attach suitable map or sketches (if needed)
80. Attach photographs of pertinent features (if available)

Placer methods of mining include the following:

1. Panning. The gold pan and its relatives, and their uses, are described in chapter 7. These are more effective for sampling and scientific study than for actual mining.

2. Sluicing. A sluice box is more effective than a pan, but requires ample water. A rocker has less capacity, but can be used with less water. A long tom is sometimes used to break clay or cementing material. The

obstructions that collect gold and other minerals are called riffles, of which many types are used (such as common or slat, pole, block, rock or stone).

3. Hydraulicking. A hydraulic giant is a water nozzle attached to the end of a large pipe that can be pivoted. A ground sluice releases water suddenly. Both devices wash gravel into a sluice box, using much water.

4. Mechanical washing. Gravel is fed into a hopper by a power shovel, dragline, scraper, or other machinery. Abundant water is needed for this sort of plant, which is also known as a dry-land dredge.

5. Dredging. A dredge is a floating boat that digs gravel from a pond by an endless string of buckets and deposits the waste material behind it. Various means are employed to recover different minerals.

Mechanical gold pans are manufactured for larger scale efforts. *(Joy Manufacturing Company)*

Sluice box. *(Jeanette Maas)*

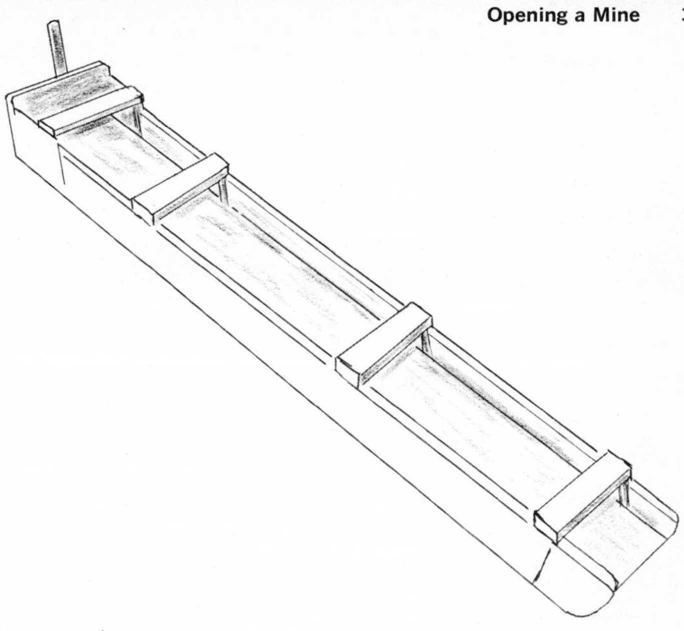

A long tom. *(Jeanette Maas)*

The Denver Gold Saver tests and samples placer deposits by means of vibrating riffles underneath the tubular scrubbing trommel. *(Joy Manufacturing Company)*

Hydraulicking in the tin deposits of Malaya. *(British Information Services)*

The Denver mechanical gold pan fits on a truck or trailer and duplicates the motion of a hand pan. *(Joy Manufacturing Company)*

A dredge recovers placer minerals by separating them from sand and gravel. This one dipped to 50 feet. *(Yuba Consolidated Industries)*

Discussing the use of explosives does not seem appropriate in this book. The *Blasters' Handbook,* sold by E. I. du Pont de Nemours & Company, Inc., and the condensed booklet *Explosives Development and Use,* obtainable free from the same firm (Wilmington, Delaware 19898), explain the subject in sufficient detail. The term *explosives* includes the following:

1. Dynamite: a detonating, or "high explosive," cap-sensitive mixture with nitroglycerin or other explosive compound.
2. Black blasting powder: a deflegrating, or "low explosive," mixture made of sulfur, charcoal, and nitrate (potassium—saltpeter—or sodium); it has a slow shearing and heaving action.
3. Pellet powder: modified "B" (sodium nitrate) black blasting powder that is pressed into cylindrical pellets 2 inches long and 1¼ to 2 inches in diameter.
4. Blasting caps: nonelectric detonators to initiate explosive charges by ignition of burning safety fuse; they are aluminum shells that are loaded with two or more explosive charges, at least one of which is a detonating charge.
5. Electric blasting caps: instantaneous or delay types of electric detonators to initiate explosive charges.
6. Detonating cord: a round, flexible cord that contains a covered center core of high explosive, which initiates charges of high explosives.

Dynamite is being rapidly replaced by ammonium nitrate—fuel oil, what is termed ANFO. Other chemicals have likewise cut deeply into the market.

As a mine is enlarged, further exploration as warranted should be carried on by these procedures:

1. Drifting along exposed stringers of ore
2. Following fractures and other openings that might contain ore
3. Following igneous contact
4. Tracing the general trend of ore bodies that have already been found
5. Following minerals of local significance; for example, pyrite that contains gold but is otherwise of no value

The Marietta continuous miner is a formidable monster, well adapted to underground operations in parallel beds. *(National Mines Service Company)*

The above exploration may be conducted by the following methods, accompanied by frequent sampling, assaying, and mapping:

1. Extending present workings
2. Driving shafts, drifts, crosscuts, raises, or winzes
3. Drilling at the surface, at the face of drifts or crosscuts, into the bottom of shafts or winzes, or on the tops of raises

BOOKS ON PRELIMINARY MINING

Much information on this subject is given in section 10 of *Mining Engineer's Handbook,* 3d edition, Robert Peele (editor), John Wiley & Sons, Inc., New York, 1941 (2 volumes). The following are also useful:

This electric drill is angling into a gold mine in Colombia. *(E. J. Longyear Company)*

Diesel locomotives are widely used in mining. *(National Mine Service Company)*

Transporting this mining drill in Cuba was slow work. *(E. J. Longyear Company)*

Blasters' Handbook, 15th edition, E. I. du Pont de Nemours & Company, Inc., Wilmington, Delaware, 1969.
Diamond Drill Handbook, J. D. Cumming, J. K. Smit & Sons of Canada Ltd., Toronto, 1951.
The Examination of Prospects, 2d edition, G. Godfrey Gunther and Russel C. Fleming, McGraw-Hill Book Company, New York, 1932.
Sampling and Estimation of Ore Deposits, C. F. Jackson and J. B. Knaebel, U.S. Bureau of Mines Bulletin No. 356, 1934.

MARKETING MINERAL PRODUCTS

Inasmuch as few small mines have sufficient reserves of ore to justify the cost of erecting a mill to concentrate it, most mining operators will ship the raw ore to a custom mill or smelter, which is one that processes ore for independent mines, perhaps in addition to its own output. Large mines may operate their own mills, but not necessarily so, the situation depending on the cost factors involved. These factors contrast the cost of building and running a mill and shipping the concentrates, on one hand, with the cost of shipping the ore to the mill and paying its charge. A small mill, furthermore, usually recovers a smaller percentage of the values than does a large one. The availability of water may prove to be the decisive factor in operating a concentrator, as indeed it has been known to be in the original mining itself; some attorneys and engineers believe that the lack of water rights is today the chief limiting element in opening a mine in western America. Adequate power must also be available. The guidance of a metallurgist who is experienced in milling is essential.

Private attempts at mineral recovery are usually unsuccessful, whether they are done in a crucible or electric furnace (not recommended) or not. Prospectors are often dissatisfied with mill and smelter payments because they fail to support assay reports; almost always has the cause been traced to poor sampling (barren rock is worthless) or a lack of understanding as to how a plant calculates its treatment charges, deductions, and penalties. Base metals, for example, are not paid for when they occur in gold or silver ore. Another fact to be remembered is that the presence of certain minerals and metals prevents the recovery of others, and certain metals (such as zinc and arsenic) are often regarded as an outright abomination. Some prospectors have the strange idea that assays somehow do not reveal the entire gold content of an ore but that the mill should be able to find it and pay for it.

Under certain circumstances (technically complicated), gold and silver ores are shipped direct to the smelter. The reduced number of mills and smelter in the United States, owing to the expansion during recent decades in foreign mining, adds a serious problem to the troubles of mining and puts a burden on the miner to seek out a custom plant that can satisfactorily handle his particular type and grade of ore. Agreement should always be reached before shipping.

On March 15, 1968, the U.S. government discontinued the purchase of gold, making all previous gold regulations obsolete. Eligible deposits for exchange were accepted by the U.S. Mint until the end of 1969. Persons or firms that hold valid government gold licenses are permitted to buy gold from private producers.

BOOKS ON MARKETING MINERALS

Mineral Facts and Problems, 4th edition, U.S. Bureau of Mines Bulletin 650, Washington, 1970 (1971).

The Marketing of Metals and Minerals, Josiah Edward Spurr and Felix Edgar Womser (editors), McGraw-Hill Book Company, New York, 1925.

Non-Metallic Minerals, 2d edition, Raymond B. Ladoo and W. M. Myers, McGraw-Hill Book Company, New York, 1951.

BOOKS ON MINING

Mining Engineers' Handbook, 3d edition, Robert Peele (editor), John Wiley & Sons, Inc., New York, 1941 (2 volumes).

Mining Geology, Hugh Exton McKinstry, Prentice-Hall, Inc., Englewood Cliffs, N.J., 1948.

Field Geology, 6th edition, Frederic H. Lahee, McGraw-Hill Book Company, New York, 1961.

Principles of Field and Mining Geology, James Donald Forrester, John Wiley & Sons, Inc., New York, 1946.

Elements of Mining, 3d edition, R. S. Lewis and G. B. Clark, John Wiley & Sons, Inc., New York, 1964.

Elements of Mining, 3d edition, George J. Young, McGraw-Hill Book Company, New York, 1932.

Introduction to Mining, Bohuslav Stoces, Pergamon Press, Oxford, England, 1954 (2 volumes).

Methods of Working Coal and Metal Mines, Seth D. Woodruff, Pergamon Press, Oxford, England, 1966 (3 volumes).

Quarrying, Opencast and Alluvial Mining, John Sinclair, Elsevier Publishing Company Ltd, Barking, England, 1969.

Underground Practice in Mining, 3d edition, Bernard Beriger, Mining Publications Ltd., London, 1947.

Mineral Processing, 3d edition, E. J. Pryor, Elsevier Publishing Company, New York, 1965.

Mining Methods and Equipment Illustrated, K. S. Stout, Montana Bureau of Mines and Geology Bulletin 63, Butte, Mont., 1967.

Handbook for Small Mining Enterprises in Montana, V. M. Sahinen et al., Montana Bureau of Mines and Geology Bulletin 39, Butte, Mont., 1964.

Mining, B. Boky, John Scott (translator), Mir Publisher, Moscow, 1967.

Mining of Mineral Deposits: A Textbook, L. Shevyakov, V. Schiffer (translator), Foreign Language Publishing House, Moscow, 1963.

SELLING A MINING CLAIM

Mining property may be up for sale because the owners either cannot operate it or do not want to. The second reason may be closely related to the first. The inability to run a mine profitably may be ascribed to one or more of the following:

1. Insufficient mineralization: an obvious reason, which usually should have been foreseen by proper exploration.
2. Declining economic conditions: nobody's fault.
3. Poor management: many factors are involved, including a falling out of partners or investors and the hiring of incompetent or dishonest people.
4. Lack of capital: should be attended to before operations are begun.
5. Disasters: fire, caving, excess water underground, lawsuits.
6. Mill construction: poor management or inexperience may lead to this being done too early, using up money that is needed for other purposes.

Theodore J. Hoover gave 10 reasons why mines shut down:

1. Too great a depth
2. Different kind of ore at depth
3. Depletion of ore
4. Reduced price of metal
5. Flooding by water
6. Caving in mine
7. Poor management
8. Conversion of property to raise capital
9. Abandoned too soon
10. Change in laws

Lacking the capital, patience, or interest to develop a mining claim oneself, the problem of disposing of it should be considered. A prompt sale may provide funds for future prospecting, further investment, or other purposes. A partnership that is entirely suitable for prospecting, moreover, may prove unsatisfactory when dealing with the complex business of mining, in which unfamiliar decisions need to be made. Most prospectors lack the highly specialized technical knowledge to carry on advanced exploration and the development of mining property.

The names of mining companies are given in the appropriate directories, as listed on page 407 of this book. A suitable buyer may be met by advertising in the mining periodicals listed on page 405. Or a mining-company scout may hear of your find and seek you out. When you make the approach yourself, nothing is more convincing than to submit samples that are truly representative, as discussed in chapter 12. Accompanying sketches of the property and its accessibility are helpful.

Certain companies devote their main attention to acquiring mineral claims, exploring and developing them more fully, and then selling them to operating companies, to which they have better access than the lone prospector. Furthermore, they can judge better the true value of a claim, for the enthusiastic prospector usually either places too high a price on what he has discovered—no outsider has any sentimental attachment to it!—or else he is too diffident to ask enough while in the august presence of a mining-company executive who has been trained to look haughty in his overstuffed chair.

You may eventually have to choose between a quick sale at a price that is probably too low, or a delayed return that will pay off gradually as mining proceeds (and the property is proved), or perhaps dying in poverty because of your excessive stubbornness. Remember that large companies may go on merrily for a longer time than you are apt to: the initials H. B. C. that stand for Hudson's Bay Company were long ago represented by Canadians to mean Here Before Christ. Unless you have had past experience or employ a mining engineer, mining geologist, or lawyer who knows how to negotiate, no one can offer you much advice in this matter of contracts. Asking for an overriding royalty, however, is almost never a mistake, and such a perpetual share in the profits can be the principal part (and the most delightful part) of one's estate. Beware of selling to a purchaser who is buying for an investment, with no intention of mining; seldom does the owner receive a fair return for his mineral rights. An accountant can suggest better ways to sell or lease mining property.

Mining property should be put into reasonably presentable condition before being offered for sale. As much of the most favorable part of a deposit as possible should be exposed, and the rest should be limited to careful sampling, while avoiding such wasteful practices as driving long tunnels, erecting mills, building roads, or buying equipment in order to "look good." Side issues can be disastrous, says C. M. Bennett. Especially should workings from which profitable ore is alleged to have been taken, and holes in which favorable testing has been done, be cleaned for examination. Whether it is developed or not, property should be marked and outlined on base maps that show road access. All legal papers (claim records, partnership agreements) should be available, together with assay reports and shipping records.

In addition to these more or less obvious aids to salesmanship, the owner should strike a balance between his demands and his expectations. An initial cash payment is not apt to be approved, nor are high royalties that suggest trying to deplete the mine too quickly. Keep in mind that the examining geologist or engineer who represents a large company must satisfy the stockholders, his superiors, and his fellow geologists and engineers; it has been said that the good opinion of one's peers is what really "keeps a man honest." Furthermore, even a nearly guaranteed profit on a particular property may not have equal appeal to all firms, which have different long-range plans and tax requirements. The high risk in mining justifies a bigger return on the investment than do most other enterprises. Property that is badly in debt or (especially) is involved in any sort of litigation is under a serious handicap. Most important, a prospective buyer wants to know as much as possible at the start, so that a report on the essential features and likely ore reserves—"proved" (blocked out), "probable" (engineering estimate), or "possible" (geologic estimate)—will be very helpful in setting into motion the negotiations that it is hoped will secure an exploration lease with an option to buy.

The Atomic Energy Commission classifies uranium ore as "measured" (tonnage and grade computed from revealed ore and detailed sampling), "indicated" (based on observation and projection), and "inferred" (based on geologic estimates).

In conclusion, a willingness to share the financial risk will almost always secure the maximum return on a mining property.

The subject of selling shares in a limited company (corporation) and additional methods of financing mines is discussed in chapter 8 of *Out of the Earth: The Mineral Industry in Canada,* by G. B. Langford (University of Toronto Press, 1954).

The following periodicals that deal with mining are of general interest and are issued commercially. Except the journals of the American Mining Congress, the American Institute of Mining, Metallurgical, and Petroleum Engineers, and the Canadian Institute of Mining and Metallurgy, they do not include the publications of organizations, nor those of various state, provincial, or national governments. There are, however, many such that are of value, and you are referred for their names, addresses, and prices to library directories such as *The Standard Periodical Directory* and *Ulrich's International Periodicals Directory.* Magazines that deal with the mineral-collecting hobby and related aspects are listed on pages 233–234.

Australian Mining
415 Bourke Street
Melbourne C.1, Australia

California Mining Journal
P.O. Drawer 628
Santa Cruz, California 95060

Canadian Mining and Metallurgical Bulletin
906-1117 St. Catherine Street
Montreal 2, Quebec

Canadian Mining Journal
Gardenvale, Quebec

Canadian Pit and Quarry
Gardenvale, Quebec

Engineering and Mining Journal
1221 Avenue of the Americas
New York, New York 10020

Industrial Minerals
46 Wigmore Street
London W.1, England

Mineral Industries
104 Mineral Sciences Building
University Park, Pennsylvania 16802

Mining in Canada
1450 Don Mills Road
Don Mills, Ontario

Mining Congress Journal
1100 Ring Building
Washington, D.C. 20036

Mining Engineering
345 East 47th Street
New York, New York 10017

Mining Journal and Mining Magazine
15 Wilson Street
London, E.C.2, England

Mining Record
718 17th Street
Denver, Colorado

National Prospector's Gazette
Segundo, Colorado 81070

Northern Miner
77 River Street
Toronto 2, Ontario

Queensland Government Mining Journal
2 Edward Street
Brisbane, Queensland

Rock Products
300 West Adams Street
Chicago, Illinois 60606

South African Mining and Engineering Journal
66 Commissioner Street
Johannesburg, South Africa

South Australia Department of Mines Mining Review
Box 38, Rundle Street P.O.
Adelaide, 5085 S.A., Australia

Western Miner
305-1200 West Pender Street
Vancouver, British Columbia

World Mining
500 Howard Street
San Francisco, California 94105

With interlibrary loans, nearly universally available in the United States, the reader can count on the assistance of his public or school librarian to obtain him virtually every item that has been published. Certain libraries emphasize material on minerals more than do others; librarians have lists of such places. The 1,010 so-called government depository libraries, situated in every state, generally have many (or most or all) of the publications of the U.S. Bureau of Mines and U.S. Geological Survey. In addition, the U.S. Government Printing Office maintains bookstores for government publications in Atlanta, Boston, Chicago, Dallas, Denver, Kansas City, Los Angeles, New York, Pueblo (Colorado), and San Francisco; others are planned for opening in federal buildings in other cities. Five stores are operated in Washington, D.C. where most orders are filled by mail by the Superintendent of Documents, zip number 20402. The U.S. Atomic Energy Commission has a list of depository libraries in which all unclassified reports are available.

MINING INDUSTRY DIRECTORIES

Following are directories of the mining and mineral industry, covering production statistics, prices, names and addresses of mines and smelters, and other information. The many useful publications of state, provincial, and federal governments are not included but can be obtained by writing to the agencies listed on pages 43 to 47.

Yearbook of American Bureau of Metal Statistics
50 Broadway
New York, New York 10004

Mines Register
525 West 42d Street
New York, New York 10036

Pit and Quarry Handbook and Purchasing Guide
431 South Dearborn Street
Chicago, Illinois 60605

Canadian Mines Handbook
77 River Street
Toronto 247, Ontario

Canadian Mining Manual
Gardenvale, Quebec

Financial Post Survey of Mines
481 University Avenue
Toronto 2, Canada

Mining Yearbook
Vintry House, Queen Street Place
London E.C.4, England

34

Mineral Processing

A wide diversity of physical and chemical processes are included in this term, which often is given as mineral dressing, ore dressing, or beneficiation.

Mineral processing embraces all the means that regulate the size of a substance, remove unwanted constituents, and improve the quality, purity, or assay grade of a commercial product; for ores, it also involves concentrating or otherwise preparing them for smelting by drying, flotation, or magnetic separation. Rocks and minerals are crushed and ground by wet or dry methods. They are separated into different kinds, and the waste (either undesired or deleterious) matter is removed. Either the valuable or worthless content may be dissolved by chemical action for further treatment.

A more highly technical knowledge is needed to make use of these methods than for the mining operations themselves, many of which were, to a limited degree, known even to primitive man. The actual chain of processes that is most suitable to a particular venture must be determined by an experienced mining engineer or metallurgist. Hence, this chapter can serve only to outline, in true handbook fashion, the principles of mineral processing, in order to acquaint the prospector and mineral collector with the sort of techniques that are in everyday use. Some means are, of course, much simpler than others; even a child can select commercial scheelite (tungsten) ore under an ultraviolet lamp.

Besides reading about them, how do you go about deciding what processes might be useful to you? Consult a skilled metallurgist, obviously. But this is expensive, and a practicing geologist ought to know if a metallurgic problem is involved and at what stage the program should be turned over to the metallurgist. The complications may, in fact, call for the rejection of the entire enterprise. Size is a major factor, however, for an answer to problems of processing can almost always be found if

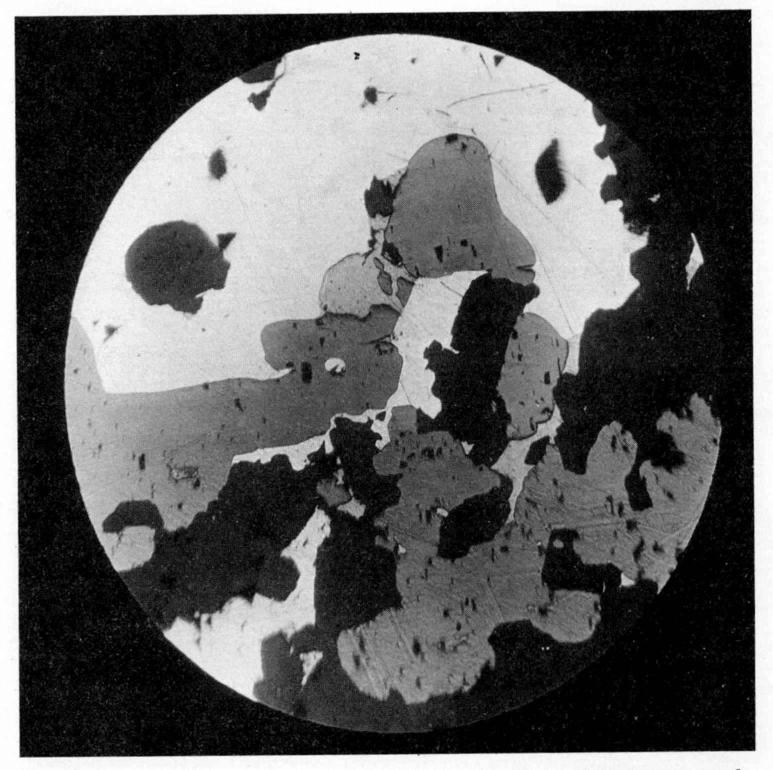

The size, shape, and associations of each mineral as seen in a polished section under a reflecting microscope are important factors in deciding the method of treating the ore. *(U.S. Geological Survey)*

the deposit is large enough, whereas a small deposit may not be able to afford much experimentation.

Outside, expert help can often be obtained without much cost—a situation that is seldom known to the beginning operator. Most railroad companies are eager to offer valuable advice, with the hope that a profitable mine will become a shipper of rock, mineral, ore, or concentrate; certain railroads maintain regular staffs for this purpose.

Manufacturing of equipment and suppliers of chemical reagents, both of which advertise in the mining periodicals that are listed on page 405, are other sources of information, perhaps to the extent of preparing a flow sheet that will be suitable for the given operation. It may ultimately prove cheaper to send the broken rock to a custom plant instead of proceeding further, but only a careful study of the kind and grade of mineral, the local conditions, and the financing that is available can suggest the best move.

Certain firms, which advertise in the mining periodicals (see page 405), specialize in selling or leasing portable equipment (some as mobile units) for concentrating ores. Used equipment has its own dealers; they, and auctioneers of such items, also advertise in the mining periodicals. Putting together miscellaneous parts can lead to costly and time-consuming operating troubles but need not be avoided altogether for that reason. The use of a furnace for smelting ore or concentrate on a small scale has not yet been worked out satisfactorily, mainly because most ores are of complex composition, and each metal that is present affects the treatment that needs to be given. Gold and silver, for instance, require copper or lead as a collector, after which the copper matte or lead bullion has to be refined. A large stock of any ore is needed, for a smelter must run continuously. Smelting was done in ancient times, of course, when the experts in this business, although they did not have a college education, were regarded as the most highly skilled of all craftsmen, as the Bible makes clear. Smelting is a job that still calls for proficiency and judgment.

Short courses of instruction in the various phases of mineral processing are given on occasion by certain of the mining schools in the United States and Canada that are listed on pages 211–214. Instruction manuals are available from the manufacturers of equipment.

Ore Reduction

Rock is always broken when it is removed, but it may have to be further reduced in size in order to be treated; this process, called ore reduction, involves breaking, crushing, and grinding. Hammers may be used (sledging, spalling, or cobbing), but a jaw crusher, or gyratory crusher, is usually employed as a primary crusher for a small mine, the rock afterward being raised to a storage bin or dropped into it. Further (secondary) reduction is obtained by some kind of mill. Stamps, or stamp mills, are of older vintage, but they are not entirely obsolete for use with hard ore (often containing much quartz) or moist and sticky

These rod mills crush ore from Butte, Montana. *(The Anaconda Company)*

The Dorr heavy-duty classifier employs reciprocating rakes and inclined-slope bed. *(Dorr-Oliver Incorporated)*

ore. Soft ore is better crushed in a Huntington-type (rolling) mill. Ball mills (using steel balls) and rod mills (using long steel rods) reduce hard ore, which can be ground to an even smaller size by a tube mill (using steel balls, flint pebbles, or hard rock). Several kinds of classifiers separate the discharge from a ball mill into oversize material and the desired product. The practices mentioned vary according to the local circumstances, and there is probably no one best method.

Waste Removal

Physical beneficiation, called concentration, has two functions: it enables complex ores to be separated for milling at a lower cost than they can be smelted; and it reduces the tonnage of waste rock, thereby saving on shipping and smelter charges and increasing plant capacity. The ores that go into a mill are known as millheads, or heads; the discarded material becomes the tailings, or tails; and the useful product is the concentrate.

Several simple processes are used to remove waste rock. One is weathering, whereby the atmosphere causes the rock to disintegrate or decay. In slack times, for example, diamond-bearing rock in South Africa was spread over the ground and allowed to weather, releasing the gem crystals at minimum cost.

Hand picking is another method, widely used for larger pieces, especially when they are not too scattered (disseminated) and have features that are easily recognized, such as the white quartz (containing gold) in dark slate at the Alaska Juneau mine. Saving desired material is called selection; discarding undesired rock is called sorting. Ore minerals may be selected before crushing (even underground) as well as after; they are often washed or sprayed after being crushed and sieved (for size) but before being picked over. After the shipping ore and waste are removed, milling ore remains.

The extent to which hand picking is profitable will depend on the particular costs of mining; lone prospectors and miners are likely to find that it pays them well to sort out waste rock before freighting ore (as high-grade, or shipping, ore) to an outside plant where charges are high. Sorted ore may be worth more per ton than mill concentrate and is less subject to loss when treated later. Hand picking, furthermore, raises the assay value above the original estimate that was based on impartial sampling. Certain mineral products—among them asbestos, barite, beryl, building stone, copper, feldspar, fluorspar, gemstones, graphite, lead, manganese, mercury, sillimanite, tin, tungsten, and zinc—are especially suitable for hand picking, particularly if they are too fragile to be worked on mechanically.

When chemical treatment is to be applied, sorting may save chemicals and perform other helpful though inconspicuous functions. Sorting is done on belt or pan conveyors, revolving or stationary tables, shaking devices, or the floor. Geiger counters and similar instruments are in use for selecting radioactive minerals, and fluorescent lamps are used for luminescent minerals.

Gravity concentration is a modification of gold panning. Both specific gravity and particle size affect the utility of this older method, which is best adapted to heavy, nonmetallic minerals. The greater the difference in specific gravity between ore and gangue, the more effective the separation. But, as to size, uniformity is necessary, or else small, heavy minerals will settle as fast as large, light ones. Some minerals tend to slime, spoiling this method for cinnabar, clay, malachite, molybdenite, serpentine, stibnite, and tungsten minerals. Jigs (see below) and vibrating tables are the most common kinds of equipment used; minerals can also be brought to settle in a rising current of water. A jig separates material by a pulsating, up-and-down current of water, which leaves them in layers according to their specific gravity. Gravity concentration is most often used in combination with other processes of treatment.

Magnetic and electrostatic concentration have a limited application to iron-bearing minerals. Several (magnetite especially) are strongly magnetic, others (chromite, ilmenite, franklinite, garnet, wolframite, tantalite) are weakly so, and some (hematite, goethite-limonite, siderite) can be made magnetic by roasting them. Magnetic minerals can also be removed from nonmagnetic minerals that have a greater value (for use in ceramics, as an example) by resorting to the same technique but in reverse.

A minor but interesting method of dry concentration is employed in the asbestos industry, where fibers of this remarkable mineral are blown away from the broken rock by strong air currents and are sucked into a vacuum device.

The sink-and-float method employs a finely ground, heavy solid (such as galena or ferrosilicon) suspended in water, making the equivalent of a heavy fluid. Still fairly new, it has been used to recover lead ore, Lake Superior iron ore, Tennessee zinc ore, and African diamonds.

Flotation

Partly physical and partly chemical in its action, flotation is the reverse of gravity concentration. With its use, heavy minerals can float while light, undesired minerals sink. Flotation has become the principal means of separating most metallic minerals and some nonmetallic ones. By selec-

One of the largest flotation installations recovers nickel in Canada. *(Denver Equipment Company)*

tive flotation—whereby appropriate reagents are used, and the alkalinity (pH) of the solution is controlled—almost any sulfide mineral can be separated from all its associates, both ore and gangue, no matter how complex. Native gold, scheelite (tungsten), and certain other nonsulfide minerals (such as graphite, which is flaky) can be isolated by flotation, but sulfides (especially copper, lead, and zinc) are the most amenable to this treatment, and silicates are the least.

Flotation is an intriguing process to watch. Finely ground rock is added to water in a series of flotation cells to make a pulp, into which special oils are mixed as frothers, collectors, and conditioners. Air is then bubbled up through the cells. The mineral particles attach themselves to the oily bubbles, forming a froth, and are skimmed off at the surface where the bubbles collapse. This method of ore treatment has to be seen to be believed.

Amalgamation

Likewise partly chemical and partly physical, amalgamation consists of alloying mercury to metallic gold and silver (either native or obtained by treatment). This method, an ancient one, is used most for placer gold but is suitable also for many oxidized deposits. Gold in pyrite or other minerals cannot be recovered by amalgamation. A plate of silvered copper is coated with mercury, and the pulp is passed over it. Or (more often) a revolving barrel or grinding mill brings mercury into contact with the wet-ground ore. The mercury holds the gold and silver, somewhat absorbing them. The precious metals are released when the amalgam is scraped off and heated to vaporize the mercury, which condenses for reuse. Amalgamation is today a subsidiary process, combined with flotation and cyanidation.

The use of corduroy blankets to trap gold has been substituted for amalgamation at the Rand (Witwatersrand) gold mines in the Republic of South Africa. Besides gold, corduroy also catches sulfide minerals and has the advantage of not causing chemical reactions as do metal plates. The minerals come to rest in the furrows between the ribs, which are farther apart than in the corduroy fabric that is used in personal clothing.

Hydrometallurgy

Wet chemical methods that result in dissolving a metal and then precipitating it by electrolytic means are classified as hydrometallurgy. Leaching and cyanidation, discussed below, belong in this category.

Leaching

Some ores can be treated by leaching them; that is, dissolving them in liquids: sometimes acids, sometimes alkalies, sometimes plain water. Sulfuric acid does a good job of leaching unoxidized copper ore except where too much limestone is present to take up the acid; adding water to some copper ore produces sulfuric acid (and often ferric sulfate) naturally. Ferric sulfate had been used with sulfuric acid in some places. A cheap and unique method for recovering copper is by allowing the copper sulfate solutions to react with scrap iron and replace it. Oxidized ores of lead and zinc are not susceptible to profitable leaching.

Ammonia solution dissolves some copper ores. Rock salt dissolves in water, as do potash, nitrate, and other soluble substances. The recovery of native sulfur by the Frasch process, whereby sulfur is melted in hot water and pumped to the surface, is a kind of distinctive mining operation that is rather different from the above examples of leaching.

Cyaniding

Gold and silver ores, especially if they lack much copper and lead, are often recovered by dissolving the precious metals in a weak solution of cyanide (sodium or calcium) of known strength. The metal is then precipitated with shavings or dust of aluminum (if lime is absent) or of zinc. After being filtered, the metal is melted (with fluxes) and cast into bars of bullion. The cyanide process works best with native gold (not too coarse), in oxidized ores, native silver, and other silver minerals that contain sulfur, antimony, or arsenic (or combinations of these elements). Many minerals, such as soluble sulfides, are not suitable. In difficult circumstances, this method may be utilized with others in a mixed treatment, and suitable flow sheets should be worked out in advance. An important advantage of cyaniding is that the product can be melted and sold without undergoing smelting. Base metals that may be present, however, cannot be recovered; but, in general, this process is less expensive than flotation.

Pyrometallurgy

Instead of a wet chemical, heat is the active agent in pyrometallurgy, which includes volatilization and smelting. This most ancient process is also the most important of the extractive techniques, being used for nearly all iron (and steel), nickel, and tin, most copper, and much gold, silver, and zinc, as well as many of the minor metals.

Volatilization

Volatilization is much used for zinc ore and for oxidized (sulfur-free) lead ore that contains gold and silver.

Smelting

As the word suggests, smelting is the process of melting ores to obtain metal. Because only a few of the largest mines operate their own smelters, concentrates are usually shipped to a custom smelter, which buys them outright and blends them with other ores for treatment.

The feed may be ordinary or selected ore, or it may consist of concentrates. High-grade copper and lead ore can be smelted without concentration. Coke, natural gas, or other fuel (such as burning pyrite in the ore) heats the blast or (nowadays) reverberatory furnace into which the feed and flux (to aid fusion) are put. The kind of flux depends on the ore (only a few are self-fluxing): limestone flux for iron, iron flux for siliceous ore, siliceous flux for pyritic concentrate. The result is to melt the ore, which separates into slag that floats and metal that sinks. A collector, such as copper or lead, is added (if it is not already present) in order to carry down any drops of molten gold.

Each metal has its own form of treatment. Copper, for example, is

Teeming a heat of open-hearth steel at Sparrows Point, Maryland. *(Bethlehem Steel Company)*

removed as matte, which is purified as blister copper. Lead becomes bullion, cast as pigs and sent to the refinery. If you understand the terms used in the buying and selling of ores, and the reasons for them, you might be able to sort and classify your mine product in such a way as to obtain a maximum return. This obviously requires experience with ores as well as an up-to-date knowledge of the marketplace. Certain ore, such as iron and zinc, is often handled on a simple basis, which can be contracted for in advance, but most ore is mixed and variable; terms vary also according to the region, the smelter, and the state of the economy. In addition, the problems of refining (discussed below) must be taken into account for whatever effect they may have on the whole matter of metal recovery.

Refining

The final separation and purification of metal that comes from the smelter are termed refining. If the smelter sends its product to a separate refinery, the cost is included in the original agreement with the mine, just as if the refining is done on the premises. In order to isolate the precious metals, to remove undesirable constituents, and to obtain metal of commercial quality, any of three kinds of refining may be carried out, as discussed below:

1. Fire refining is most often applied to copper that does not contain gold or other metals; copper matte is converted to blister copper. Silver, gold, and antimony are removed from lead bullion by this method.

2. Electrolytic refining involves the dissolving of metal (usually copper) in an acid solution through which electric current is passed. The metal is redeposited in a pure state while the impurities remain in the sludge or slime—but this may contain a valuable amount of gold or other metals. The platinum metals are recovered in this way. Bismuth may be recovered electrolytically from refined lead. Zinc ore is apt to be treated by an electrolytic process rather than by the older method of smelting.

3. Chemical refining is accomplished by a number of methods in which chemical reactions play a part rather than heat or electricity.

Inasmuch as most small metal mines sell crude ore or concentrates rather than finished metal, or else sell cyanided gold and silver, their dealings are ordinarily with a custom smelter. Apart from the precious and common base metals, however—which are quoted in mining periodicals (pages 405–406)—many mineral products are bought on rather rigid specifications, and the market is much more limited. For some materials, in fact, an open market does not exist, but negotiations have to be conducted on an individual basis. Information regarding the preparation and marketing of mineral products that will be of special value to the prospector and small-mine operator is given in chapters 18 to 32, where the minerals and metals are discussed separately.

BOOKS ON ORE TREATMENT

Much information on this subject and further references are given in sections 31 through 33 of the *Mining Engineers' Handbook*, 3d edition, Robert Peele (editor), John Wiley & Sons, Inc., New York, 1941 (2 volumes).

Part 7

Data Lists and Tables

Extensive tables of use in the mineral industries are given in the following books:

Mining Engineers' Handbook, 3d edition, Robert Peele (editor), John Wiley & Sons, Inc., New York, 1941 (2 volumes).

Examination and Valuation of Mineral Property, 4th edition, Charles H. Baxter and Roland D. Parks, Addison-Wesley Publishing Company, Inc., Reading, Mass., 1957.

THE CHEMICAL ELEMENTS

Natural Elements	*Symbol*	*Natural Elements*	*Symbol*
Aluminum	Al	Neodymium	Nd
Antimony	Sb	Neon	Ne
Argon	A	Nickel	Ni
Arsenic	As	Niobium (columbium)	Nb
Barium	Ba	Nitrogen	N
Beryllium	Be	Osmium	Os
Bismuth	Bi	Oxygen	O
Boron	B	Palladium	Pd
Bromine	Br	Phosphorus	P
Cadmium	Cd	Platinum	Pt
Calcium	Ca	Potassium	K
Carbon	C	Praseodymium	Pr
Cerium	Ce	Protactinium	Pa
Cesium	Cs	Radium	Ra
Chlorine	Cl	Radon	Rn
Chromium	Cr	Rhenium	Re
Cobalt	Co	Rhodium	Rh
Columbium (niobium)	Nb	Rubidium	Rb
Copper	Cu	Ruthenium	Ru
Dysprosium	Dy	Samarium	Sm
Erbium	Er	Scandium	Sc
Europium	Eu	Selenium	Se
Fluorine	F	Silicon	Si
Gadolinium	Gd	Silver	Ag
Gallium	Ga	Sodium	Na
Germanium	Ge	Strontium	Sr
Gold	Au	Sulphur	S
Hafnium	Hf	Tantalum	Ta
Helium	He	Tellurium	Te
Holmium	Ho	Terbium	Tb
Hydrogen	H	Thallium	Tl
Indium	In	Thorium	Th
Iodine	I	Thulium	Tm
Iridium	Ir	Tin	Sn
Iron	Fe	Titanium	Ti
Krypton	Kr	Tungsten (wolfram)	W
Lanthanum	La	Uranium	U
Lead	Pb	Vanadium	V
Lithium	Li	Wolfram (tungsten)	W
Lutecium	Lu	Xenon	Xe
Magnesium	Mg	Ytterbium	Yb
Manganese	Mn	Yttrium	Y
Mercury	Hg	Zinc	Zn
Molybdenum	Mo	Zirconium	Zr

COMMON CHEMICAL COMPOUNDS OF IMPORTANCE TO THE MINERAL INDUSTRIES

Compound	Formula
Acetic acid	CH_3COOH
Aluminum hydroxide	$Al(OH)_3$
oxide	Al_2O_3
Ammonium carbonate	$(NH_4)_2CO_3 \cdot H_2O$
chloride (sal ammoniac)	NH_4Cl
hydroxide (ammonia)	NH_4OH
molybdate	$(NH_4)MoO_4$
oxylate	$(NH_4)_2C_2O_4 \cdot H_2O$
Antimony oxide	Sb_2O_3
(tri-) sulfide	Sb_2S_3
Arsenic (tri-) sulfide	As_2S_3
Barium carbonate	$BaCO_3$
chloride	$BaCl_2$
hydroxide	$Ba(OH)_2$
oxide	BaO
sulfate (blanc fixe)	$BaSO_4$
Bismuth iodide	BI_3
Cadmium sulfide	CdS
Calcium carbonate	$CaCO_3$
fluoride	CaF_2
oxide (quicklime)	CaO
sulfate (plaster of paris)	$CaSO_4$
Carbon dioxide	CO_2
monoxide	CO
tetrachloride	CCl_4
Cobalt (ous) nitrate	$Co(NO_3)_2 \cdot 6H_2O$
Copper oxide	CuO
Dimethylglyoxime	$CH_3C(NOH)C(NOH)CH_3$
Ethyl alcohol (ethanol)	C_2H_5OH
Ferric hydroxide	$Fe(OH)_3$
Ferrous carbonate	$FeCO_3$
Hydrochloric acid	HCl
Hydrogen chloride	HCl
fluoride	HF
peroxide	H_2O_2
sulfide	H_2S
Iron (di-) sulfide	FeS_2
Lead carbonate (white lead)	$PbCO_3$
chloride	$PbCl_2$
(mon) oxide (litharge)	PbO
sulfate	$PbSO_4$
sulfide	PbS
Lithium fluoride	LiF

Compound	Formula
Magnesium carbonate	$MgCO_3$
oxide	MgO
phosphate	$Mg_3(PO_4)_2 \cdot 4H_2O$ and other formulas
Manganese carbonate	$MnCO_3$
(di-) oxide	MnO_2
Molybdenum (di-) sulfide	MoS_2
Nickel (mon) oxide	NiO
Nitric acid	HNO_3
Permanganic acid	$HMnO_4$
Potassium acid sulfate	$KHSO_4$
cyanide	KCN
ferricyanide	$K_3Fe(CN)_6$
Potassium ferrocyanide	$K_4Fe(CN)_6 \cdot 3H_2O$
hydroxide (caustic potash)	KOH
iodide	KI
nitrate (saltpeter)	KNO_3
permanganate	$KMnO_4$
Silicon carbide (Carborundum)	SiC
(di-) oxide (silica)	SiO_2
Silver chloride	$AgCl$
nitrate (lunar caustic)	$AgNO_3$
sulfide	Ag_2S
Sodium bicarbonate	$NaHCO_3$
(tetra) borate (borax)	$Na_2B_4O_7 \cdot 10H_2O$
carbonate (soda)	Na_2CO_3
chloride (salt)	$NaCl$
cyanide	$NaCN$
fluoride	NaF
hydroxide (caustic soda)	$NaOH$
phosphate	Na_3PO_4 and other formulas
(mono) sulfide	Na_2S
Stannic oxide	SnO_2
Stannous chloride	$SnCl_2$
Strontium carbonate	$SrCO_3$
Sulfur dioxide	SO_3
Sulfuric acid	H_2SO_4
Titanium dioxide	TiO_2
Tungsten dioxide	WO_2
trioxide	WO_3
Uranium oxide	U_3O_8
Water	H_2O
Zinc oxide (zinc white)	ZnO
sulfide	ZnS

WEIGHTS AND MEASURES

LENGTH

Metric. The meter is the unit of length:

1 meter = 10 decimeters = 100 centimeters = 1,000 millimeters = 0.20 rods (approx.) = 1.09 yards (approx.) = 3.28 feet (approx.) = 39.37 inches.

1 kilometer = 1,000 meters = 0.62 miles (approx.) = 3,280.83 feet,

1 centimeter = 0.39 inches (approx.).

To convert from meters, etc., multiply by the appropriate numbers above.

Common English. The yard is the unit of length:

1 yard = 3 feet = 36 inches = 0.18 rods (approx.) = 0.91 meters (approx.).

1 foot = 12 inches = 0.30 meters (approx.).

1 inch = 2.54 centimeters (approx.).

1 statute mile = 1,760 yards = 5,280 feet = 1,609 meters (approx.).

To convert from yards, etc., multiply by the appropriate numbers above.

Other common measures of length:

1 fathom = 6 feet; 1 hand = 4 inches; 1 span = 9 inches.

1 furlong = 40 rods = 220 yards = 660 feet; 1 rod = 1 pole = 1 perch; 1 league = 3 miles.

Surveyors:

1 link = 7.92 inches = 0.66 feet.

1 rod = 25 links = 16.5 feet.

1 chain = 100 links = 66 feet = 4 rods.

1 furlong = 40 rods = 0.125 miles.

1 mile = 320 rods = 80 chains = 8 furlongs.

Latin America:

1 Vara = 0.80 to 1.1 meters, according to country.

AREA

Common English. The square yard is the unit of area:

1 square yard = 9 square feet.

1 square foot = 144 square inches.

1 square rod = 30.25 square yards.

1 acre = 43,560 square feet = 4,860 square rods.

1 square mile = 640 acres.

To convert from square yards, etc., multiply by the appropriate numbers above.

VOLUME

Common English. The cubic yard is the unit of volume:

1 cubic yard = 27 cubic feet = 46,656 cubic inches.

1 cubic foot = 1,728 cubic inches.

To convert from cubic yards, etc., multiply by the appropriate numbers above.

CAPACITY

Metric. The liter is the unit of capacity:

1 liter = 10 deciliters = 100 centiliters = 1,000 milliliters = 1,000 cubic centimeters = 33.8 fluid ounces (approx.) = 2.11 pints (approx.) = 0.26 gallons (approx.) = 61 cubic inches (approx.).

1 gallon = 231 cubic inches (United States) = 277 cubic inches (approx., British) = 0.13 cubic feet (approx.) = 4 quarts = 8 pints.

1 quart = 2 pints.

1 pint = 16 fluid ounces.

1 cubic foot = 7.48 gallons (approx.).

To convert from liters, etc., multiply by the appropriate numbers above.

MASS

Metric. The gram is the unit of mass or weight:

1 gram = 10 decigrams = 100 centigrams = 1,000 milligrams = 0.04 ounces avoirdupois (approx.) = 0.03 ounces troy (approx.) = 15.43 grams (approx.).

1 kilogram = 1,000 grams = 2.20 pounds (approx.) = 32.15 ounces troy (approx.) = 35.27 ounces avoirdupois (approx.).

1 metric ton = 2,204.62 pounds avoirdupois = 35,273.96 ounces avoirdupois.

1 grain = 64.80 milligrams (approx.).

1 carat = 200 milligrams = 0.2 gram = 3.09 grains (approx.).

Avoirdupois. The pound is the unit of weight:

1 long ton = 2,240 pounds (British ton).

1 short ton = 2,000 pounds (used for United States mineral industries except where the metric system is used solely).

1 pound = 16 ounces = 7,000 grains.

1 ounce = 437.5 grains.

1 gram = 0.04 ounces (approx.).

Troy. The pound is the unit of weight for precious metal (and drugs):

1 pound = 12 ounces = 5,760 grains.

1 ounce = 20 pennyweight = 480 grains = 155 carats.

1 pennyweight = 24 grains.

To convert troy to avoirdupois, multiply pounds by 0.82 (approx.) and ounces by 1.10 (approx.). To convert avoirdupois to troy, multiply pounds by 1.22 (approx.) and ounces by 0.91 (approx.).

Latin America:

1 libra = 256 adarmes = 16 onzas = 459 (approx.) to 1,000 grams, according to country.

1 cajon = 2,300 (approx.) to 2,761 (approx.) kilograms, according to country.

1 arroba = 25 libras.

1 quintal = 100 libras = 4 arrobas.

1 carga = 300 libras = 12 arrobas = 3 quintals.

1 tonelada = 2,000 libras = 80 arrobas = 20 quintals.

TEMPERATURE

Centrigrade (C.) 0° at freezing point of water, 100° at boiling point.
Fahrenheit (F.) 32° at freezing point and 212° at boiling point.

To convert Centigrade to Fahrenheit, multiply by 1.8 and add 32. To convert Fahrenheit to Centigrade, subtract 32 and divide by 1.8.

If below freezing, subtract 32, instead of adding, to get Fahrenheit; add 32, instead of subtracting, to get Centigrade. Celsius is the same as Centigrade.

ELECTRICAL

Ampere (amp.): The unit of rate of flow of electric current.

Ohm: The unit of electric resistance.

Voltage (volt): The unit of electric pressure.

Watt: The unit of electric power =1 ampere × 1 volt. 1,000 watts =1 kilowatt; 746 watts =1 horsepower.

Kilowatt-hour: The unit of power sale,
1 kilowatt for 1 hour.

OIL GRAVITY

The A.P.I. (American Petroleum Institute) scale is used in the United States.

GRAVITY SCALES AND EQUIVALENT WEIGHTS

Gravity			
A.P.I.	Baumé	Specific gravity	Weight, pounds per gallon
10	10.0	1.0000	8.328
20	19.9	0.9340	7.778
30	29.8	0.8762	7.296
40	39.7	0.8251	6.870
50	49.6	0.7796	6.490
60	59.5	0.7389	6.151
70	69.4	0.7022	5.845
80	79.3	0.6690	5.568
90	89.2	0.6388	5.316
100	99.1	0.6112	5.085

ACIDITY AND ALKALINITY

Litmus paper turns red for acid, blue for alkaline.
Phenolpthalein solution turns crimson in alkaline cyanide solution.
Indicator paper for pH: 7 is neutral, below 7 is acid, above 7 is alkaline.

ALTITUDE

Sea-level pressure is 14.7 pounds per square inch, =29.9 inches of mercury.

Each 1,000-foot rise in altitude reduces the pressure about 0.5 pound and the boiling point of water 2°.

Heat-Color Scale

980 °F. . . . . Red
1,530 °. . . . . Cherry red
1,790 °. . . . . Dark orange
2,050 °. . . . . Yellow
2,280 °. . . . . White

Calculations

Area of circle = diameter × diameter × 0.7854 (about 4/5)
Circumference of circle = diameter × 3.1416 (about 3 1/7 or 22/7)
Diameter of round tank = circumference × 0.3183 (about 1/3)
Cubic content of round tank = diameter × diameter × 4/5 × depth (in feet)
Cubic content of square or rectangular tank = length × depth × width (in feet)
Gallon capacity of tank = cubic content × 7.5
Pipe capacity is increased four times by doubling its diameter

Water

Gallons per minute = cubic feet per second × 449.
Tons per day = gallons per minute × 6.
Hydraulic pressure = head in feet × 0.43 = pounds per square inch.
Specific gravity = 1; sea water = 1,028 (approx.).
Rain: 1 inch per acre = 100 long tons = 27,000 gallons.
1 acre-foot = 1 acre × 1 foot depth = 43,560 cubic feet = 325,850 gallons.

Part 8

Glossary of Terms

These terms are carefully selected for their utility from the many thousands that apply to prospecting, mining, geology, and mineralogy. The repetition of terms that are defined or discussed in the text and appear in the index is purposely avoided. Words of general use, likely to be found in almost any English dictionary, are also not included. In addition to the glossary and index, see the geologic time scale (page 274) and the lists and tables of data in Part 7.

The two most extensive glossaries of terms in the categories represented here are the following:

A Dictionary of Mining, Mineral, and Related Terms, compiled and edited by Paul W. Thrush and the Staff of the Bureau of Mines, U.S. Government Printing Office, Washington, D.C., 1968.

Glossary of Geology, American Geological Institute, Washington, D.C., 1972.

See also the many entries in *Gems, Minerals, Crystals, and Ores: The Collector's Encyclopedia,* by Richard M. Pearl, Odyssey Press, Inc., New York, 1964.

Aa: Rough, clinkery lava.
Abyssal: Pertains to rock formed deep within the earth.
Accessory mineral: A minor mineral in an igneous rock.
Acid: A chemical compound that can react with a base to form a salt. (The complete definition is more technical.)
Acidic: A not very exact term, best applied to igneous rocks containing more than two-thirds silica; in this sense, the opposite of **basic**; unlike **intermediate**, and most like **silicic.** Acidic rocks (such as granite) are typically light colored and light in weight.
Acre-foot: The amount of water that covers 1 acre to a depth of 1 foot, or 43,560 cubic feet.
Acmite: A mineral of the pyroxene group.
Actinolite: A mineral of the amphibole group.
Adamantine: A brilliant luster like that of diamond.
Adamellite: Either quartz monzonite or granodiorite.
Adit: A tunnel open at one end and used for mining or drainage.
Adobe: A deposit of clay, silt, and usually lime, used to make sun-dried brick. (Other definitions are less common.)
Adularia: A variety of orthoclase feldspar.
Aegirine, aegirite: A mineral of the pyroxene group.
Aeolian: Refers to wind deposition.
Agglomerate: A volcanic rock composed of large fragments embedded in ash.
Aggregate: Loose or broken rock or slag added to cement to make concrete or mortar.
Air leg: A cylinder operated by compressed air and attached to a rock drill in order to press it into a hole.
Air shaft: A shaft used to ventilate a mine.
Alaskite: A light-colored granite having less than 5 percent dark minerals.
Alkali: One of the chemical elements that include cesium, lithium, sodium, and potassium; also ammonia; similar to **base** but less inclusive. Also, a salt left on the surface by evaporation in an arid or semiarid region.
Alkalic: Often the same as **alkaline,** but also applied to various other uses, especially an igneous rock containing much potassium or sodium.
Alkali flat: An arid or semiarid basin or plain where salts have accumulated by evaporation of the water in which they were dissolved.
Alkaline: Basic, but also has other meanings. Often the same as **alkalic.**
Alkaline earth: One of the oxides of barium, calcium, or strontium, or sometimes beryllium, magnesium, or radium.
Alliaceous: The odor of garlic, caused by arsenic.
Alloy: A mixture of a metal and another metal or a nonmetal.
Alloy steel: A steel containing certain metals of especial value in alloying.
Alluvium: Loose clay, silt, sand, or gravel carried and deposited by running water.
Alnoite: A dark igneous rock occurring in dikes.
Alumina: Aluminum oxide, Al_2O_3.
Aluminate: A mineral or other compound of aluminum, another metallic element, and oxygen.
Amalgam: An alloy of mercury and one or more other metals, usually gold or silver. It may be natural or artificial.

Ambergris: A waxy secretion from a sperm whale. Not of mineral origin or use.

Ammonia: A light gas, NH_3, but also refers to ammonium hydroxide, NH_4OH, a liquid reagent.

Amorphous: Noncrystalline.

Amortize: To pay off a debt by advance financing.

Amphibole: A group of silicate minerals.

Amygdale, amygdule: A rock cavity containing secondary minerals.

Amygdaloid: A rock in which the cavities are lined or filled by later minerals. Also, the secondary minerals themselves.

Amygdaloidal: Pertaining to an amygdaloid rock.

Analcime: A mineral of the zeolite group.

Andesine: A member of the plagioclase series of feldspar minerals.

Angle brace: A brace that is used to hold mine timber in place.

Anhydrous: Without water.

Anneal: To improve metal by heating and cooling it. There are also other kinds of annealing.

Annual labor: Assessment work done to hold a mining clam.

Anorthite: A member of the plagioclase series of feldspar minerals.

Anthophyllite: A series of minerals of the amphibole group.

Anticline: A rock upfold having opposite limbs dipping away from the axis.

Antimonate: A mineral or other compound composed of antimony, oxygen, and other elements.

Antimonide: A mineral or other compound composed of antimony and a more metallic element or group of elements.

Apache tear: A rounded lump of obsidian.

Aphanitic: Applies to a fine-grained rock.

Apophyllite: A silicate mineral.

Apophysis: An offshoot from a rock or vein.

Aqua regia: A mixture of three parts hydrochloric acid and one part nitric acid, used mostly to dissolve gold or platinum.

Aqueous rock: A sedimentary rock deposited by or in water.

Aquifer: A water-bearing rock.

Arenaceous: Sandy.

Argentiferous: Containing silver.

Argillaceous: Clayey; like **lutaceous** and **pelitic** but more common.

Arrastre: A circular, rock-lined pit in which broken ore is crushed by heavy stones attached to horizontal poles and dragged around the pit.

Arroyo: A steep-walled gulley that is dry most of the time; but different areas use the term variously.

Arsenate: A mineral or other compound of arsenic and oxygen, AsO_4, AsO_3, or As_2O_7.

Arsenical: Pertaining to arsenic.

Arsenide: A mineral or other compound composed of arsenic and a more metallic element or group of elements.

Arsenious: Pertaining to arsenic.

Assay foot: The assay value multiplied by the number of feet across which the sample was taken.

Assay grade: The percentage of an element or compound determined in a sample.
Assay inch: The assay value multiplied by the number of inches across which the sample was taken.
Assay split: A compromise value between buyer's assay and seller's assay.
Assay ton: An amount of ore assayed so that the number of milligrams of bullion corresponds to the number of troy ounces of bullion per ton of ore.
Assay value: The value of gold or silver, in ounces per ton of ore, as assayed. (Other, similar meanings are also used.)
Assay wall: The limits of profitable mining as shown by assays.
Assessment work: Annual labor done to hold a mining claim.
Astringent: Having a puckering taste.
Attitude: The position in space of a rock body or structure, including strike, dip, and plunge.
Augite: A mineral of the pyroxene group.
Aureole: Halo or surrounding zone.
Auric: Pertaining to gold in its state of highest combining power.
Auriferous: Containing gold.
Aurous: Pertaining to gold in its state of lowest combining power.
Axinite: A silicate mineral.
Azimuth: Horizontal angle between a given line and a true meridian.
Back: A hole drilled at the tip of a round to outline the opening; also, the upper side of a vein. (Also many other uses.)
Backsight: A backward reading in a surveying traverse.
Baffle: A deflector in a cyaniding box (or similar usage).
Banket: A deposit of conglomerate, usually containing gold or copper.
Bar: A bank of sand or gravel at the mouth of a river, in a stream, or along the shore. (Also other uses.)
Bar diggings: Workings in a sand or gravel bank.
Base: A chemical compound that can react with an acid to form a salt. (The complete definition is more technical.)
Basenite: A dark lava rock, usually porphyritic.
Basic: Refers to igneous rock having a low content of silica and a high content of iron and magnesium. Basic rocks (such as periodotite) are typically dark and heavy.
Battery: A group of stamps for crushing and pulverizing ore. (There are other uses.)
Bedding: Layering of rocks.
Bedding plane: Boundary between two layers of rock.
Bedrock: Solid rock, either exposed or covered by loose material.
Bench: A terrace or ledge.
Beneficiation: Concentration and treatment of ore.
Biotite: A mineral of the mica group.
Bit: Usually, the cutting end of a drill.
Blanket deposit: A flat deposit of ore, longer and wider than thick.
Blind shaft: One that is not open to the surface.
Blind vein: One that does not reach the surface.
Block: A large, angular fragment of volcanic origin.
Block caving: A mining method in which blocks of unsupported ore break off by gravity.

Bomb: A large, rounded fragment of volcanic origin.
Bonanza: A find of rich ore.
Booming: The accumulation and sudden discharge of water in placer mining.
Borate: A mineral or other compound of boron and oxygen, BO_3.
Borehole: An exploratory hole made in prospecting.
Boss: A dome-shaped mass of igneous rock exposed by erosion. (There are other uses.)
Bostonite: A fine-grained, sugary variety of syenite.
Breast: The face of a mine working. (Also other descriptive uses.)
Breccia: A rock consisting of fragments.
Bromide: A mineral or other compound composed of bromine and another element.
Bronzitite: A coarse-grained igneous rock composed mainly of bronzite.
Bulkhead: A tight partition for protection against gas, fire, and water in a mine. (There are other, similar uses.)
Button: A globule of metal remaining in an assaying cupel or crucible after firing.
Bytownite: A member of the plagioclase series of feldspar minerals.
Calcareous: Containing calcium carbonate; limy. (Other definitions are less common.)
Calcic: Containing calcium.
Calcification: Petrifaction by calcium carbonate.
Calcine: To heat, burn, or roast; also, the products obtained.
Caliche: Cemented sediment; also, the calcium carbonate cement; also, evaporite deposits, especially Chilean nitrate.
Camptonite: A dark igneous rock occurring in dikes.
Cancrinite: A mineral of the feldspathoid group.
Cap rock: Barren rock believed to overlie productive mineralization. (Also related usages.)
Carat: A metric unit for weighing gems, equal to 200 milligrams and divided into 100 points. Different from **karat.**
Carbonaceous: Containing organic carbon or coal.
Carbonate: A mineral or other compound of carbon and oxygen, CO_3. (Also refers to rocks—such as limestone—that are composed mainly of carbonate minerals.)
Carbon steel: Steel made only of iron and carbon.
Carborundum: An artificial abrasive and grinding material, silicon carbide. Different from **corundum,** a mineral.
Cave-in: Collapse of mine workings.
Cavestone: Mineral deposits in a cavern.
Caving: Breaking ore by producing collapse.
Cement: A natural bonding material of mineral matter deposited by underground solution; also, a construction material made from limestone. (There are other uses.)
Chabazite: A mineral of the zeolite group.
Chimney: An ore shoot or other vertical structure, natural or manmade.
Chloride: A mineral or other compound of chlorine and one or more other elements.
Chlorite: A series of silicate minerals.
Chromic: Pertaining to chromium in its state of intermediate combining power.

Chromous: Pertaining to chromium in its state of lowest combining power.

Chrysocolla: A silicate mineral.

Chute: A passage for the descent of ore. (Also various others uses.) Not the same as **shoot.**

Cinder: A volcanic fragment of medium size. A **cinder cone** is composed of cinders and scoria.

Clastic: Broken, fragmental.

Clay gouge: Clay on the wall of a vein.

Claystone: A sedimentary rock composed of clay-sized particles.

Cleavable mass: An irregular mineral specimen showing good cleavage throughout.

Clinoenstatite: A mineral of the pyroxene group.

Clinohypersthene: A mineral of the pyroxene group.

Clinometer: An instrument for measuring vertical angles.

Clinozoisite: A mineral of the epidote group.

Cobaltic: Pertaining to cobalt in its state of highest combining power.

Cobaltous: Pertaining to cobalt in its state of lowest combining power.

Cohesion: The ability of a mineral to hold together.

Collar: The top of a mine shaft. (Also other usages.)

Color: A particle of gold.

Columbate: A mineral or other compound composed of niobium (columbium), another metal, and oxygen.

Columnar jointing: Breakage of rock into columns by cooling or drying.

Compact: Fine grained and solid.

Compound: A definite combination of two or more chemical elements.

Concentrator: Mill in which ore is concentrated.

Concretion: A lump of cementing material in a sedimentary rock.

Conformable: Applies to continuous deposition between successive layers.

Cordierite: A silicate mineral.

Core: A rod-shaped portion of rock removed by a drill bit.

Correlate: To match rock formations that were once continuous or are of the same age or otherwise related.

Cortlandite: An igneous rock variety of peridotite.

Creek claim: A mining claim (especially in Oregon) that includes the bed of a stream.

Crested: Occurring in tabular groups arranged in ridges.

Crib: A mine structure built of timber arranged like a log cabin. (Also other uses.)

Cropping: Rock exposed at the surface.

Crossbedding: Layers of sediment or sedimentary rock that lie at angles to the main strata.

Crosscut: A horizontal passage in a mine, at right angles to the main tunnel or adit, or crossing a vein. (Also various other uses.)

Crosscutting: Extending a horizontal mine working across a vein.

Cross-section: A slice or plan at right angles to the trend or strike.

Crucible: A furnace hearth; also a heat-resistant receptacle for melting or calcining materials.

Crude: In a natural state before treatment.

Cummingtonite: A series of minerals of the amphibole group.

Cupel: A small cup used in assaying; also the hearth of a small furnace used in assaying.

Cupola: A dome on a body of igneous rock; also, a kind of metallurgical furnace.

Cupric: Pertaining to copper in its state of highest combining power.

Cuprous: Pertaining to copper in its state of lowest combining power.

Cut: A hole drilled to make the first blast in a series of explosions. (Also many other uses.)

Cut-and-fill stoping: A mining method in which ore is removed from the back and the space is filled with waste rock; there are various types.

Dana system: A classification of minerals based on chemistry and crystallography.

Datolite: A silicate mineral.

Decrepitate: To fly to pieces when heated.

Deflagrate: To catch fire.

Dense: Fine grained and tightly arranged.

Denudation: Weathering and erosion of rock surface.

Detrital: Consisting of loose or cemented fragments.

Detritus: Loose sediment or rock waste, often with organic material.

Diallage: A variety of monoclinic pyroxene marked by front parting.

Diallagite: A coarse-grained igneous rock composed mainly of diallage.

Diaphaneity: Degree of transparency.

Differentiation: Separation of different rock types from a single molten body.

Diopside: A mineral of the pyroxene group.

Dip: The angle at which a vein or tabular body is inclined from the horizontal.

Disseminated: Scattered throughout.

Divining rod: A forked piece of tree branch (or manmade substitute) used to detect water, oil, or a mineral deposit; same as **dowsing rod, wiggle stick.** Von Bernewitz defined it: "Keep clear of such frauds."

Downcast: A shaft carrying fresh air down to underground workings.

Drag: Bending a rock near a fault; also, rock fragments along a fault surface. (Also various other uses.)

Drift: A horizontal passage in a mine, usually following a vein.

Drifter drill: A heavy drill mounted on a column or bar carriage.

Drift gravel: Alluvial gold or tin on slate or granite, covered by basalt.

Drifting: Extending a horizontal mine working, especially along a vein. (Also other meanings.)

Driving: Extending a horizontal mine working, especially along a vein. (Also other uses.)

Ductile: Can be drawn into wire.

Dump: The waste pile of a mine.

Earthy: Fine grained and powdery; also, having a dull luster.

Easer: A hole drilled around the first blast (the **cut**) to enlarge it; same as **relief.**

Eclogite: An igneous or metamorphic rock composed of garnet and pyroxene.

Efflorescence: A coating of crystals deposited by evaporation. (Also other, related meanings.)

Elasticity: The ability of a mineral to snap back after being bent.

Electrum: A mineral alloy of gold and silver.

Element: A chemical substance that cannot be changed into any other substance by ordinary chemical means. The table of known **chemical elements** is given in Part 7 of this book.

Emulsion: A mixture of oil and water.

Endline: One of the two boundary lines of a mining claim that cross the surface trend of the vein.

Enlarger: A hold drilled to enlarge the opening.

Enstatite: A mineral of the pyroxene group.

Entry: An adit or slightly inclined shaft in a mine. (Many terms are used for various kinds.)

Epidote: A group of silicate minerals; also, a member of the group.

Epigenetic: Refers to a mineral deposit that was formed later than the enclosing rock.

Epithermal: Applies to a mineral deposit formed at low temperature and shallow depth.

Eruptive: Extrusive.

Essential mineral: A major mineral in a rock.

Essexite: A coarse-grained igneous rock related to gabbro.

Etched: Surface corroded by chemical action.

Evaporite: A mineral or rock deposited by the evaporation of mineral-bearing (lake or sea) water.

Exploit: To utilize a mining property completely.

Exposure: Outcrop of igneous rock.

Face: A flat surface on a crystal; also, the working surface of a mine. (Also many other uses.)

False set: A temporary structure of timber used before a permanent set.

Fault breccia: Broken rock on a fault surface.

Fault gouge: Soft material on a fault surface.

Feldspathoid: A group of silicate minerals that includes nepheline, sodalite, analcime, leucite, hauynite, noselite, nepheline, cancrinite.

Felsic: Applies to igneous rocks containing much feldspar or feldspathoid minerals and light colored. Also refers to these minerals, as well as to quartz and muscovite mica, which are all light colored and have a moderate to high content of silica.

Felsitic: Applies to a fine-grained, light-colored rock.

Ferrate: A mineral or other compound composed of iron, another metallic element, and oxygen.

Ferric: Pertaining to iron in its state of highest combining power.

Ferromagnesian: Containing iron and magnesium and hence typically dark and heavy.

Ferrous: Pertaining to iron in its state of lowest combining power.

Ferruginous: Containing iron.

Fire: To ignite an explosive.

Fissile: Easily split.

Fissure: An extensive crack in rock.

Fissure vein: An ore vein in a rock crack.

Flexibility: The ability of a mineral to be bent and stay bent.

Fluoride: A mineral or other compound of fluorine and one or more other elements.

Fluvial: Applies to streams.

Flux: A substance that reduces the melting temperature of another. (Also other usages.)

Foliation: A parallel arrangement in a metamorphic rock. (Also other definitions, often conflicting.)

Footwall: The lower side of a fault. (Also related meanings.)

Foresight: A foreward reading in a surveying traverse.

Form: Includes all crystal faces that are similar. (Also other, less specific definitions.)

Foyaite: A coarse-grained igneous rock; the same as **nepheline syenite.**

Friable: Crumbly.

Fumarole: A smoke hole of volcanic origin.

Gangue: Worthless minerals associated with ore.

Gash vein: A small, shallow vein. (Also other definitions of more specialized use.)

Geode: A hollow nodule, usually lined or filled with crystals.

Geosyncline: Trough in which sediment accumulates and later becomes a system of folded mountains.

Geyserite: A deposit of silica at geysers and hot springs.

Glacial drift: Material carried and deposited by a glacier.

Glance: Added to names of minerals that have a bright luster.

Glauconite: A silicate mineral.

Glaucophane: A mineral of the amphibole group.

Glistening: Describes minerals that have a bright but diffuse luster.

Glucinum: The former name for beryllium.

Gophering: Haphazard prospecting or working. (Also more specific uses.)

Gouge: Soft material on the wall of a vein.

Grain: A distinct particle in a rock.

Granular: Composes of individual grains of about equal size; also grainlike.

Graphic: Applies to an intergrowth of two minerals, resembling ancient writing.

Graphic granite: Pegmatite having a pattern like ancient writing.

Graphitic: Containing graphite or other carbon.

Graywacke: Sandstone having considerable dark minerals.

Greasy: Applies to the luster of oil.

Greenstone: Greenish igneous or metamorphic rock of various kinds; in Canada, especially ancient lava.

Grit: Usually a coarse, angular sand or sandstone.

Grizzly: A metal grating for screening rock or ore to size; also a protection device in a mine.

Gulch: A narrow ravine or valley, perhaps deep.

Gully: A small ravine, usually in soft material.

Gumbo: Clayey soil that turns to sticky mud when wet. (Also specialized local uses.)

Gyratory: Operating in an off-centered manner, as certain rock crushers.

Habit: The shape of a mineral, whether imitative, massive, or in crystal form.

Hackly: Applies to a jagged fracture.

Halide: A mineral or other compound of a metallic element and one of the halogen elements (bromine, chlorine, fluorine, iodine).

Halogen: One of the chemical elements that include bromine, chlorine, fluorine, and iodine.
Hanging wall: The upper side of a fault. (Also related meanings.)
Hardpan: Cemented gravel or glacial clay. (Also other, similar uses.)
Harzburgite: An igneous rock variety of periodotite; the same as **saxonite.**
Headframe: A structure built over a mine shaft for various operating purposes.
Hedenbergite: A mineral of the pyroxene group.
Heulandite: A mineral of the zeolite group.
Hornblende: A mineral of the amphibole group.
Hornfels: A fine-grained, often spotted metamorphic rock.
Horse: A body of waste rock in a vein or mine. (Also other usages.)
Hydrate: A mineral or other compound containing water.
Hydrothermal: Applies to hot solutions from a magma.
Hydrous: Containing water.
Hydrous oxide: A mineral or other compound containing water.
Hydroxide: A mineral or other compound containing hydroxyl (OH), which yields water when heated.
Hygroscopic: Capable of absorbing moisture from the atmosphere.
Hypabyssal: Shallow-depth intrusive processes and bodies.
Hypersthene: A mineral of the pyroxene group.
Hypersthenite: A coarse-grained igneous rock composed mainly of hypersthene.
Hypogene: Refers to rising solutions.
Hypothermal: Applies to a mineral deposit formed at considerable temperature and depth.
Idocrase: A silicate mineral.
Ijolite: A coarse-grained igneous rock.
Incline: A sloping shaft, tunnel, drift, or other opening in a mine.
Inclusion: Foreign matter enclosed within a mineral.
Incrustation: A crust or coating.
Interbedded: Occurring between two strata.
Intermediate: Applies to rocks lying between those that are acidic and those that are basic. Their composition is matched by an intermediate color tone (darkness-lightness) and specific gravity.
Intermediate level: A horizontal working starting from a mine raise rather than a main shaft.
Iodide: A mineral or other compound composed of iodine and another element or group of elements.
Iridescence: A play of rainbow colors.
Ironstone: A sedimentary rock having a high percentage of iron minerals.
Jaspilite: A sedimentary rock composed of red jasper and black hematite in alternate layers.
Karat: One-twenty-fourth part of pure gold in artificial alloys; thus, 24K is pure gold, and 12K is half gold and half other alloying metals. Different from **carat.**
Keratophyre: A kind of porphyritic rock.
Kersantite: A dark igneous rock occurring in dikes.
Labor: Annual assessment work.
Lacustrine: Applies to lakes.

Lamellar: Occurring in layers of scales or plates.
Lamprophyre: Dark igneous rock occurring in dikes.
Lapilli: Volcanic fragments of medium size.
Lateral: A horizontal mine working. (Also other usages.)
Laterite: Red soil and rock typical of tropical weathering.
Laumontite: A mineral of the zeolite group.
Lawsonite: A silicate mineral.
Lens: A pillow-shaped or lenticular body of ore or rock.
Lenticular: Pillow or lens shaped.
Lepidolite: A mineral of the mica group.
Leptothermal: Applies to a mineral deposit formed just above intermediate temperature and depth.
Lessee: One who leases land or other property; **leaser** is incorrect.
Lessor: The grantor of a lease.
Leucite: A mineral of the feldspathoid group.
Leucocratic: Describes light-colored igneous rock.
Leucophyre: A porphyritic rock of light color.
Level: A horizontal mine opening at a given elevation. (Also other meanings.)
Lifter: A hole drilled at the bottom of an opening to be exploded after the cut and relief holes. (Also other uses.)
Lime: Usually means calcium carbonate (as in **limy**) or calcium. Properly, lime is quicklime, CaO.
Liparite: Rhyolite.
Lithology: The study of rocks, especially sedimentary rocks.
Litmus paper: Treated paper used to test whether a solution is acidic (turns red) or alkaline (turns blue).
Loam: Crumbly topsoil composed of clay, silt, sand, humus, or combinations of these. (Also other, special uses.)
Location: A mining claim; also, fixing the legal boundaries of a mining claim. (Also other uses.)
Longwall caving: A mining method in which rock is removed in a continuous line from flat beds.
Lutaceous: Clayey; like **argillaceous, pelitic.**
Lutite: A fine-grained sedimentary rock.
Macroscopic: Visible to the unaided eye.
Mafic: Containing a high percentage of magnesium and iron.
Magmatic water: Water derived from cooling magma.
Magnesia: Magnesium oxide.
Malleable: Capable of being hammered.
Manganate: A mineral or other compound composed of manganese, another metallic element, and oxygen.
Manganic: Pertaining to manganese in its state of higher combining power.
Manganous: Pertaining to manganese in its state of lower combining power.
Manto: A blanket-shaped or cylindrical body of ore.
Massive: Without orientation or distinctive form; also, thickly bedded.
Matrix: Rock surrounding a mineral or ore. (Also other uses.)
Meager: Applies to a rough feel.

Megascopic: Visible to the unaided eye or with a hand lens.

Melanocratic: Describes dark igneous rock.

Melaphyre: A porphyritic rock of dark color.

Mercuric: Pertaining to mercury in its state of highest combining power.

Mercurous: Pertaining to mercury in its state of lowest combining power.

Mesh: The opening in a sieve or screen; also, the size of opening. (Also related usage.)

Mesothermal: Applies to a mineral deposit formed at intermediate temperature and depth.

Metal: One of the chemical elements or alloys characterized by good conductivity and typically opaque and lustrous. (The distinction from a nonmetal is technical, and there is also a chemical definition, but the familiar metals are readily recognized. See also **road metal.**)

Metallic: Containing a metallic element, or having a metallic luster.

Metasomatism: The metamorphic exchange of mineral matter for barren rock.

Meteoric water: Water from the atmosphere and surface sources.

Miarolitic cavity: A large cavity in igneous rock, especially pegmatite.

Milling ore: Ore that must be concentrated before being shipped.

Miners' right: A permit to prospect in Australia or to mine in California.

Minette: A dark igneous rock occurring in dikes.

Monchiquite: A dark igneous rock occurring in dikes.

Monitor: A nozzle or jet used in hydraulicking; a kind of mine car; a sampling or measuring device.

Monument: A permanent object marking a boundary of land, including a mining claim.

Mother lode: A principal vein system in a large area. The Mother Lode is in California.

Muck: Loose, broken, useless, or mixed debris in a mineral deposit or mine.

Mucker: A mine shoveler.

Mudstone: A sedimentary rock composed of mudlike particles.

Muriatic acid: Hydrochloric acid.

Native: Applies to a mineral that occurs as an element.

Natrolite: A mineral of the zeolite group.

Nepheline: A mineral of the feldspathoid group.

Nickelic: Pertaining to nickel in its state of highest combining power.

Nickeliferous: Containing nickel.

Nickelous: Pertaining to nickel in its state of lowest combining power.

Nitrate: A mineral or other compound of nitrogen and oxygen, NO_3.

Nodule: A lump or concretion.

Nonmetallic: Not containing a metallic element; also, a luster other than metallic.

Nugget: A lump of mineral, usually water-worn and metallic.

Ocherous: Occurring in earthy, powdery form.

Oölite: A small, rounded body.

Option: An agreement to buy.

Orbicular: Applies to a rounded structure in rock.

Orebody: A vein and closely surrounding rock.

Ore shoot: A rich part of a steep vein.

Organic: Of plant or animal origin.

Orogeny: Mountain making.

Overburden: Uncoated material, either loose or solid, that lies above a mineral deposit.

Oversaturated: Refers to igneous rocks containing enough silica to form quartz (which is pure silica) in addition to other minerals.

Oxide: A compound of oxygen and a metallic element.

Oxidize: To combine with oxygen.

Oxysulfide: A mineral or other compound composed of a metal, sulfur, and oxygen.

Pahoehoe: Smooth, ropy lava.

Pantellerite: A rock similar to rhyolite.

Paragenesis: An association of minerals and the order in which they were formed.

Paragonite: A mineral of the mica group.

Parting: A soft layer between a vein and the wall rock. (Also many other uses.)

Pay dirt: Placer material of commercial value.

Pay gravel: Placer material of commercial value.

Payload: Useful material taken from a mine.

Pay ore, pay rock: Ore of commercial value.

Pay shoot: The part of a deposit composed of ore of commercial value.

Pay streak: The rich part of a gold placer, or the profitable part of a vein.

Pearly: Describes the iridescent luster of pearl and most cleavage surfaces of minerals.

Pectolite: A silicate mineral.

Pelitic: Clayey; like **argillaceous, lutaceous.**

Persistent: Extensive, as applied to a vein or orebody.

Petrography: The description and classification of rocks.

Phaneritic: Consisting of crystals visible to the unaided eye.

Phlogopite: A mineral of the mica group.

Phosphate: A mineral or other compound of phosphorus and oxygen, PO_4.

Phyllite: A metamorphic rock intermediate between schist and slate.

Picrite: Olivine-rich diabase.

Piezoelectricity: Electricity or polarity caused by pressure.

Pig: An ingot cast of metal. (Also other descriptive usages.)

Pigeonite: A mineral of the pyroxene group.

Pillar: An area of rock that is left to support a mine.

Pillar caving: A mining method in which ore is broken in a series of tall rooms, leaving pillars that later collapse.

Pinched out: Describing a vein that narrows to disappearance.

Pipe: A tubular-shaped, mineralized body. (Also other descriptive usages.)

Pitch: The angle at which the axis of a body is inclined from the horizontal.

Plagioclase: A series of minerals of the feldspar group.

Plumbic: Pertaining to lead in its state of highest combining power.

Plumbous: Pertaining to lead in its state of lowest combining power.

Plutonic: Pertains to deep geologic activity.

Point: A metric unit for weighing gems, equal to 2 milligrams and 1 percent of a carat.

Porosity: The percentage of open space in rock or soil.

Potash: Potassium oxide, K_2O.

Potassic: Containing potassium.

Precipitate: A solid that forms when one solution is added to another; also, to cause such a precipitate to form.

Prehnite: A silicate mineral.

Property: A characteristic or quality.

Propylitization: A kind of rock alteration.

Prospect: An unproved but promising mineral property.

Protore: Primary mineralization that failed to be enriched into ore.

Pseudomorph: A mineral having the shape or crystal form of a different substance.

Puddingstone: Conglomerate.

Pyritization: Petrifaction by iron sulfide.

Pyroclastic: Broken by volcanic action.

Pyrometasomatism: Contact metamorphism.

Pyroxene: A group of silicate minerals.

Qualitative analysis: A chemical determination of the kinds and nature of chemical elements that are present in a substance. A **qualitative test** is made for one or a few elements.

Quantitative analysis: A chemical determination of the percentages of the chemical elements that are present in a substance. A **quantitative test** is made for one or a few elements.

Quartzose: Containing mostly quartz.

Raise: A vertical or inclined opening driven upward from a mine level. (Also other uses.)

Raker: A hole drilled to clean out and enlarge the first blast (the cut) in a series of explosions.

Reagent: A chemical substance used for testing, assaying, flotation, or other purposes.

Reef: A lode or vein in Australia; bedded deposits in South Africa; barren shale in African diamond mines. (Other geologic uses elsewhere.)

Reef rock: Bedrock of slate or sandstone of a certain age in Australia.

Refractory: Refers to a complex ore difficult to treat; also, a material having a high melting point.

Relief: A hole drilled around the first blast (the cut) to enlarge it; same as **easer.** (Also other uses.)

Replacement: Substitution of one mineral for another.

Resinous: Applies to the luster of resin.

Riebeckite: A mineral of the amphibole group.

Rise: British for **raise.**

Road metal: Broken stone (or other material) used in building roads.

Room and pillar: A mining method that leaves supporting columns of rock.

Round: A series of blasts. (Also other uses.)

Royalty: Payment to a mine owner, based on production.

Rudite: A fragmental sedimentary rock.

Run-of-mill: Ore selected for mill treatment.

Run-of-mine: Unsorted ore or coal taken from a mine.

Rusty gold: Free gold that is coated with iron oxide or other substance that hinders amalgamation.

Saddle: A rock structure shaped like an anticline.

Saline: Salty.

Sanidine: A mineral of the feldspar group.

Saturated: Refers to igneous rocks containing enough silica to form certain **saturated minerals** but not enough for quartz. Also, rock and soil in which the pore spaces are filled with water.

Saxonite: An igeous rock variety of peridotite; the same as **harzburgite.**

Scaly: Flaky.

Scapolite: A series of silicate minerals.

Schiller: A bronxy, iridescent luster.

Schistose: Refers to a banded rock that splits readily.

Scoria: A rough, irregular fragment of porous lava or volcanic debris. **Scoriaceous** pertains to scoria.

Sectility: The ability of a mineral to be cut with a knife.

Selenide: A mineral or other compound composed of selenium and a more metallic element or group of elements.

Selvage: A layer of clay or other soft rock along the wall of a vein. (Also other, similar uses.)

Sericite: Fine, flaky muscovite mica.

Shonkinite: An igneous rock related to syenite.

Shoot: A rich part of a steep vein. (Also other uses.) Not the same as **chute.**

Shrinkage stoping: A mining method in which ore is removed in slices from the bottom upward. (Other names are also used.)

Shuttle car: A trackless vehicle used to transfer raw material to the main transportation system.

Sideline: One of the two boundary lines of a mining claim that parallel the surface trend of the vein.

Silica: Silicon dioxide, SiO_2. Quartz is composed of silica. Do not confuse with **silicon** (a chemical element) or **silicate** (a class of minerals).

Silicate: A mineral or other compound of silicon and oxygen, in which tetrahedrons of SiO_4 are linked in chains (single or double), sheets, isolated units, or three-dimensional networks.

Siliceous: Containing silica.

Siliceous sinter: A deposit of silica at hot springs and geysers.

Silicic: Containing a high percentage of silica; the opposite of **basic,** unlike **intermediate,** and most like **acidic.** Silicic rocks (such as granite) are typically light colored and light in weight (specific gravity).

Silicification: Petrifaction by silica.

Silky: Describes the luster of fibrous minerals.

Siltstone: A sedimentary rock consisting of silt-sized particles.

Sinker drill: A hand drill used to drill holes downward.

Skarn: Rock formed by contact metamorphism.

Slaty cleavage: A parallel splitting in metamorphic rock.

Slickensides: Scratches and polish on a fault surface.

Sludge: Cuttings and mud made by a bit. (Also other meanings.)
Sluice box: A long trough used to recover placer minerals.
Slusher: A scraper or dragshovel loader.
Soda: Applies to sodium or several sodium compounds.
Sodalite: A mineral of the feldspathoid group.
Solfatara: A sulfur-emitting hole of volcanic origin.
Sourdough: An experienced prospector.
Spall: To break ore with a heavy hammer. (Also other uses and spellings.)
Specular: Mirrorlike.
Sphene: A silicate mineral.
Splendent: Shiny.
Stalactite: Iciclelike deposit formed by dripping water.
Stalagmite: A rising deposit formed by dripping water.
Stannic: Pertaining to tin in its state of highest combining power.
Stannous: Pertaining to tin in its state of lowest combining power.
Staurolite: A silicate mineral.
Stilbite: A mineral of the zeolite group.
Stockwork: A group of small veins mixed as a unit, or solid mass of ore.
Stope: The place where ore is being mined. (Also various other uses.)
Stoper drill: A small drill used in narrow openings, mainly overhead.
Stratified: Deposited in strata, beds, or layers.
Stratigraphy: The study of layered rocks.
Streak plate: A piece of unglazed porcelain, used to mark the streak of a mineral.
Striated: Scratched or grooved.
Striations: Fine parallel lines on a crystal.
Strike: The compass direction of a vein or tabular body.
Stringer: A narrow vein.
Strip: To remove overburden from useful material.
Structure: The general arrangement of rock bodies; also, the larger features of a rock body.
Stull: A timber prop in a mine. (Also other uses.)
Subaerial: Refers to geologic processes or products that occur on land surface.
Subaqueous: Refers to geologic processes or products that occur beneath a body of water.
Sublevel caving: A mining method in which thin blocks of ore are caved by undermining.
Submetallic: Describes a luster that is not quite metallic.
Sulfantimonide: A mineral or other compound composed of sulfur and antimony.
Sulfarsenide: A mineral or other compound composed of sulfur and arsenic.
Sulfate: A mineral or other compound of sulfur and oxygen, SO_4
Sulfide: A mineral or other compound of sulfur and one or more other elements.
Sulfite: A compound of sulfur and oxygen, SO_3.
Sulfosalt: A mineral or other compound composed of sulfur, a metallic element, and a semimetallic element.
Sump: A pit for collecting water. (Also other uses.)
Supergene: Refers to descending solutions.
Surficial: On the surface of the earth.

Syncline: A rock downfold having opposite limbs dipping toward the axis.

Syngenetic: Refers to a mineral deposit that was formed at the same time as the enclosing rock.

Tabular: Occurring (as a mineral) in flat plates or (as a mineral deposit) in a flat bed.

Tachylite: Basaltic glass.

Tactite: A metamorphic rock occurring between an intrusive body and a carbonate rock.

Talus: Slide rock at the bottom of a cliff.

Tantalate: A mineral or other compound composed of tantalum, another metallic element, and oxygen.

Tarnish: Surface discoloration on a mineral or other substance.

Telethermal: Applies to a mineral deposit formed at very low temperature and very shallow depth.

Telluride: A mineral or other compound composed of tellurium and a more metallic element or group of elements.

Tenacity: The ability of a mineral to hold together.

Tenor: Percentage or (for the precious metals) ounces per ton of metal in an ore.

Tephra: Explosive material from a volcanic eruption.

Tephrite: A dark lava rock, usually porphyritic.

Texture: The smaller features of a rock body as shown by an individual specimen.

Thallic: Pertaining to thallium in its state of highest combining power.

Thallous: Pertaining to thallium in its state of lowest combining power.

Theralite: A coarse-grained igneous rock of the alkali-syenite group.

Thermoluminescence: Glowing when heated.

Thorite: A silicate mineral.

Till: Loose, unsorted glacial material.

Tillite: A sedimentary rock consisting of unsorted glacial material.

Tinguaite: A variety of phonolite occurring in dikes.

Titanate: A mineral or other compound composed of titanium, another metallic element, and oxygen.

Titaniferous: Containing titanium.

Titanite: A silicate mineral now called sphene.

Tool steel: Carbon and alloy steel suitable for tools that work metal or wood.

Top-slice caving: A mining method in which horizontal slices of ore are removed from the top down. (Other names are also used.)

Tough: Flexible without being brittle.

Translucent: Passing light through a thin edge.

Tremolite: A mineral of the amphibole group.

Trend: Strike, direction, or bearing of a vein, rock body, or other structure; also a variation in grade of ore.

Triboluminescence: Glowing when rubbed or scratched.

Trim: A hole drilled to outline the final opening.

Trommel: A revolving sieve.

Tuff: Rock consisting of volcanic ash.

Tundra: A treeless but mossy plain in the far north.

Twin crystal: Two or more crystals intergrown in a definite way.

Ultrabasic, ultramafic: Applies to igneous rock that contains less than 45 percent silica.

Umpire assay: An assay made to settle a dispute.

Unakite: A metamorphic rock containing green, pink, and white minerals.

Unconformity: A time gap between adjacent rock formations.

Unctuous: Greasy feeling.

Undersaturated: Refers to igneous rocks that contain some or all unsaturated minerals (such as feldspathoid minerals), which have a deficiency of silica.

Unit: The amount of metal or compound in a ton of 1 percent ore. (Also other meanings.)

Unsaturated: Applies to minerals (such as feldspathoid minerals) that have a deficiency of silica.

Upcast: A shaft carrying stale air up to the surface.

Uranate: A mineral or other compound composed of uranium and oxygen.

Uranic: Pertaining to uranium in its state of highest combining power.

Uranous: Pertaining to uranium in its state of lowest combining power.

Vadose zone: The part of the earth's crust that lies above the water table.

Vanadate: A mineral or other compound of vanadium and oxygen, VO_4 or VO_3.

Vanadic: Pertaining to vanadium in its state of higher combining power.

Vanadous: Pertaining to vanadium in its state of lower combining power.

Vesicle: A gas cavity in igneous rock.

Vesicular: Pertaining to a rock made porous by gas cavities.

Vesuvianite: A silicate mineral now called idocrase.

Viscosity: The stickiness of a fluid.

Vitreous: The luster of glass.

Vitrophyre: Porphyritic volcanic glass.

Volatile: Readily evaporated or heated to a gas.

Volcanic ash, volcanic dust: Small particles of volcanic origin.

Volcanic breccia: Broken rock of volcanic origin.

Volcanism: Volcanic activity.

Vosgesite: A dark igneous rock occurring in dikes.

Vulcanism: Volcanic and magmatic activity.

Water table: The boundary between water-saturated soil or rock (below) and percolating water (above).

Weathering: Changes due to the atmosphere, moisture, plants, and organisms.

Websterite: An igneous rock variety of pyroxenite.

Wehrite: An igneous rock variety of peridotite.

Wildcat: A risky venture in mining or oil-well drilling in an untested area.

Winze: A vertical or inclined opening driven downward from a mine level.

Wollastonite: A silicate mineral.

Working face: The place in a mine where work is being done.

Xenolith: A foreign body within igneous rock.

Yellow cake: Uranium concentrate after milling.

Zeolite: A group of silicate minerals.

Zirconate: A mineral or other compound composed of zirconium, another metallic element, and oxygen.

Zoisite: A mineral of the epidote group.

Zone of aeration: The part of the earth's crust lying above the water table.

Zone of enrichment: The part of the earth's crust lying below the water table.

Zone of flowage: The deeper part of the earth's crust, where rock moves plastically rather than breaks.

Zone of fracture: The shallower part of the earth's crust, where rock breaks rather than moves plastically.

Zone of oxidation: The shallower part of the earth's crust, where minerals have been oxidized.

Zone of saturation: The part of the earth's crust lying below the water table.

Zone of weathering: The shallower part of the earth's crust, where oxidation and hydration take place.

Index

Abrasive sand, 375
Abrasives, 326, 360–363
  siliceous, 361–362
  soft, 363
Absorption magnetometer, 143
Absorption spectrograph, 153
Achroite, 356
Acicular habit, 218
Acid test, 229
Acidity data, 426
Actinide rare earths, 339
Actinolite, 227
Adamantine luster, 220
Adit, 390
Adularia, 292
Adverse claim, 197
Aerial photography, 4, 81–84, 138
  books on, 83–84
Aerial survey, 4
Aeromagnetic map, 143
Africa, 6, 299, 351, 358, 413
  (*See also* South Africa and specific countries)
Agate, 354, 355
Agatized wood, 354
Agricola (Georg Bauer), 14
Aids to prospectors, 27–47
Air mattress, 87
Airborne magnetometer, 143
Airborne survey, 124, 143
Airplane, transportation by, 7, 18, 53, 54
Ajo, Arizona, 299
Alabama, 41, 211, 310
Alabaster, 376
Alaska, 15, 18, 29, 31, 35, 42, 93, 109, 110
  claims in, 173, 178, 187, 196, 198–200
  mineral deposits in, 210, 235, 279, 280, 288, 292, 295, 353
  schools in, offering degrees related to mining, 211
  state agencies in, 43
  transportation in, 48–50, 52, 53
Alaska Juneau mine, 412
Alberta, 32, 46, 202, 212, 213
Albite, 134, 227, 352
Aldan, region of Siberia, 352
Alexander, Bill, 18
Alexander, Dee, 18
Alexander, John, 18
Alexandrite, 351
Alkali industry, 371
Alkali lakes, 371
Alkaline test, 229
Alkalinity data, 426
Allanite, 332, 339, 340
Allophane, 365
Alluvial deposits, 267, 277
  (*See also* Placers)
Alluvium, 109
All-terrain vehicles (ATV), 8, 55–57

Almaden, Spain, 341
Almandite, 227, 353, 361
Alnoite, 352
Alpha rays, 123
Alteration of wallrock, 273
Altitude data, 426
Aluminite, 136
Aluminum, 154, 164, 275, 416
  associations of, 298, 321, 324–326, 328, 333, 344
  test for, 227, 231
Aluminum myth, 276
Alunite, 302, 381
Amalgamation, 414–415
Amazonstone, 109, 352
Amber, 349
Amblygonite, 239, 372
American emery, 361
American Federation of Mineralogical Societies, 23
American Geological Institute, 212
American Mining Congress, 182
American mining law, 172–201
Amethyst, 354
Amherst, Ohio, 390
Amosite, 378
Amphibole, 240, 353, 378
Amphibolite, 240, 317
Amygdaloidal, 218
Anaconda Company, 170
Analysis, chemical, 164, 165
Anatase, 329
Andalusite, 366, 368–369
Andes Mountains, 49
Andesite, 240, 295
Andradite, 353
Angel's Camp (California), 12
Anglesite, 134, 136, 294, 303, 304
Anhydrite, 246, 302, 371, 376
Animals:
  burrowing, 109
  pack, 49
Anomaly, 138, 146, 334, 335
Anorthosite, 240, 329
Anthills, 163
Anthony, Leo Mark, 103, 114
Anthophyllite, 378
Antimonite, 145
Antimony, 34, 111, 140, 149, 150, 164, 247, 276, 320, 342–343, 416, 418
  coatings, 228
  flame color of, 227
  test for, 229
Antlerite, 300, 302
Apatite, 221, 231, 239, 328, 358
Apex, law of, 184–186, 201
Aplite, 239
Aquamarine, 350
Arabian Peninsula, 50
Aragonite, 133–136, 218, 231
Arbuckle Mountains of Oklahoma, 273
Archimedes, 223
Area, data on, 424
Argall, George O., Jr., 49
Argentina, 204
Argentite, 220, 293
Arizona, 7, 30, 42, 275, 387
  claims in, 173, 187, 199, 201
  mineral deposits in, 210, 295, 299, 336, 359
  schools in, offering degrees related to mining, 211
  state agencies in, 43
Arkansas, 43
  claims in, 173, 187
  mineral deposits in, 210, 352
  state agencies in, 43
Arkose, 243–244, 334
Arsenates, test for, 232
Arsenic, 150, 164, 276, 333, 343, 416
  coatings, 228
  flame color of, 227
  test for, 229
Arsenopyrite, 111, 292
Asbestos, 33, 143, 377, 378–379, 412, 413
  geology, mineralogy, and petrology, 236, 240, 248, 256
Asbestos mitt, 91
Asia, 306
  Central, 49
  (*See also* specific countries)
Asiatic ladle, 112
Aspen, Colorado, 50
Asphalt, 331, 337
Assaying mining claims, 157–166, 386
Assessment work, 183, 193–195
Atacama Desert of Chile, 10
Atacamite, 300, 302
Atomic-absorption spectrophotometer, 154
Atomic-absorption spectroscopy, 153
ATV (all-terrain vehicles), 8, 55–57
Auger-drill sample, 159
Austin, Texas, fossils from, 245
Australia, 6, 8, 13, 20, 49, 53, 108, 111, 171, 258, 267, 277, 278
  mineral deposits in, 210, 288, 294, 306, 313, 330, 339, 345, 353
  mining laws in, 203, 204
  (*See also* specific states)
Automobiles, 54–57, 97, 98
Autunite, 132, 134, 135, 331, 332
Aventurine, 354
Axe, two-handed, 67
Azurite, 231, 300

Background count, 123–125, 141
Baddeleyite, 345, 369
Bag, sample, 63
Ball, Sydney H., 14, 26
Ball clay, 365
Ball mill, 412
Ballas, 361
Balsley, Howard, 29
Banded habit, 218
Barrier lode, New South Wales, Australia, 294
Barbados, 204
Barite, 112, 145, 209, 279, 366, 380, 412
  luminescence, 132, 135, 136
  metallic associations of, 292, 301, 313
Barium, 164
  flame color of, 227
  test for, 231
Barrows, Jim, 169
Barylite, 328
Basalt, 240, 295, 352
Basalt porphyry, 241
Base line, 173–175
Base metal, 294, 334, 345, 416
Basic dike, 256
Basic igneous rock, 296, 297, 317, 329, 351, 353, 354, 376, 379, 380
Basic ore, 248
Bastnaesite, 339, 340
Batea, 112
Bateman, Alan M., 257, 381
Bates, Robert L., 172
Batholith, 254, 255, 258–259
Bauer, Georg (Agricola), 14
Bauxite, 33, 136, 276, 321, 324–326, 344, 363, 366, 369, 381
Bead tests, 122, 132, 228
Bench marks, 78
Bench placer, 278–280
Bendigo, Australia, 267, 288
Beneficiation, 386, 408
Benitoite, 135
Bentonite, 365, 366, 381
Bergendahl, M. H., 292
Bertrandite, 328
Beryl, 239, 327, 328, 350, 412
Berylometer, 126, 127, 327
Beta rays, 123
Beta-uranophane, 135, 332
Betafite, 323
Big Blue mine at Quartzburg, California, 12
Bingham, Utah, 293, 299, 303, 305, 319
Bioluminescence, 130
Biotite, 237, 239, 247, 302
Biotite schist, 247
Birmingham, Alabama, 310
Bisbee, Arizona, 299
Bismite, 343
Bismuth, 34, 111, 164, 343–344, 418
  coatings, 228
Bismuth flux, 225
Bismuthinite, 343
Bitter lakes, 371
Bittern, 326
Bitumens, 377
Black-band ore, 311
Black lead, 368
Black light, 129–136
Black sand, 109, 110, 243, 283
Black shale, 331, 350
Blackjack, 305
Bladed habit, 218
Blasting, 397
Blende, zinc, 305
Blind River area of Ontario, 19
Blister copper, 417
Bloodstone, 354
Blowouts, 276
Blowpipe methods, 224–232
Blowpiping, books on, 232–233
Blue asbestos, 379
Blue copper carbonate, 300
Blue rock, 358
Boats, transportation by, 50–53, 55
Boehmite, 325
Bolivia, 204, 293, 307, 320
Bonanza, 115
Bonanza vein, 235
Books:
  on blowpiping, 232–253
  on geochemical prospecting, 156
  on geology, 283
  on geophysical prospecting, 148
  on luminescence, 136
  on maps, 81
  on marketing minerals, 401
  on metallic mineral deposits, 346
  on mining, 401–402
  on nonmetallic mineral deposits, 381–382
  on ore treatment, 418
  on petrology, 248–249
  on preliminary mining, 398–399
  on radioactive materials, 127
Boots worn in prospecting, 88
Boracite, 372
Borate deposits, 172
Borates, 371–372
  test for, 231
Borax, 135, 357, 366, 369, 372
Bornite, 220, 300–302
Borocalcite, 372
Boron, 155
  flame color of, 227

Boron flux, 225
Boronatrocalcite, 372
Bort, 361
Botryoidal habit, 218
Bournonite, 216
Bowls, heavy plastic, 91
Boyle, R. W., 295
Braden, Chile, 299
Braggite, 296
Brannerite, 331, 332
Brass, 305, 308
Braunite, 314
Brazil, 204, 288, 312, 313, 339, 345, 361
Bread pan, 91
Breccia, 243, 291, 295, 351
Brine, 326, 359, 371–373
British Columbia, 7, 20, 21, 30–32, 43, 46, 49, 202, 212, 213, 293, 303–306
British Guiana, 111
Brochantite, 300, 302
Broken Hill, New South Wales, Australia, 306
Bromine, 372
  test for, 230
Bronze, 308
Bronzite, 227
Brookite, 111, 329
Brown, Harrison, 24
Brown mica, 379
Brown rock, 358
Brucite, 135, 326, 369
Brunton compass, 61, 62
Buckeye sandstone, 390
Building materials, 374–377
Building stone, 374–375, 412
Building-stone placer, 193
Bunker Hill mine in Idaho, 12, 28, 29, 186
Bunsen burner, 224–225
Bureau of Land Management, 199, 200
Burgin, Lorraine, 200
Burma, 204, 353
Burros, 5, 50
Butler, G. Montague, 274–276
Butler, Jim, 12
Butte, Montana, 269, 299, 313, 411

Cable tool drilling bits, 67
Cadmium, 33, 305, 344
Cairngorm, 354
  flame color of, 227
Calcareous tufa, 246
Calcite, 145, 218, 222, 231, 242, 244
  luminescence, 132–136
  metallic associations of, 291, 295
  nonmetallic associations of, uses for, 350, 353, 363, 381
Calcium, test for, 231
Calaverite, 289, 292
California, 4, 11, 12, 15, 28, 42, 54, 110, 112, 114, 169, 172, 208, 227, 277, 386
  claims in, 173, 185, 187
  mineral deposits in, 210, 280, 288, 293, 295, 320, 339, 340, 352, 353, 372
  schools in, offering degrees related to mining, 211
  state agencies in, 43
California Highway Patrol, 98
California pan, 112
California Tertiary gravel, 108
Calomel, 134
Cambrian Period, 274
Camels, 49
Camp Bird mine in Colorado, 16, 167, 207
Campbell, Colin, 9
Camping, 85–88
Can opener, 91
Canada, 6, 10, 13, 16–18, 20, 21, 30–32, 48–53, 63, 65, 79, 80, 83, 96, 97, 111, 118, 144, 150, 171, 210, 212–214, 258, 267, 410, 414
  land subdivision in, 176–178
  mineral deposits in, 295, 306, 330, 331, 335, 352
  mining laws in, 176, 181, 201–204
  schools in, offering degrees related to mining, 212–213
  (*See also* specific provinces)
Canadian Shield, 50, 335
Candles, 87, 91, 97
Canoes, transportation by, 50–52
Canon City, Colorado, 103
Capacity, data on, 425
Capillary habit, 218
Caprok, 262
Carbon (carbonado), 361
Carbonaceous matter, 331, 334
Carbonate minerals, rocks, 263, 275, 306, 332, 380
Carbonates, test for, 230, 231
Carbonatite, 339, 352
Carboniferous periods, 274
Carborundum, 361
Cariboo in western Canada, 16
Carlin mine in Nevada, 5, 292
Carnallite, 359
Carnelian, 354
Carnotite, 322
Cascade Mountains, 104
Cassiterite, 110, 111, 216, 277, 278
  metallic associations of, 306–308, 323, 327, 329, 345
  test for, 230

Cathode rays, 130
Cat's-eye, 350
Cavity, 252, 262
Cavity filling, 258, 288, 293, 299, 303, 305, 307, 318, 320, 321
Celestite, 135
Cement, 236, 241, 244, 246, 326, 375–376
Cement gravel, 280
Cenozoic era, 274
Central City, Colorado, 15
Ceramic materials, 364–366
Cerargyrite, 294
Cerium, 338
Cerro de Pasco, Peru, 299, 303, 305
Chalcedony, 135, 354, 355
Chalcocite, 145, 300, 302
Chalcopyrite, 111, 115, 145, 227, 292, 300–302, 315
Chalk, 246, 363
Chalybite, 311
Chamosite, 311
Channel sample, 159
Charcoal, 225
Charcoal tests, 228–229
Chattanooga Formation, 331
Chellson, Harry C., 6, 25
Chemical analysis, 164, 165
Chemical compounds, 422–423
Chemical elements, 421
Chemical refining, 418
Chemicals, 370–373
Chemiluminescence, 130
Chernyshov Museum of Geological Prospecting in Leningrad, 18
Chert, 246, 334, 361
Chessylite copper, 300
Chile, 10, 204, 299, 359
China, 118, 204, 320, 353
China clay, 363, 365
China stone, 366
Chinese miners and prospectors, 14
Chip sample, 159
Chisels, 66
Chlorapatite, 134
Chlorine:
  flame color of, 227
  test for, 230
Chlorite, 116, 220
Chromates, test for, 232
Chromatography, 153
Chromite, 33, 110, 111, 140, 145, 277, 279, 317, 318, 329, 351, 369, 413
Chromium, 110, 140, 149, 165, 208, 240, 248, 256, 275
  bead test for, 228
  metallic associations of, 296, 309, 315–319, 329, 345
Chromium diopside, 351, 352
Chrysoberyl, 318, 350–351
Chrysocolla, 300, 302
Chrysolite, 326
Chrysolite claim, 168
Chrysoprase, 354
Chrysotile, 378
Chuquicamata, Chile, 299
Churn-drill sample, 159
Cinnabar, 111, 112, 277, 340, 341, 413
Citizenship, 196
Citrine, 354, 356
Claim map, 80, 178
Claim mining, 386, 402–406
Claim selling, 402–406
Claim staking, 7, 167–204
Classifier, 411, 412
Clawson, Marion, 200
Clay, 115, 242, 244, 262, 281, 413
  metallic associations of, 299, 302, 313, 321, 325, 327
  nonmetallic associations of, 363–366, 369, 375, 377, 380, 381
Clay ironstone, 311
Clay ore, 326
Cleavage, 220
Climax, Colorado, 266, 318, 319
Clinton ore, 310
Closed tube, 225
Closed-tube tests, 229
Clothing, 88–89
Coal, 122, 145, 218, 321, 331, 334, 344, 368
Coast Ranges, 297
Coatings, 228–229
Coats, 88
Cobalt, Ontario, 168, 323
Cobalt, 16, 33, 115, 122, 140, 149, 154, 155, 165, 240
  bead test for, 228
  bloom, 115, 323
  metallic associations of, 296, 309, 315, 318, 322–323, 333
  test for, 226
Cobbing, 411
Coeur d'Alene district of Idaho, 12, 186, 268, 303–306
Coffeepot, 91
Coffinite, 331, 332
Cogwheel ore, 216
Colemanite, 135, 136, 372
Colleges offering degrees related to mining, 211–212
Colloform habit, 218
Colombia, 204, 296, 350, 399
Color, 115–116, 220
Colorado, 5, 11, 12, 15, 16, 28, 29, 31, 42, 50, 97, 103, 104, 106–108, 110, 112, 114, 117, 125, 167, 168, 171, 207, 209, 240, 266, 271, 275
  claims in, 173, 178, 187, 201

Colorado *(Cont.)*:
mineral deposits in, 210, 288, 293, 303, 305, 318, 319, 333
schools in, offering degrees related to mining, 211
state agencies in, 43
Colorado Mining Association, 23, 182
Colorado Plateau, 155, 209, 274, 321, 322, 332–334, 336
Colorimetric tests, 151
Colorimetry, 154
Columbite, 111, 239, 327
Columbite-tantalite, 323
Columbium, 33, 165, 323, 339
Column chromatography, 153
Columnar habit, 218
Comb structure, 261
Common varieties, 197
definition of, 236
Compass, 61, 140
Comstock, Henry, 170
Comstock Lode, Nevada, 293
Concentrates, 412
Concentrating ores, 410
Concentration of ores, waste removal by, 412
Concentric habit, 218
Concessions, prospecting, 201
Conchoidal fracture, 220
Concrete, 236, 241
Concretion, 218
Conglomerate, 242, 243, 280, 292, 334, 335
Congo, 111, 204, 299
Connecticut, 44, 210
Conservation, 22, 23
Contact, 109, 251–252
Contact metasomatism (metamorphism), 253, 258, 288, 293, 296, 299, 305, 310, 318, 320, 354, 358, 377
Contact print, 122
Contact-print test, 332
Contour, index, 78
Contour interval, 70, 78
Contours, rules of, 77
Cooking, 89–95
Cooper and Peck (firm), 29
Cooperite, 296
Copper, 18, 33, 110, 116, 164, 298–303, 320, 321, 344, 345, 370
associations of, 294, 296, 305, 314, 322, 323, 332, 333, 336, 343
bead test for, 228
flame color of, 227
geochemical and geophysical methods, 123, 140, 141, 145, 149, 154–156
geology, mineralogy, and petrology, 235, 240–242, 267, 277
Copper *(Cont.)*:
mining and milling, 410, 412, 414–418
test for, 228, 231
Copper glance, 300
Copper pyrites, 300
Coquina, 246
Coral, 218, 243, 245, 246, 349, 369
Corduroy, 415
Core drill, 68
Cores, 160, 161
Cornish stone, 366
Cornwall, England, 14
Cornwall, Pennsylvania, 310, 389
Cortez mine in Nevada, 5, 292
Corundum, 33, 109, 111, 134, 216, 221, 239, 350, 361
Cosmic rays, 123, 125
Cost concept, 24
Costa Rica, 204
Country rock, 115, 253, 254, 269
Courses in prospecting, 30–32, 410
in Canada, 212–213
in Great Britain, 214
in United States, 211–212
Covellite, 220, 300, 302
Cowcill (prospector), 9
Creede, Colorado, 207
Creek placer, 278
Creekology, 117
Cretaceous Period, 274, 294
Cripple Creek, Colorado, 5, 10, 11, 15, 97, 114, 167, 168, 207, 209, 240, 271, 275
Cristobalite, 353
Crocidolite, 379
Crocombe, Tommy, 107
Crocus, 363
Cross, Jules, 312
Cross, W. R., 9
Crosscut, 390
Crushed rock, 276
Crushed stone, 236, 237, 244, 375
Crusher, 411, 412
Cryolite, 136
Cryptomelane, 314
Crystallography, 215
Crystals, 215
optical, 381
Cuba, 315, 400
Cuprite, 111, 145, 300, 302
Cuproscheelite, 134, 135
Cups, 91
Curie, Marie, 29
Custom plant, 410
Cut-and-fill stoping, 391
Cyaniding and cyanidation, 415–416
Cylindrite, 307

Dacite, 240, 291
Dake, H. C., 25, 96
Dark ruby silver, 293
Data list and tables, 419–427
Daughter element, 123
Davidite, 331, 332
Death Valley, California, 227
Decrepitate, 227
Delaware, 44, 169
Demantoid, 353
Dendrites, 219
Dendritic habit, 218
Density, 222, 223
Descloizite, 322
Desert placer, 278, 279
Desert prospecting, 106
Desert varnish, 106
Devonian Period, 274
Diabase, 240, 291
Diamond, 10, 11, 33, 110, 351–352
  geology, mineralogy, and petrology, 216–218, 221, 240, 242, 248, 279
  industrial, 360–361
  luminescence, 130, 135
  processing, 412, 413
Diamond drill, 18, 68, 69, 160
Diamond mortar, 224
Diaspore, 325, 366, 369
Diatomaceous earth (diatomite), 362, 366, 369, 377, 381
Dickite, 365
Dickson, Benny, 18
Dike (dyke), 254–256, 282, 291, 304, 351, 352, 380
Dimension stone, 236, 237, 239, 240, 243, 244, 247, 248, 374, 375
Diodorus Siculus, 10
Diopside, 351, 352
Diorite, 239, 240, 291, 295, 302
Dip angle, 147
Dip needle, 140, 141
Directories, mining industry, 406–407
Discovery of vein or lode, 191–192
Dish towels, 91
Dishcloths, 91
Dishpan, 91
Disseminated minerals, 253, 254, 258, 277, 296, 301, 302, 314, 317, 412
Disseminated replacement, 258, 288, 293, 299, 303, 305, 319
Divergent habit, 218
Divide, Montana, 266
Divine Gulch, California, 11
Divining rod, 276
Dogs, 117
Dogtooth spar, 218
Dole, Hollis M., 8
Dolerite, 240
Dolomite (dolostone), 134, 136, 145, 231, 246, 248, 263, 325–327
  nonmetallic association of, uses of, 357, 369, 375–377, 379, 380
Dominican Republic, 204
Double boiler, 91
Double jack, 66
Drainage, 114
Dredging, 394, 397
Dreyer, Robert M., 17
Drifting, 390, 398
Drifting pick, 66
Drill steel, 66
Drilling, 109, 387, 388
Drilling bit, 67, 161
Drilling machine, 159, 160
Drive-pipe method, 162
Drusy habit, 220
Dry-bone ore, 306
Dulong, 112
Dumortierite, 366, 368, 369
Dunite, 240, 296, 315, 379
Du Pont company, 171
Dynamite, 397
Dysprosium, 338

Earth, studying the, 207–213
Eclogite, 351
Ecuador, 204
Egypt, 14, 204
Elastic, 220
Eldorado uranium and silver mine, 208
Electrical data, 426
Electrical methods, 145–147
Electrolytic refining, 418
Electromagnetic method, 146–147
Electron-beam magnetometer, 143
Electrostatic concentration, 413
Elements, chemical, 421

Elliott, Thomas, 20
El Salvador, 204
Eluvial placer, 278
Ely, Northcutt, 203, 204
Ely, Nevada, 299
Emanometry, 127
Emerald, 350
Emery, 361
Emission spectrograph, 153
Emma mine in Utah, 170
Emmons, Samuel F., 209
Emmons, William H., 275
Enargite, 300, 343
En echelon veins, 271
Engine sand, 375
England, 14
Enstatite, 351

Eocene Epoch, 274
Eolian placer, 278, 279
Epidote, 112, 302
Epoch, geologic, 274
Epsomite, 135, 357
Equilibrium effect, 125
Equipment:
  cooking, 89–91
  prospecting, 58–84
Era, geologic, 274
Erbium, 339
Erythrite, 323
Eschynite, 323
Eucryptite, 134, 372
Europe, 313
  (*See also* specific countries)
European ore, 325
Europium, 338
Euxenite, 323, 340
Evaporation, 258, 358
Exfoliate, 227
Exploration assistance, 32–42
Exploration claim, 183
Explosives, 69, 97, 397
Exposure, 108
Extralateral rights, 181–186, 201
Extrusive rock, 257

Fahlore, 294, 300
False bedrock, 281
Faulk, Terry, 6
Faults, 109, 139, 262–264, 276
Feel of minerals, 223
Feldspar, 109, 134, 292, 302, 327, 352, 363, 366, 412
  geology, mineralogy, and petrology, 221, 227, 237, 239, 240, 243, 247, 262
Feldspathoid, 353
Felsite, 291
Felted habit, 220
Fennimore, James, 170
Ferberite, 111, 320, 321
Fergusonite, 111, 323, 340
Ferrimolybdite, 319
Ferroalloy metals, 309–323
Ferrosilicon, 413
Fertilizers, 357–359
Fibrous habit, 220
File, 225
Filiform, 220
Fillers, 380
Filter sand, 375
Filters, 381
Finland, 118
Fire clay, 365, 369
Fire refining, 418
Fire sand, 375
Firearms, 88
First aid, 98
Fishing tackle, 88
Fissionable source material, 198
Fissure, 260, 269
Flake mica, 379
Flame spectroscopy, 154
Flashlight, 87
Flawn, Peter T., 24, 172, 203
Fletcher, Andrew, 305
Flexible habit, 220
Flin Flon, Manitoba, 299, 305
Flint, 362
Float, 108–110
Flood gold, 278
Florida, 44, 173, 187, 210, 277, 331, 339, 345, 358
Flotation, 408, 413–415
Flow sheet, 410, 416
Fluorescence, 127, 154, 228, 232, 319, 321, 332, 337, 340, 367, 372, 413
  (*See also* Luminescence; Ultraviolet)
Fluorescent bead, 122
Fluorescent minerals, 129–136
Fluorite (fluorspar), 33, 109, 132, 140, 165, 327, 412
  geology, mineralogy, and petrology, 222, 239, 241
  luminescence, 132, 134–136
  nonmetallic associations of, uses of, 366–368, 381
Fluxgate magnetometer, 143
Fly bar, 87
Folds, 139, 262, 264
Foliated, 220
Food, 91–95
Fool's gold, 289, 292
Foraminifer, 246
Forceps, 225
Forks, 91
Formanite, 323
Formation, 267
Fossils, 242, 245, 246
Founding sand, 369
Fowler, M. Gordon, 14
Fracture, 220
France, 18
Franckeite, 307
Franklinite, 111, 140, 305, 314, 318, 329, 413
Frasch process, 415
Free miner's certificate, 171
Freibergite, 294
Fresnillo, Mexico, 293
Friction tape, 88
Frohberg, M. H., 208
Frost action, 109

Frostbite, 97
Fryer Hill in Colorado, 28
Frying pan, 91
Fuels, 5
Fuller's earth, 363, 365, 381
Fumarole, 371
Furnace:
  blast, 416
  reverberatory, 416
Furnace sand, 369, 375
Fusibility, 226–227

Gabbro, 239–240
Gad, 66, 68, 159
Gadolinite, 111, 339, 340
Gadolinium, 338
Galena, 109, 111, 112, 145, 227, 294, 303, 304, 413
Galenite, 303
Galicia, Spain, 10
Gallium, 305, 344
Gamma-ray spectrometry, 125
Gamma rays, 123
Gangue, 260, 301, 413, 414
Ganister, 369
Garden of the Gods in Colorado, 125
Gardner, E. D., 6
Garnet, 111, 112, 215, 227, 279, 302, 329, 351–353, 361, 362, 413
Garnet schist, 247
Garnierite, 315
Gas, natural, 139
Geiger counter, 6, 18, 64, 119–128, 141, 413
Gem scoop, 69
Gems, 110, 130, 239, 277, 280, 327, 349–356, 412
Gemstones (*see* Gems)
General Mining Laws, 186–198
Geobotanic prospecting, 149, 155
Geochemical prospecting, 114, 149–156
  books on, 156
Geode, 219, 220
Geologic agencies, state, 43–46
Geologic maps, 79
Geologic time scale, 274
Geology, 250–283
  books on, 283
Geophysical methods, 7, 18, 54
Geophysical prospecting, 137–148
  books on, 148
Georgia, 44, 210
Germanium, 305, 344
Ghana, 111, 313
Gibbsite, 325
Gilpin County, Colorado, 106
Gilsonite, 377
Glacier deposits, 278, 280
Glass:
  blue, 225
  volcanic, 241
Glass cutter, 225
Glass sand, 375
Glass tube, 225
Glasses, safety, 69
Glauberite, 135
Globular habit, 220
Glory holing, 391
Gloves, 89
Gneiss, 247, 252, 291
Goethite, 140, 311, 312, 413
Goggles, safety, 69, 97
Gold, David P., 352
Gold, 5, 10, 34, 64, 110–114, 157, 164, 168, 287–292, 370
  coarse, 109
  geochemical and geophysical methods, 141, 144, 149, 153–156
  geology, mineralogy, and petrology, 207, 209, 239–242, 247, 254, 275, 277–281
  metallic associations, 293, 294, 296, 300, 301, 303, 305, 308, 314, 315, 323, 343, 345
  mining and milling, 398, 410, 412, 414–416, 418
  phantom, 157
  price of, 5
  sweating of, 10
  test for, 228
Gold Coast, 111
Gold-digging ants, 10
Gold farm, 180
Gold pan, 63, 102, 159, 350
Gold quartz, 354
Golden beryl, 350
Gophering, 385, 391
Gossan, 115–116, 267, 294, 301, 306
Gouging, 385, 391
Grab method, 162
Grab sample, 159
Grand Canyon, Arizona, 359
Granite, 114, 116, 145
  geology and petrology, 237, 239, 240, 252, 254
  metallic associations of, 291, 295, 319, 320, 327, 339, 343, 345
  nonmetallic associations of, 349–382
Granite gneiss, 247
Granite porphyry, 241
Granodiorite, 239, 240, 302, 343
Grants, New Mexico, 19
Granular, 220
Graphic tellurium, 289, 294
Graphite, 33, 145, 146, 217, 368, 412, 414

Gravel, sand, 354, 360
Gravel-plain placer, 278
Gravimetric methods, 144–145
Gravity, specific, 222, 223
Gravity concentration, 413
Gravity meter, 8, 144, 145
Gravity scales, 426
Gray copper ore, 294, 300
Greywacke, 292
Great Bear Lake, Canada, 208, 295
Great Britain, courses in mineralogy, schools offering, 214
Great Diamond Hoax, 350
Great Dyke, Rhodesia, 256, 318
Great Lakes, 311, 352
Greece, 18
Greek emery, 361
Green copper carbonate, 300
Greenockite, 344
Greensand, 357
Greisen, 307
Grid survey, 145, 155
Griddle, 91
Grill, charcoal, 90
Grossularite, 353
Groundmass, 233, 241
Grub hoe, 67
Grubstaking, 27–30
Guano, 359
Guatemala, 204
Guffey, Colorado, 103
Gulch placer, 278
Gummite, 332
Gump, Richard, 118
Guyana, British Guiana, 111
Gypsite, 376
Gypsum, 134, 135, 222, 227, 236, 243, 246
  nonmetallic associations of, uses of, 357, 358, 371, 375–377, 381

Habit, 216, 218–220
Hachures, 78
Hackmannite, 134
Hafnium, 345–346
Haiti, 204
Halite, 134, 246
Halloysite, 365
Halos, 127, 302
Hamme, North Carolina, 320
Hammer, 64, 225
Hanksite, 135
Hard rock, 358
Hardness, 110, 220–222
Hardness points, 225
Haroernis, Captain, 14
Hatchet, 67
Hats, 88
  hard, 97
Hausmannite, 314
Hauynite, 353
Hawaii, 42, 44, 254
Hawk's-eye, 354, 379
Hayes, George Gibbon, 169
Heads, 412
Health and safety, 96–99
Heat-color scale, 427
Heater, 88–89
Heavy minerals, 110–114
Helicopter, 8, 53
Heliotrope, 354
Helium, 198
Hematite, 111, 112, 140, 145, 220, 242, 292, 311, 312, 318, 329, 361, 377, 413
Hemimorphite, 134, 227, 305
Henderson, Charles W., 104
Herodotus, 10
Hessite, 292, 294
Hiddenite, 354
High-level gravel, 280
Hirschhorn, Joseph, 19
Holmes, Elisha, 11
Holmium, 339
Homestake, South Dakota, 288
Honduras, 204
Hong Kong, 204
Hook, George P., 28
Hoover, Herbert, 104, 181
Horizontal loop, 147
Horn silver, 294
Hornblende, 112, 237, 239, 240, 247
Hornblende gneiss, 247
Hornblende schist, 247
Hornblendite, 240
Horseflesh ore, 300
Horses, 49–50
Horsetail structure, 269
Hoskin, Arthur J., 117
Howlite, 134
Huebnerite, 314, 320
Hydraulicking, 394
Hydrocarbon, 331, 377
Hydrogen, test for, 229
Hydrometallurgy, 415
Hydrothermal, 258, 296, 299, 307, 313, 315, 318, 320, 321
Hydrothermal fluid, 254
Hydrozincite, 135, 136
Hypersthene, 240

Iceland spar, 381
Idaho, 12, 28, 42, 44, 211
  claims in, 173, 183, 186, 187

Idaho *(Cont.)*:
  mineral deposits in, 210, 268, 293, 294, 303–306, 320
Igneous rocks, 235–241, 267, 273, 275
  metals and minerals in, 296, 302, 310, 314, 317, 339, 341, 343, 351–354, 362
Illinois, 44, 210, 211
Illite, 365
Ilmenite, 110–112, 140, 277, 279, 351, 352, 413
  associations of, 318, 328, 329, 345
Ilsemannite, 319
Inclination, 141
Index contour, 78
India, 10, 204, 277, 289, 313, 339, 345, 368
Indian land, 187
Indiana, 44
Indicator, 150
Indicator plant, 107, 149, 155
Indicolite, 356
Indium, 305, 344
Indonesia, 111, 112, 204
Industrial diamond, 360–361
Infrared mapping, 138
Injected rock, 253, 258, 318, 320
Insects, 97
Insulation, 377
Intrusive rock, 252–257, 275, 291, 310, 320, 339
Intumesce, 227
Iodine, test for, 230
Iowa, 44
Ireland, 18, 171
Iridium, 296, 297
Iridosmium, 111
Iris quartz, 354
Iron, 33, 110, 116, 140, 165, 181, 309–312, 321, 322, 364
  bead test for, 228
  geology, mineralogy, and petrology, 239–241, 246, 275, 276
  metallic associations of, 296, 303, 305, 322, 326, 328, 329, 333
  mining and milling, 413, 415–417
  test for, 228, 231
Iron hat (*see* Gossan)
Itaberite, 310

Jackson, George A., 112
Jade, 109, 118, 353
Jadeite, 353
Jamaican ore, 325
Jamesonite, 343
Japan, 204
Jasper, 354, 355
Jasperized wood, 354
Jaspilite, 310
Jefferson, Thomas, 173
Jet, 349
Jig, 413
Joints, 252, 260, 262
Jolly balance, 222
Jones, John, 9
Jurassic Period, 274, 333

Kainite, 359
Kalgoorlie, Western Australia, 114
Kambalda nickel deposit, 9
Kansas, 42, 44, 187, 210
Kaolin (kaolinite), 365
Kaolinization, 115
Katanga, Congo, 299
Kellogg, Noah, 12, 29
Kennecott copper mine in Alaska, 235
Kentucky, 44, 210, 211
Kern County, California, 372
Kernite, 372
Kettle, 91
Kiangsi, China, 320
Kidney ore, 31
Kimberley, British Columbia, 11, 293, 306
Kimberlite, 240, 351, 352
Kirkland Lake, Ontario, 288
Kiruna, Sweden, 310
Klondike, 16, 20
Knives, 88, 91
Knowledge, 4, 5
Kolar, India, 288
Korea, 111, 320
  South, 368
Koschmann, A. H., 292
Kramer, 172, 372
Krennerite, 289, 292
Krumlauf, Harry E., 385, 388
Kunzite, 354
Kyanite, 33, 279, 366, 368–369

La Bine, Gilbert, 208
Labrador, 16, 311, 312
Labradorite, 352
Laccolith, 256, 257
Ladder vein, 268
Laizure, C. McK., 111
Lake Superior, 140, 299, 310, 311, 413
Lakeshore copper deposit in Arizona, 7
Lamellar, 220
Laminated, 220
Lamp, 224
Land pebble, 358
Land plaster, 358

Lander, Wyoming, 243
Lang, A. H., 17, 21, 49, 59, 96, 202
Langbeinite, 359
Langford, G. B., 405
Lantern, 87
Lanthanides (lanthanons), 339
Lanthanum, 338
Laos, 204
Lapis lazuli, 353
La Sal Mountains, 15
Laterite, 315–317, 325
Latin America, 210
Latite, 240
Lava, 235–237, 240, 254, 256, 260, 273
Lazurite, 353
Leaching, 275, 415
Lead, 122, 164, 303–304, 320, 321
  associations of, 294, 298, 305, 333, 343–345
  coatings, 228
  geochemical and geophysical methods, 140, 149, 155, 156
  geology, mineralogy, and petrology, 247, 268, 275
  mining and milling, 410, 412–417
  test for, 228
Leadhillite, 134
Leadville, Colorado, 28, 106, 107, 168, 209, 293, 303, 305
Leadville Blue Limestone, 108
Lean-to, 87
Leasing, 7, 34, 172, 188, 193, 195, 236, 358, 359, 372
Length, data on, 424
Lens, magnifying, 62, 63
Lepidolite, 366, 372, 379
Leucoxene, 329
Light ruby silver, 293
Lighting, camp, 87
Lightweight metals, 324–329
Lignite, 331
Lime, 165, 244, 357, 358, 369
  (*See also* Calcium)
Limestone, 114, 218, 231, 244–245, 248, 263, 275, 415, 416
  metallic associations of, 301, 302, 304, 306, 320, 325–327
  nonmetallic associations of, 350, 354, 357, 358, 367, 369, 370, 375, 378, 381
Limonite, 140, 220, 242, 306, 311, 312, 326, 333, 377, 413
Lincoln, Abraham, 12
Lindgren, Waldemar T., 257
Linked vein, 271
Lithium, 122, 327, 366, 379
  flame color of, 227
Lithium minerals, 372
Lithosphere, 252
Litmus paper, 226, 229
Little Cottonwood Canyon in Utah, 170
Little Pittsburg mine in Colorado, 28
Livingstonite, 341
Llamas, 49
Llangynnog, Wales, 10
Location, claim, 187, 189, 191, 192
Location notice, 192
Locomotive, 399
Lode, 108, 115, 171, 172, 281, 320
Lode claim, 190, 191, 196
Lode replacement, 258, 288, 293, 299, 303, 305
Lodestone, 140, 223, 311, 312
Long tom, 395
Lopolith, 257
Lorraine-Luxembourg, 310
Los Pilares, Mexico, 268
Lost mines, 17
Louisiana, 44, 173, 178, 187, 210
Luminescence, 123
  books on, 136
  (*See also* Fluorescence; Phosphorescence; Ultraviolet)
Luminescent minerals, 129–136
Luster, 220
Lutetium, 339
Luxembourg, 310
Lynn Lake, Manitoba, 315
Lyon Mountain, New York, 310

Maclure, William, 209
McGee, John, 18
McKelvey, V. E., 113
McKinstry, Hugh E., 276, 312
McLaughlin, Patrick, 170
McLeod, Norman, 12
Magazines, 233–234
Magma, Arizona, 299
Magmatic concentration, 258, 288, 296, 299, 317, 318, 320, 321
Magmatic dissemination, 258, 296, 318
Magmatic rock, 252, 254, 260, 310, 352
Magnesia, 369
Magnesite, 145, 248, 326, 357, 363, 366, 369, 376, 381
Magnesium, 165, 246, 324, 326–327, 376, 380
  test for, 231
Magnesium ilmenite, 351, 352
Magnesium mica, 379
Magnet, 62, 225
Magnetic balance, 141
Magnetic declination, 141
Magnetic iron ore, 311
Magnetic methods, 140, 145

Magnetic separation, 408
Magnetism, 123, 222, 223, 227, 333, 351, 413
Magnetite, 110-112, 413
geochemical and geophysical methods, 140, 141, 145, 147
geology, mineralogy, and petrology, 222, 223, 277, 279, 283
metallic associations of, 311, 312, 318, 329
nonmetallic associations of, 361, 363, 377
Magnetometer, 140-144
Magnifying glass, lens, 62, 225
Maine, 44, 210
Malachite, 231, 300, 302, 413
Malaya, 396
Malayasia, 204, 277, 278, 307, 320
Malleable, 110, 220
Mammillary, 220
Manganapatite, 134
Manganese, 33, 116, 163, 165, 246, 279, 412
bead test for, 228
geochemical and geophysical methods, 122, 140, 145, 149, 150
metallic associations of, 295, 309, 312-314, 316, 328, 329
Manganite, 314
Manitoba, 32, 46, 50, 202, 212, 299, 305, 315
Mantle, 254
Manufacturing materials, 378-382
Map case, 61
Maps, 60-61, 70-81
books on, 81
claim, 80
geologic, 79, 138
topographic, 74-79
Marble, 231, 248, 263, 350
Marketing minerals, 400-401
books on, 401
Martinez, Paddy, 19
Maryland, 44, 210, 417
Mass, data on, 425
Mass effect, 125
Massachusetts, 44, 210, 211
Massive habit, 215
Massive replacement, 258, 288, 293, 299, 303, 305
Matches, 91
Matte, 417
Mattress, air, 87
Meanders, 280, 281
Measures, weights and, 424
Measuring tape, 62
Mechanical concentration, 258, 288, 296, 307, 317, 320
(*See also* Placers)
Mechanical processes, 257
Mediterranean bauxite, 325
Meerschaum, 381
Melaconite, 300
Mercury, 34, 116, 132, 150, 154, 165, 248, 340-341, 412, 414, 415
test for, 229
Mercury-vapor lamp, 133
Merensky Reef, South Africa, 296
Mertie, John B., Jr., 112
Mesozoic era, 274
Meta-autunite, 134, 331, 332
Metacinnabar, 341
Metal detecting, 147, 148
Metallic mineral deposits, books on, 346
Metallic minerals, 287-346
Metallurgical materials, 367-369
Metals:
ferroalloy, 309-323
lightweight, 324-329
minor, 342-346
nonferrous, 298-308
precious, 287-297
radioactive, 330-337
rare earth, 338-340
Metamorphic rocks, 291, 313, 314, 317, 320, 350, 352-354, 361, 362, 368, 379, 380
geology, mineralogy, and petrology, 235, 237, 246-248, 252, 254, 258, 267, 273
Metatorbernite, 331, 332
Metatyuyamunite, 331
Metes and bounds, 177
Mexico, 204, 210, 268, 293, 299, 303, 305
Meyerhoff, Howard A., 12
Meyerhofferite, 136
Miami, Arizona, 299
Miarolitic cavity, 262
Mica, 33, 115, 116, 122
geology, mineralogy, and petrology, 220, 237, 239, 244, 247, 262
metallic associations of, 302, 307, 327
nonmetallic associations of, uses of, 351, 372, 377, 379-381
Mica schist, 247
Micaceous habit, 220
Micaceous iron ore, 311
Michigan, 44, 140, 210, 211, 312
Microcline, 109, 352, 366
Microlite, 323
Milbourne, Gordon, 18
Milky quartz, 354
Mill:
ball, 412
grinding, 415
rod, 411

Mill *(Cont.)*:
  rolling, 412
  tube, 411
Mill City, Nevada, 320
Mill site, 191, 193, 196, 197
Millhead, 412
Milling, 122
Milling ore, 412
Mills, uranium, 336
Mina Ragra, Peru, 321
Mine, opening a, 385–407
Mine safety, 97
Mineral deposits, classification of, 257–258
Mineral dressing, 408
Mineral processing, 408–418
Mineral surveyor, 195, 196
Mineral wool, 377
Mineralogical Society of Arizona, 98
Mineralogy, 215–234
Miners Basin, Utah, 14
Miner's certificate, 171
Miner's license, 171
Miner's pan, 112
Miner's permit, 171
Mines, provincial departments of, 46–47
Minette ore, 310
Mining, 122
  books on, 401–402
    preliminary, 398–399
Mining claim, 167–204
  definition of, 186
  selling, 402–406
Mining districts, 15
Mining industry directories, 406–407
Mining laws, 172–204
  in Australia, 203
  in Canada, 201–203
  in United States, 172–201
  world, 203–204
Minnesota, 44, 140, 141, 210, 211, 312
Minor metals, 342–346
Miocene Epoch, 274
Misch metal, 340
Mississippi, 45, 173, 187
Mississippian Period, 274
Missouri, 45, 210, 211, 294, 304
Mock, H. Byron, 385
Mohs scale, 220
Moil, 66, 159
Mojave Desert, 98
Molding sand, 375
Molybdates, test for, 232
Molybdenite, 111, 145, 302, 319, 320, 368, 345, 413
Molybdenum, 19, 33, 115, 165
  bead test for, 228
  coatings, 228
  flame color of, 228
Molybdenum *(Cont.)*:
  geochemical and geophysical methods, 130, 150, 155
  geology, mineralogy and petrology, 237, 239, 266, 275
  metallic associations of, 302, 309, 318–320, 333, 336, 343–345
  test for, 229
Monazite, 33, 110–112, 126, 231, 239, 277, 279, 283, 329, 337, 339, 340, 345
Montana, 18, 42, 45, 173, 187, 210, 211, 242, 264, 266, 269, 299, 313, 331, 336, 411
Montmorillonite, 365
Monumenting, 192
Monzonite, 239, 240, 302, 343
Moonstone, 352
Morgan (prospector), 9
Morganite, 350
Morion, 354
Morrison in the Colorado Plateau, 274
Morro Vehlo, Brazil, 288
Mortar, 375
  and pestle, 67, 159, 224
Mosquito netting, 87
Moss agate, 354
Mossite, 323
Mother Lode, California, 114, 186, 208, 277, 288
"Mother lode," 110
Motor vehicles, 54–57
Mountain Pass, California, 339
Muck, 109
Mud, 244
Mudstone, 334
Mulberger, Henry C., 6
Mullite, 368
Multiple use, 197, 198
Muscovite, 237, 379–381
Muscovite schist, 247
Mwadui, Tanganyika, 22

Nacozari, Mexico, 268, 299
Nacrite, 365
Nails, 88
Nambia (South West Africa), 352
National monuments, 187
National parks, 187
Native bismuth, 343
Native copper, 277, 300, 302
  (*See also* Copper)
Native gold, 218, 281, 288, 289, 345, 414, 416
  (*See also* Gold; Placers)
Native mercury, 341
  (*See also* Mercury)

Native platinum, 295, 297
(*See also* Platinum)
Native silver, 293, 416
(*See also* Silver)
Native sulfur, 358, 370, 415
(*See also* Sulfur)
Nazibia, South-West Africa, 352
Nebraska, 42, 45, 173, 187
Neck, volcanic, 257
Nelsonite, 329
Neodymium, 338
Nepheline, 134
Nepheline syenite, 239, 240
Nephrite, 353
Nevada, 5, 7, 12, 42, 170, 171
claims in, 173, 183, 201
mineral deposits in, 292, 293, 299, 320, 353
schools in, offering degrees related to mining, 211
state agencies in, 45
New Brunswick, 46, 202, 212
New Caledonia, 315
New Guinea, 6, 53, 111
New Hampshire, 45, 210
New Jersey, 45, 132, 210, 305
New Mexico, 19, 31, 42, 243
claims in, 173, 187, 188
mineral deposits in, 210, 299, 319, 381
schools in, offering degrees related to mining, 211
state agencies in, 46
New South Wales, Australia, 203, 294, 306
New York, 45, 210, 211, 310
New Zealand, 111, 353
Newfoundland, 47, 202, 212, 213, 310
Newitt, Harry R., 107
Nicaragua, 204
Nichrome wire, 225
Nickel, 33, 110, 115, 122, 127, 137, 140, 155, 165, 240, 241, 248, 414, 416
bead test for, 228
bloom, 115
metallic associations of, 296, 309, 314–316, 318, 319, 322, 323, 333
Niobium, 323
Niter, 359
Nitrates, 357, 359, 381, 415
test for, 231
Nodular land rock, 358
Nolan, Thomas B., 7
Nome, Alaska, 279
Nonferrous metals, 298–308
Nonmetallic mineral deposits, books on, 381–382
Nonmetallic minerals, 349–382
Nontronite, 365
Noranda, Quebec, 288, 299
Norite, 240
North America, 313, 352
(*See also* specific countries)
North Carolina, 45, 207, 210, 320
North Dakota, 42, 45, 173, 187, 210, 331, 336
Northward pull, 143
Northwest Territories, 47, 201, 202
Norway, 18
Nova Scotia, 32, 47, 203, 212
Nuclear energy (*see* Radioactivity)
Nuclear-precision magnetometer, 143
Nye, Bill, 168

Obsidian, 241
Ocher, 377
Odor of minerals, 223
Office of Indian Affairs, 187
Office of Minerals Exploration, 32–42
Ogdensburg, New Jersey, 305
Ohio, 45, 211, 390
Oil, petroleum, 122, 139, 218, 321, 334, 358
Oil gravity, 426
Oklahoma, 42, 45, 187, 210, 273
Oligocene Epoch, 274
Oligoclase, 352
Olivine, 240, 326, 351, 354
Olivine basalt, 240
Ontario, 10, 19, 32, 47, 50, 93, 104, 178, 203, 212, 214, 289, 296, 305, 312, 315, 323
Onyx, 354
Oolitic habit, 220
Opal, 134, 135, 353
Opalized wood, 353
Open tube, 225
Open-tube tests, 229
Opening a mine, 385–407
Ophir mine, 170
Optical crystals, 381
Ordovician Period, 274
Ore dressing, 408
Ore reduction, 410–412
Ore shoot, 276
Ore treatment, books on, 418
Oregon, 42, 45, 173, 187, 210
Oriental topaz, 350
Orientation, map, 77
O'Riley, Peter, 170
Orpiment, 343
Orthite, 339
Orthoclase, 221, 227, 302, 352, 366
Osmium, 296, 297
Ouray, Colorado, 16, 288

Outcrop, 108-110, 114, 115, 269, 275, 276, 386
Oven rack, 91
Ovens, 90
Over, Edwin W., Jr., 110
Overburden, 108
Oxidation and enrichment, 258, 293, 299, 311
Oxides, test for, 230
Oxidizing flame, 226
Oxygen, test for, 229

Pachuca, Mexico, 293
Pacific Ocean, 240
Pack animals, 49
Packsack, 87
Paeonia, 10
Pakistan, 204
Paleocene Epoch, 274
Paleozoic Era, 274
Palladium, 296
Pan:
  bread or roll, 91
  gold, 63, 102, 159
Panama, 204
Pandermite, 372
Panel sample, 159
Panning, 63, 110–114, 393–396, 413
  (*See also* Gold pan)
Pao, Venezuela, 311
Paper chromatography, 153, 154
Paraguay, 204
Park, Charles F., 24
Park City, Utah, 293, 303, 305
Parka, 88
Parral, Mexico, 293
Parting, 220
Patent, claim, 178, 183, 189, 196, 197
Pathfinder element, 150
Patronite, 322
Paul, Rodman W., 4
Peacock ore, 300
Pearce, Richard, 10, 208, 251
Pearceite, 293
Pearl, 349
Pearly luster, 220
Pectolite, 134, 135
Peele, Robert, 25, 69, 93
Pegmatite, 110, 237–239, 254, 261, 262
  metallic associations of, 318, 320, 327, 335, 339
  nonmetallic associations of, uses of, 350, 356, 361, 366, 369, 372, 380
Pennant, Thomas, 10
Pennsylvania, 45, 117, 210, 211, 310, 389
Pennsylvania Period, 274
Pentlandite, 111, 315
Percussion-drill sample, 159–160
Peridot, 354
Peridotite, 208, 240, 248, 296, 315, 317, 351, 354, 379
Period, geologic, 274
Perlite, 377
Permian Period, 274
Perovskite, 323, 329
Peru, 204, 299, 303, 305, 321, 322
Pestle, mortar and, 67, 159, 224
Petalite, 372
Petrified Forest National Park, 355
Petrified wood, 354
Petroleum, oil, 122, 139, 218, 321, 334, 358
Petroleum placer, 193
Petrology, 235–249
  books on, 248–249
Petsamo, U.S.S.R., 315
Petzite, 289, 292, 294
Pewter, 308
Pfleider, Eugene P., 390
Phantom gold, 157
Phenacite (phenakite), 328
Phenocrysts, 237
Philippines, 6, 204, 315
Philipsburg, Montana, 313
Phonolite, 240, 275
Phonolite myth, 275
Phosgenite, 134
Phosphate rock, 122, 165, 321, 331, 357, 358, 369, 381
  test for, 231
Phosphor, 123
Phosphorescence, 129, 132, 337
Phosphorescent minerals, 129–136
Phosphorus flame, 227
Photographs:
  oblique, 82
  trimetregon, 82
  vertical, 82
Photography, aerial, 81–84
  books on, 83–84
Pick, drifting, 66
Pigments, 376–377
Pigs, lead cast as, 417
Pike, James A., 98
Pikes Peak, Colorado, 109, 112
Pillar stoping, 391
Pine Creek, California, 320
Pipe, 313, 351
Pipe clay, 363
Pisolitic habit, 220
Pitchblende, 331, 332
Placer claims, 191, 193
Placers, 108–114, 171, 172, 257, 258, 267, 277-283, 288, 289, 291, 292,

Placers, *(Cont.)*:
296, 297, 321, 322, 329, 350–353, 356, 361, 414
Plagioclase, 237, 240, 352
Plants, 117, 155, 156, 276
Plasma, 354
Plaster, 236, 246, 375
Plates, 91
Platinum, 34, 110–112, 114, 140, 165, 418
geology, mineralogy, and petrology, 208, 240, 248, 256, 275–277, 279–281
metallic associations of, 287, 296–297, 315, 318, 319, 345
Platinum group metals, 34, 281, 296–297
Playa, 359, 371, 372
Pleistocene Epoch, 274
Pliers, 88
Pliocene Epoch, 274
Plumbago, 368
Plumose habit, 220
Polarity, 140
Polarography, 154
Poling boat, 52
Pollucite, 344
Polybasite, 293
Porcupine district of Ontario, 10
Porphyry, 116, 237, 239, 241, 291, 295, 320, 351
Porphyry copper, 241, 254, 262, 294, 301, 302, 319, 320, 345
Porter, William W., II, 25
Poseidon mine in Australia, 8
Postmasburg, South Africa, 313
Potash and potassium, 227, 276, 357, 359, 366, 371, 372, 415
Potassium flame, 227
Potassium mica, 379
Potosí, Bolivia, 293
Powderhorn case, 171
Powellite, 130, 135, 136, 319, 320
Power drill, 69
Prase, 354
Praseodymium, 338
Precambrian Era, 254, 274, 291, 294, 335
Precious metals, 287–297
Priceite, 372
Prime meridians, 174
Prince Edward Island, 47, 203, 213
Principal meridian, 173–176
Prismatic habit, 220
Processing, mineral, 408–418
Projection, map, 74
Promethium, 338
Promoters, 170
Prompt-neutron-capture—gamma-ray method, 127, 137, 154
Prospect, where to, 3, 102–108
Prospecting:
in Africa, 6
art of, 3–25
assistance, 32–42
in Australia, 6, 8
basic, 102–118
in Canada, 6
conditions in, 4
courses on, 30–32
equipment, 58–84
history of, 6–22
knowledge in, 3
laws, 178–204
license, 171
literature on, 25–26
in New Guinea, 6
in Philippines, 6
profitability of, 6
science of, 3–25
Prospectors:
aids to, 27–47
employment for, 29, 30
organizations of, 29
syndicates of, 30
Prospector's pick, 64, 65
Proton-precession magnetometer, 143
Proustite, 293
Provincial departments of mines, 46–47
Provisions, 91–95
Prudent-man doctrine, 182, 188
Pry bar, 68
Pseudocleavage, 220
Psilomelane, 313, 314
Public-land states, 173, 175
Puerto Rico, 45
Puget Sound, 104–106
Pumice, 241, 361, 362, 377
Purple copper ore, 300
Pyramidal, 220
Pyrargyrite, 111, 293
Pyrite, 112, 115, 116, 145, 247, 260, 398, 415, 416
metallic associations of, uses of, 289, 292, 300–302, 306, 311, 319, 335
nonmetallic associations of, 353, 358, 370
Pyrite halo, 262
Pyrochlore, 323
Pyrolusite, 313
Pyrometallurgy, 416
Pyromorphite, 304
Pyrope, 351, 352
Pyrophanite, 329
Pyrophyllite, 366, 380
Pyroxene, 239, 240, 351, 353, 356
Pyroxenite, 240, 296, 317, 380
Pyrrhotite, 111, 140, 145, 222, 292, 296, 311, 315

Quartering, 162
Quartz, 33, 109, 110, 112, 114–116, 135, 145
  geology, mineralogy, and petrology, 208, 216, 221, 227, 239, 242, 247, 254, 260, 262, 268, 277, 283
  metallic associations of, 291, 301, 302, 307, 328, 329, 344
  nonmetallic associations of, uses of, 354, 356, 360–362, 369, 381
  processing, 411, 412
Quartz cat's eye, 354
Quartz crystal, 34
Quartz diorite, 144, 239, 240
Quartz Hill district in Montana, 266
Quartz monzonite, 141, 239, 240, 291, 302, 327
Quartz porphyry, 291
Quartzburg, California, 12
Quartzite, 114, 247, 291, 368, 369
Quaternary Period, 274
Quebec, 32, 47, 50, 178, 203, 213, 288, 299, 311
Queensland, 111, 203
Quicklime, 375
Quicksilver, 340

Radar, 138
Radiated habit, 220
Radioactive materials, literature on, 127
Radioactive metals, 330–337
Radioactive minerals, 6, 119–128, 239, 322, 323
  (*See also* Thorium; Uranium)
Radioactive source materials, 198
Radioactivity, 119–128, 252, 295, 327, 351
Radiometry, 154
Radium, 330
Radon, 127
Rainbow quartz, 354
Raincoat, 88
Rand, the (Witwatersrand), South Africa, 144, 288, 331, 415
Ranges, 173–177
Rare-earth metals, 338–340
Rare-earth minerals, 132
Rare earths, 34, 165
Rasorite, 372
"Raspberry" (miner), 12
Ray, Arizona, 299
Realgar, 343
Recent Epoch, 274
Red copper ore, 300
Red iron ore, 311
Red ocher, 311
Reducing flame, 226
Reduction, ore, 410–412
Reed, Conrad, 207
Refining, 417–418
Reflection survey, 139
Refraction survey, 139
Refractories, 326, 364
Refractory materials, 367–369
Remote sensing, 137
Reniform habit, 220
Replacement, 258, 272, 288, 293, 299, 302, 303, 305, 310, 314, 315, 320, 321, 341, 376, 380
Representation, 201
Republic of South Africa, 11, 380, 415
  (*See also* South Africa)
Residual concentration, 258, 289, 311, 313, 315, 317, 321
Residual deposits, 310, 365, 380
Residual placer, 278
Residual process, 257
Residual weathering, 324
Resin jack, 305
Resinous, 220
Resistivity, 145
Resuing, 391
Reticulated habit, 220
Rhenium, 336, 345
Rhode Island, 46, 210
Rhodesia, 111, 256, 299, 317, 318
Rhodium, 296, 297
Rhodochrosite, 314
Rhodonite, 134, 314
Rhyolite, 240, 295
Richter, Frank, 7
Rickard, T. A., 12, 25, 167
Ridge, John D., 170, 386
Riebeckite, 379
Riffled pan, 112
Riffles, 281, 282
Rinkenberger, Richard K., 30
Rio Algom-Rio Tinto organization, 29
Rio Tinto, Spain, 299
Ripple marks, 243
Riprap, 236
Rische, August, 28
River pebble, 358
River placer, 278
Rivets, 88
Road metal, 236, 375
Roasting, 228
Robie, Edward H., 203
Rock(s), 235–249
  blue, 358
  brown, 358
  definition of, 235
  extrusive, 257
  hard, 358

Rock(s) *(Cont.)*:
igneous, 236–241
intrusive, 254–257
metamorphic, 246–248
nodular land, 358
sedimentary, 242–246
soft, 358
white, 358
Rock crystal, 354, 381
Rock gypsum, 376
Rock hammer, 64, 65
Rock salt, 415
Rock wool, 377
Rocky Mountains, 334, 390
Rod mill, 411
Roedder, Edwin, 209
Rogers, "Lovely," 12
Rogers, M. J., 95
Roll-type uranium, 335
Rolling mill, 412
Roof pendant, 253
Room and pillar stoping, 391
Roots, Robert W., Jr., 113
Rope, 88
Roscoelite, 322, 379
Rose quartz, 239, 354
Rosin jack, 305
Rottenstone, 362
Rouge, 363
Rowboats, 52
Rubellite, 356
Rubicelle, 359
Ruby, 109, 130, 134, 216
star, 350
Ruby blende, 305
Ruby copper, 300
Ruby silver, dark, 293
Ruby zinc, 305
Rules of contours, 77
Russia (*see* Soviet Union)
Ruthenium, 296, 297
Rutile, 34, 110, 111, 277, 279
metallic associations of, 323, 328, 329, 345

Sack, 87
Saddle reef, 267
Safety, 96–99
Safety glasses and goggles, 69, 97
Safety shoes, 97
Sagenite, 354
St. Peters Dome, Colorado, 110
Saline placer, 193
Salt, 371
rock, 415
Salt dome, 371
Salting a mine, 168–170
Salts, evaporite, 122
Samarium, 338
Samarskite, 323, 340
Sample bags, 63
Sample kit, 59
Sampling, 157–167, 375, 386
San Francisco (Mexico), 293
San Juan Mountains, 16, 108
Sand, 109, 110, 244, 362, 369
foundry, 369
furnace, 369
and gravel, 354, 360, 375, 389
Sandong, Korea, 320
Sandstone, 109, 110, 243, 244, 247, 322, 334, 335, 345, 362, 369
Sanitation, 96–97
Santa Eulalia, Mexico, 293, 303, 305
Santa Rita, New Mexico, 299, 319
Saponite, 365
Sapphire, 109, 216
star, 350
Sarawak, 204
Sard, 354
Sardonyx, 354
Saskatchewan, 30, 32, 47, 203, 213
Sassolite, 372
Satin spar, 376
Saturation exploration, 139
Saw, 88
Scale, map, 74
Scandinavia, 244
(*See also* specific countries)
Scandium, 339
Scheelite, 111, 112, 129, 130, 132, 135, 136, 295, 320, 321, 409, 414, 415
Schist, 110, 247, 271, 275, 291, 329, 350, 368
Schoepite, 332
Scintillation counter, 6, 7, 18, 64, 119–128, 327
Scoop, gem, 69
Scotia nickel deposit, 8
Scouring pad, 91
Scrap mica, 379
Scree, 109
Screen, color, 225
Searles Lake, California, 372
Secondary enrichment, 267, 275
Sectile, 220
Sections, land, 173–177
Sedimentary minerals and rocks, 299, 310, 313, 314, 325, 326, 334, 352, 354, 358, 362, 364, 366, 371, 372, 376, 380
geology, mineralogy, and petrology, 235, 242–246, 254, 258, 273, 275
Segregations, 258, 296, 317
Seismic methods, 139

Selected sample, 159
Selection, 412
Selenite, 376, 381
Selenium, 34, 153, 315, 333, 345
  coatings, 228
  flame color of, 228
Self-potential, 145
Selling mining claims, 402–406
Senarmontite, 343
Sepiolite, 136, 365, 381
Sericite, 302
Sericitization, 115
Serpentine, 135, 144, 208, 236, 248, 296, 315, 317, 351, 378, 379, 413
Sewing kit, 89
Shaft, 388, 390
Shale, 244, 247, 262, 291, 292, 321, 331, 350
Shear zone, 270
Sheet, 87
Sheet mica, 379
Sheeted vein, zone, 271
Shelter half, 87
Shipping ore, 412
Shirley, Ken, 8
Shirts, 88
Shoes, 88
  safety, 97
Shovels, 67
Shrinkage stoping, 391
Siberia, 49, 111, 352
  (*See also* Soviet Union)
Siderite, 145, 231, 311, 312, 413
Sienna, 377
Sierra Nevada, 277, 297
Sieve, 69, 159, 412
Silesia, 155
Silica, 165, 325, 349, 353, 360, 361, 364, 369, 381, 416
Silicates, test for, 230, 232
Siliceous abrasives, 361–362
Silicification, 114, 115
Silicified wood, 354
Silky luster, 220
Sill, 255, 256, 351
Sillimanite, 366, 368–369, 412
Sillimanite group, 368–369
Silt, 244
Siltstone, 244
Silurian Period, 274
Silver, 34, 111, 164, 292–295, 320
  associations of, 287, 296, 303–305, 314, 315, 322, 323, 333, 336, 343, 345
  geochemical and geophysical methods, 122, 149, 150, 153, 154, 156
  geology, mineralogy, and petrology, 239–242, 267, 275
Silver *(Cont.):*
  mining and milling, 410, 414–416, 418
  test for, 228, 230
Silver glance, 293
Silver Snooper, 154, 295
Sinai Peninsula, 14
Single jack, 66
Sink-and-float, 413
Sketching materials, 62
Slate, 110, 247, 273, 291, 412
Sledging, 66, 411
Sleeping bag, 87
Sloane, Howard N., 25
Sloane, Lucille L., 25
Sluice box, 393, 394
Smelting, 386, 408, 410, 412, 416–418
Smithsonite, 231, 305, 306
Smoky quartz, 109, 354
Snakes, 97
Snell, Charles, 29
Snider, 18
Snowshoes, 57
Soap, 91
Soapstone, 380
Socks, 88
Sodalite, 134, 135
Sodium, 276, 372
  flame color of, 227
Soft abrasives, 363
Soft rock, 358
Soil-testing kit, 152
Solvent-extraction techniques, 153
Sonora, Mexico, 268
Sorting, 412, 413
South Africa, 11, 13, 60, 144, 288, 296, 313, 317, 331, 339, 412
  (*See also* Republic of South Africa)
South America, 49, 112, 180, 293
  (*See also* specific countries)
South Australia, 203
South Carolina, 46, 210
South Dakota, 42
  claims in, 173, 187
  mineral deposits in, 210, 288, 311, 336
  schools in, offering degrees related to mining, 211
  state agencies in, 46
South Korea, 204, 368
South-West Africa, 352
Soviet Union, 18, 112, 296, 313, 315, 317, 352
Spain, 10, 299, 341
Spalling, 411
Sparrows Point, Maryland, 417
Spathic iron, 311
Spatula, 91
Specific gravity, 222, 223, 283, 413

Specific ion activity, 154
Specimen bag, 63
Spectrograph, 153
  X-ray, 153
  X-ray fluorescent, 154
Spectrophotometer, 154
  atomic-absorption, 154
Spectroscope, 153
  atomic-absorption, 153
Specular iron, 311
Specularite, 311
Sperrylite, 296, 353
Sphalerite, 134, 145, 305, 306, 344
Sphene, 112, 329
Spinel, 111, 134, 354, 361
Spodumene, 134, 239, 354–356, 366, 372, 373
Sponge, artificial, 91
Sponge ore, 326
Spontaneous polarization, 145
Spoons, 91
Spot tests, 154
Square-set timber stoping, 391
Staking claims, 167, 204
Stalactitic, 220
Stamp, 411
Stamp mill, 411
Stannite, 111, 307
Star of South Africa, 11
Star ruby, 350
Star sapphire, 350
Stassfurtite, 372
State geologic agencies, 43–46
Staurolite, 217, 329
Staurolite schist, 247
Steatite, 380
Steel and steel alloy, 309–323, 326, 330, 364, 367–369, 416, 417
Steel galena, 34
Steen, Charles, 170
Steep Rock Lake, Ontario, 312
Stellated, 220
Stephanite, 294
Stibiconite, 343
Stibiocolumbite, 323
Stibiopalladinite, 296
Stibiotantalite, 323
Stibnite, 227, 342, 343, 413
Stock, 254–255, 260
Stockraising Homestead Act, 195
Stolzite, 135, 136
Stone, building, 374–375, 412
Stoping, stull, 391
Stout, Koehler, S., 388
Stoves, 89–90
Stratton, Winfield Scott, 10
Streak, 220, 221
Stream placer, 278, 279
Stream tin, 307
Stripping, 389, 391
Stromeyerite, 293
Strontianite, 135
Strontium, 369
  flame color of, 227
  minerals, 373
  test for, 231
Structures, 139, 258–272
Stull stoping, 391
Sublimate, 228–229, 258
Submerged placer, 280
Sudan, 14
Sudbury, Ontario, 16, 296, 315
Sulfates, test for, 232
Sulfide enrichment, 275
Sulfur, 34, 122, 140, 150, 165, 357–358, 370–371, 415, 416
  test for, 229, 230
Sullivan, British Columbia, 303, 305
Summit cupola, 254, 255
Sunglasses, 97
Sunshine, Idaho, 293
Sunshine, effect of, on gems, 130
Sunstone, 352
Surinam ore, 325
Survey meter, 120
Surveys, 195–196
Sweden, 18, 118, 310
Syenite, 145, 239, 240
Sylvanite, 289, 292, 294
Sylvite, 359
Symbols, map, 72–77
Synthetic ore, 313

Tabor, Horace A. W., 28, 168
Tabular habit, 220
Taconite, 311
Tactite, 320, 321
Tailing, 412
Talc, 34, 222, 248, 366, 380
Talus, 109
Tanganyika, 22
Tantalite, 239, 327, 413
Tantalite-columbite, 323
Tantalum, 34, 165, 323, 339
Tanzanite, 218
Tape:
  friction, 88
  measuring, 62
Tapiolite, 323
Tarnish, 220
Tarp, 87
Tarpaulin, 87
Tasmania, 111, 203
Taste of minerals, 223

Teallite, 307
Telluride, 291, 292
Tellurium, 34, 289, 315, 345
  coatings, 228
  flame color of, 228
Temperature data, 426
Tenacity, 220
Tennantite, 294, 300
Tennessee, 46, 210, 306, 358, 413
Tenor, 117
Tenorite, 300
Tents, 86
Terbium, 338
Terlinguaite, 135
Terrace placer, 280
Tertiary basins, 335
Tertiary gravel, 108, 208, 278, 280
Tertiary Period, 274, 291, 325, 333
Testing, 157–166, 386
Tetrahedrite, 294, 300
Texas, 42, 46, 141, 210, 245
Texture of rock, 236, 237
Thailand, 204
Thallium, 305
Thaumasite, 136
Thenardite, 134, 135
Thermoluminescence, 130, 154
Thomas Mountains in Utah, 107
Thomson, Don W., 182
Thorianite, 277, 337
Thorite, 111, 277, 337
Thorium, 4, 6, 34, 110, 119–128, 132, 165, 202, 329, 330–337, 339, 345
Thucolite, 332
Thulium, 338, 339
Tiers (of townships), 173
Tiger's-eye, 354, 379
Time scale, geologic, 274
Timmins, Ontario, 104, 305
Tin, 34, 110, 114, 149, 165, 306–308
  associations of, 295, 298, 319, 320, 323, 327, 329, 343, 345
  coatings, 228
  geology, mineralogy, and petrology, 216, 237, 239, 275, 278, 280
  mining and milling, 412, 416
  test for, 228
Tincal, 372
Tinstone, 230, 307
Tintic, Utah, 293, 299, 303, 305
Titanium, 110, 122, 140, 154, 165, 240, 279, 324, 325, 328–329, 333, 339, 345
  bead test for, 228
Tobago, 204
Tonolite, 240
Tonopah mine in Nevada, 12
Topaz, 109, 221, 239, 307, 356, 366, 368, 369
Topaz *(Cont.)*:
  Oriental, 350
Topographic maps, 70–80
Topography, 109
Torbernite, 132, 331, 332
Toughness, 110
Tourmaline, 239, 302, 308, 356, 381
Towels, dish, 91
Townships, 173–177
Trace element, 150
Trachyte, 240, 295
Trail buggies, 18
Transportation, 18, 48–57
Transvaal, 111
Traprock, 240
Traverse, 116–117, 125, 143
Travertine, 246
Tremolite, 134, 136, 379
Trenching, 387, 388
Trespassing, 22
Triassic Period, 274, 333
Triboluminescence, 130
Trinidad, 204
Tripoli, 362
Trona, 136
Trousers, 88
Tube mill, 411
Tubes, glass, 225
Tufa, 246
Tungstates, test for, 230, 232
Tungsten, 18, 110, 114, 129, 130, 132, 165, 409, 412–414
  geology, mineralogy, and petrology, 237, 239, 259, 275, 279
  metallic associations of, 295, 308, 309, 314, 319–321, 343, 345
Tunnel claim, 192
Tunnel site, 191, 192
Turbidimetry, 154
Turkey, 18, 204, 317
Turkish enery, 361
Turquoise, 14, 231, 356
Twain, Mark, 19–20, 103, 289
Twenhofel, William H., 280
Tyler, Stanley, 312
Tyuyamunite, 322, 331, 332

Ulexite, 136, 372
Ultrabasic rock, 296, 297
Ultraviolet lamp, 226
Ultraviolet light, 129–136
Umber, 377
Underclothing, 88
Underground caving, 391
United States, 13, 17, 30, 31, 111, 155, 171, 181, 258, 267, 292, 304, 306, 326, 330, 331, 333–336, 381, 410
  (*See also* specific states)

U.S. Bureau of Land Management, 178, 186
land offices, 199–200
U.S. Bureau of Mines, 130, 365
field offices, 163
U.S. Forest Service, 186
regional offices, 199–200
Universities offering degrees related to mining, 211–212
Unpatented claims, 189
Upper Cave Springs Wash area, 29
Urad, Colorado, 318
Ural Mountains, 209, 277, 296
Uranates, test for, 232
Uraninite, 111, 331, 332
Uranium, 3, 4, 6, 7, 17, 19, 29, 34, 110, 116, 119–128, 178, 181, 202, 208, 330–337, 405
associations of, 321, 323, 339, 345
bead test for, 228
geochemical and geophysical methods, 132, 134, 135, 149, 155
sampling, testing, and assaying, 158, 160, 163, 165
Uranophane, 135, 332
Uranothorite, 339
Uranotite, 332
Uruguay, 204
U.S.S.R. (*see* Soviet Union)
Utah, 14, 15, 42, 46, 107, 170
claims in, 173, 178, 187
mineral deposits in, 210, 293, 299, 303–305, 319, 336
schools in, offering degrees related to mining, 211
state agencies in, 46
Ute Pass, Colorado, 117
Utensils, cooking and eating, 90–91

Valentite, 343
Vanadates, test for, 232
Vanadinite, 111, 304, 322
Vanadium, 165, 181, 309, 321–322, 333, 336, 379
bead test for, 228
test for, 230, 231
Vancouver Island, 18
Van Niekerk, Schalk, 10
Van Wagenen, Theodore F., 203
Veins, 108, 115, 251, 254, 260, 264, 269–272, 276, 281, 282, 368, 369, 376, 380
metallic associations of, 289, 291, 292, 301, 302, 306, 313, 314, 320, 333, 335
Venezuela, 204, 311
Vermiculite, 377, 379
Vermont, 46, 210
Vertical force, 143
Vertical loop, 146
Vibrating table, 413
Victoria (Australia), 203, 267
Victoria mine in Canada, 180
Viet Nam, 204
Virginia, 46, 210, 211, 329
Visor twin, 216, 308
Vitreous luster, 220
Volatilization, 416
Volcanic ash, 381
Volcanic dust, 362
Volcanic glass, 241
Volcanic neck, 257–260, 275, 291, 295, 327, 334, 340
Volume, data on, 424
Von Bernewitz, Max W., 25, 28, 49, 54, 93, 210
Vugs, 260

Wabana, Newfoundland, 310
Wad, 313, 326
Wales, 10
Wallrock, 273, 276
Wallrock alteration, 373
Walsh, Thomas, 167, 207
Wardell, William W., 50
Wardner, Jim, 12
Warren, Harry V., 20, 118
Washbasin, 91
Washington, George, 173
Washington state, 42, 211
claims in, 173, 187
mineral deposits in, 210, 333
schools in, offering degrees related to mining, 211
state agencies in, 46
Waste removal, 412–413
Water bucket, 91
Water data, 427
Watkins, T. H., 25
Wavellite, 135
Weathering, 365, 377
Weights and measures, 424
Wells, John H., 278
Wernerite, 134
Wesselton diamond mine in the Republic of South Africa, 11
West Virginia, 46, 210, 211
Western Australia, 109, 144, 168, 203, 208
Weston, William, 167
Wet tests, 229–232
White arsenic, 343
White mica, 379
White run, 283
White rock, 358
White Sands of New Mexico, 243

Whiting, 363
Willemite, 111, 132, 135, 305
Williamson, John T., 22
Windbreaker, 88
Winter, Aaron, 227
Wire, 88, 225
Wisconsin, 46, 140, 210, 212
Withdrawals, 187
Witherite, 135, 136, 231, 380
Witwatersrand (the Rand), South Africa, 144, 288, 331, 415
Wolff, Ernest N., 18
Wolframite, 111, 277, 278, 308, 320, 321, 413
Wollastonite, 134–136, 377
Wonderstone, 380
Wood:
  agatized, 354
  jasperized, 354
  opalized, 353
  petrified, 354
  silicified, 354
Wood-tin, 230, 307
Wool, mineral (rock), 377
World mining laws, 203–204
World prospecting laws, 203–204
Wrapping paper, 91
Wulfenite, 304, 319
Wyoming, 42, 243
  claims in, 173, 187
  mineral deposits in, 210, 333, 335, 353, 371
  schools in, offering degrees related to mining, 212
  state agencies in, 46

Xenotime, 111, 339, 340
X-ray fluorescence, 127
X-ray fluorescence spectrography, 154
X-ray spectrograph, 153
X-ray spectrographic analysis, 165
X-rays, 130

Yakutia region of Siberia, 352
Yellow copper ore, 300
Yellow Pine, Idaho, 320
Yellow sand, 283
Yerrington copper mine in Nevada, 7
Youngberg, Elton A., 333
Ytterbium, 339
Yttrium, 339
Yukon Territory, 14, 16, 20, 32, 47, 201, 202

Zambia, 155
Zeschke, Gûnter, 111
Zinc, 164, 305–306, 320, 321
  associations of, 294, 295, 298, 314, 322, 344
  coatings, 228
  geochemical methods, 122, 140, 149, 150, 155, 156
  geology, mineralogy, and petrology, 247, 267, 276
  mining and milling, 412–418
  test for, 231
Zinc blende, 305
Zincite, 111, 305
Zircon, 110–112, 134, 135, 277, 279
  metallic associations of, 329, 332, 345, 346
  nonmetallic associations of, uses of, 351, 356, 369
Zirconia, 366
Zirconium, 345–346, 356
Zoning, 262, 265